Late Holocene Indigenous Economies of the Tropical Australian Coast

An archaeological study of the Darwin region

Patricia Mary Bourke

BAR International Series 2340
2012

Published in 2016 by
BAR Publishing, Oxford

BAR International Series 2340

Late Holocene Indigenous Economies of the Tropical Australian Coast

ISBN 978 1 4073 0923 1

BAR Publishing is the trading name of British Archaeological Reports (Oxford) Ltd. British Archaeological Reports was first incorporated in 1974 to publish the BAR Series, International and British. In 1992 Hadrian Books Ltd became part of the BAR group. This volume was originally published by Archaeopress in conjunction with British Archaeological Reports (Oxford) Ltd / Hadrian Books Ltd, the Series principal publisher, in 2012. This present volume is published by BAR Publishing, 2016.

Printed in England

BAR
PUBLISHING

BAR titles are available from:

BAR Publishing
122 Banbury Rd, Oxford, OX2 7BP, UK
EMAIL info@barpublishing.com
PHONE +44 (0)1865 310431
FAX +44 (0)1865 316916
www.barpublishing.com

Preface

This monograph is an edited version of my PhD thesis, which was submitted to the Northern Territory University (now Charles Darwin University) in 2000. References post-dating this time are not included. The monograph presents a study of Indigenous economies in traditional Larrakia country, the Darwin coastal region of northern Australia, during the Late Holocene period. Subsistence and settlement patterns of this period are revealed through archaeological investigation of shell mounds, which dominate the study area and have long been a topic of scholarly interest both internationally and in Australia. Addressed are cultural, environmental and taphonomic aspects of mound formation and the implications of inter and intra-midden variability for interpretations of chronological change in hunter-gatherer economic systems, particularly with regard to theories of Holocene intensification in the Australian literature. I find that these mounds are not the remains of a specialised marine economy focused on off-shore large sea creatures as found for some mounds on the east coast of Australia, but of a generalised, flexible coastal economy utilising resources of the nearshore estuarine systems, coastal plains and hinterland. In this study I explore the question of why people built mounds of shell and why they stopped this practice that had continued for millennia. I suggest that the answer lies in the intimate interaction between the hunter-gatherer inhabitants and their environment, played out according to the particular historical cultural behaviour and to perceptions of the cumulative environmental and social changes that occurred during this period on the north Australian coast.

Acknowledgements

I would like to give special thanks to members of the Larrakia community, traditional owners of this country, including Pauline Baban, Tina Baum, Richard Barnes, Russel Cubillo, Terry Fox, Donna Jackson, Alfie and Jon May, Robert and Wes Mills, Bill Risk, Lorraine Williams and Victor Williams, for permission to carry out this research, field assistance, and sharing some of their wealth of knowledge. For assistance in the field and laboratory, I thank my intrepid volunteers Hugh Bland, Joanna Caramelia, Andrew Collis, Chris Crassweller, Colin De La Rue and his wife Kathy for bringing references on European use of shell middens to my attention, and Alison Farrell, Adam Ford, Jason Kneebone, Cathy McCarthy, Bruce Mumford, Judy Opitz, Kate Peake, Patricia Puig, and Wayne Roddam.

I am grateful to my supervisors, Ian Walters and Peter Hiscock, who contributed greatly with support and advice in their respective areas of expertise, including the mundane activities of identifying faunal and stone artefact remains, as well as the more theoretical aspects of the study. Thanks to Norma Richardson of the Museums and Art Gallery of the Northern Territory (MAGNT) for first encouraging me to undertake this project, and to other MAGNT staff for assistance - Fiona Mowat, for access to the sites register and Helen Larson and Dirk Megirian for identification of faunal remains. I am also indebted to Richard Willan at MAGNT, who gave freely of his time and expertise, for identification of mollusc taxa and knowledge on their habitats. Thanks to Stephen Sutton of the Heritage Branch, who assisted with organising permits and in innumerable practical ways, Robin Gregory for help tracking down documents, Dave Howe for practical help and Peter Brocklehurst and Bart Edmeades for access to mangrove and land unit maps. Map data was provided courtesy of Northern Territory Government Department of Infrastructure, Planning and Environment.

My thanks for logistical support and funding to the Australian Institute of Aboriginal Studies (AIATSIS) to Graeme Ward of AIATSIS for advice and support, The Northern Territory University (NTU) Internal Faculty Grants and the Australian Institute of Nuclear Science and Engineering (AINSE), for radiocarbon determinations. Thanks too to Alan Hogg of Waikato Radiocarbon laboratory for assistance with my enquiries on radiocarbon dating. I would like to thank the library staff at NTU, Cathy Robertson and Elaine Glover for bibliographic support, and Evi Adams and Hemali Seneviratne for administrative support. I am grateful for the Australian National University (ANU) National Visiting Scholarship, which facilitated library research and useful discussions with colleagues at ANU, including Sue O'Connor, Doreen Bowdery, Peter Bellwood, Judith Cameron, John Chappell, Geoff Hope, Rhys Jones, Betty Meehan and Annie Clarke. I thank Colin Woodroffe for discussions on geomorphology and sending me a copy of his Shoal Bay paper, David Bowman for information on *Callitris* and scrubfowl mounds and Rod Kennett for information on turtles.

In my undergraduate years Chris Healey and David Mearns helped develop my appreciation that Archaeology is firmly rooted in Anthropology, and I am forever indebted to Wendy MacDonald for talking me out of throwing in the towel years ago. My friends and colleagues at NTU contributed to fostering an academic atmosphere rich in ideas and moral support vital to the long process of research and writing a thesis. In particular I would like to thank Chris Crassweller, Daryl Guse, Silvano Jung, Scott Mitchell, Fiona Mowat, Judy Opitz, Marc Oxenham, Patricia Puig and Sally Brockwell who, along with Peter Hiscock, Ian Walters and Matthew Fegan, proof-read various drafts of the thesis. I am also most grateful to Adam Black for his excellent work in formatting this monograph. Finally, I thank my family for emotional and practical support over the years, and especially my partner Matthew Fegan, for many hours assisting with maps and figures.

Contents

Figures

Plates

Tables

Introduction: An archaeological study of the Darwin coastal region

Introduction

Mounds of shell have long intrigued archaeologist and layperson alike, both in Australia and overseas. In particular the large shell mounds of Northern Australia have been the object of scholarly scrutiny since the inception of archaeology as a discipline in Australia. In the first half of the 20th century excavations were undertaken into the large shell mounds at Weipa (Wright 1963, 1971) and Milingimbi (Warner 1958:455; McCarthy and Setzler 1960:230-250) to determine the nature of these deposits, their origin and antiquity. The focus of early analyses though, was generally on the stone artefacts recovered from the mounds rather than the shell itself. But it is since Bailey's (1975a, 1977) seminal work on the 500 or so shell middens and large mounds at Weipa on Cape York Peninsula, that shell deposits have increasingly featured in broad archaeological interpretations of past Indigenous coastal economies in Australia. Studies in the 1980s, on the shell mounds at Princess Charlotte Bay by Beaton (1985), and on shell middens and mounds at Aurukun by Cribb (1986a; Cribb *et al.* 1988), and their respective interpretations, of a time-lag in occupation, and of domiculture, are well known in the archaeological literature.

Over the past decade there has been an expansion in Australian coastal research in general (see Hall & McNiven 1995, 1999; O'Connor 1996; O'Connor and Sullivan 1994a), and in northern Australia (Hiscock and Kershaw 1992; Sullivan *et al.* 1994; Veth and Hiscock 1996). Topical issues of this research have been the significance of the late appearance of shell mounds, in particular with regard to their Pleistocene heritage, questions on the anthropogenic origin of mounds, and regional investigations into the nature of coastal economies and the relationship between landscape and midden sites. Recent studies on shell midden and mound sites have been undertaken in North Queensland (Bailey 1994; 1999; Bailey *et al.* 1994; Barker 1991, 1999; Cribb 1996), on the coast and islands of the Northern Territory (Burns 1994; Clarke 1994; Hiscock 1996, 1997, 1999; Hiscock and Hughes in press; Hiscock and Mowat 1993; Hiscock *et al.* 1992; Mitchell 1993, 1994a; Mowat 1995; Roberts 1991, 1994) and along the north-west Australian coast as far south as the Pilbara (Bradshaw 1995; O'Connor 1990, 1996, 1999b; O'Connor and Sullivan 1994b; O'Connor and Veth 1993; Veitch 1994, 1996, 1999).

These studies have shown the dominant component of most north Australian mounds to be *Anadara granosa* shell, a marine bivalve mollusc with an ecological range from northern Australia to Japan. Today this mollusc is commercially cultivated in the Indo-Pacific region, particularly in Thailand and Malaysia (Broom 1985; Mason 1991:271). The presence of numerous shell mounds composed predominantly of *Anadara granosa* on the coasts of northern Australia, Malaysia, Sumatra, northern Vietnam (Bellwood 1985:171), Thailand (Mason 1991) and Japan (Suzuki 1986) attests to the economic importance of this roughbacked cockle in Indo-Pacific prehistory. Most of the shell mounds in Southeast Asia have been destroyed through utilisation as a source of lime (Bellwood 1985:162, 171). The large numbers of *Anadara* mounds recorded along the north Australian coast, at Princess Charlotte Bay (Beaton 1985), Weipa (Bailey 1977, 1994), Aurukun (Cribb 1986) around Milingimbi in Arnhem Land (McCarthy and Setzler 1960; Peterson 1973; Roberts 1994), in the Pilbara and Kimberley regions (O'Connor and Sullivan 1994b; Veitch 1996) and most recently in the Darwin region for this project (Figure 1.1) are the most abundant surviving remains of this type of archaeological site in the Indo-Pacific region. They represent an important cultural resource from which a great deal can be learnt about prehistoric human interaction with a dynamic late Holocene landscape.

Accordingly, this study aims, through archaeological investigation of open sites – mainly *Anadara* shell mounds - to determine the subsistence and settlement patterns of the people who occupied the Darwin region of the north Australian coast during the late Holocene period. The study describes and reports on the results of analysis of excavations of eight open archaeological sites on the coast of two different estuaries in the Darwin region - three shell deposits at Darwin Harbour, and five shell and earth deposits at Hope Inlet, Shoal Bay (Figure 1.1). Within the broad aim of this study are a number of specific objectives:

- To investigate the problem of contemporaneity of sites in the archaeological record.
- To explore the wider implications with regard to interpretations of sites based on stratigraphy, through data obtained on shell midden formation and taphonomic post-depositional processes.
- To reconstruct as far as possible the palaeoenvironment during the late Holocene and investigate the relationship between the inhabitants and their environment, including their

response to environmental change (following Erlandson 1994).

- To compare data from a number of stratified sites with potentially contrasting and complementary deposits within the same local area (Bailey 1993:15; McBryde 1974:16-17), across the different microenvironments.

- To analyse and interpret these data within the context of sites in the local area and wider north Australian coastal region, and across Australia as a whole.

- To compare detailed site histories, identify parallel trends and separate site specific and local factors from large-scale trends representative of a region (following Frankel 1993:31).

- To compare the data with ethnographic data and models on the nature of subsistence and settlement economies in the region.

- To address methodological problems as identified by Bird and Frankel (1991a, 1991b) on the manner in which different site types and depositional contexts are considered significant and the way meaning is applied to these.

Environmental context is considered a very important aspect of this study, considering the dynamic nature of Holocene climate, environmental and ecological variability in the tropics of Australia (Woodroffe 1993:166, and see below). This study covers the most recent phases in environmental history – the transitional and freshwater phases – identified in the region by Woodroffe *et al.* (1985; Clark and Guppy 1988; Hope *et al.* 1985). My theoretical orientation could be described as what Kent (1987:518) refers to as "cultural materialism/ecology", where the economy and technology of a past forager society are deemed to be mediated through the environment. From this perspective the relationship between humans and the environment is deemed interactive rather than deterministic.

The extent of environmental effects on human history is a long-standing topic in interpretations of culture change in archaeology. But in the particular historical socio-political context of our time it appears acceptable to explain early human history in terms of evolution and environment, but not recent, in particular late Holocene, human history, even though in Australia at least, these were foraging peoples (Crumley 1994:2-3; Rowland 1999:21). Yet the environment in the torrid tropics becomes very real even in the technological 1990s, when the four wheel drive is bogged up to the axles and the nearest tree is a kilometre in the distance, water supplies are running low, a wasp sting on the face is swelling and the hot shower at the end of a hard day tramping through the mangrove mud seems a long way away. In the tropical Top End, still remote to most modern white Australians who congregate on the southern coasts, a monsoonal climate means heat and humidity, which is extreme in the last few months of every year during the build-up to wet season rains when water is scarce and plant growth minimal (Jones 1980:107-9). Most of the white population, living in cities such as Darwin, is insulated from the effects of climate in everyday life by airconditioning, piped water from dams and food trucked up from south in refrigerated supermarkets. Past foraging peoples, however, would not have been insulated from the variability of this monsoonal climate and this will be discussed in Chapter Two.

The study area

Selection of the research area for this project was determined by a number of factors. Firstly, my interest in the archaeology of the Darwin area was stimulated by my Honours research, which revealed how little was known of the area and hinted that many more archaeological sites may exist in the wider Darwin region. This interest, encouraged by Norma Richardson, then curator of Archaeology at the Museum and Art Galleries of the Northern Territory, together with the logistical benefit of working close to home, without long fieldwork periods away from my young daughters, led me to undertake a PhD thesis on the archaeology of the Darwin coastal region.

The study area lies at the western edge of the Van Diemen Gulf, catchment area of the Adelaide, Mary, and Alligator Rivers, and covers 2195 square kilometres that makes up the catchment area of the Beagle Gulf (Figure 1.1). Here the main rivers of Darwin Harbour, the Elizabeth and Blackmore, and the small Howard River at Hope Inlet flow into the Beagle Gulf. Initially the research design aimed to determine the number and nature of sites within this region, and a combination of purposive and random surveys were undertaken to this purpose (see Chapter Four). Due to the sheer numbers of sites found within a few kilometres of the coast, investigations were then concentrated in these coastal zones around Darwin Harbour and Shoal Bay, 25km to the northeast.

Context of the research

In the past, archaeologists have tended to base interpretations and broad ranging models of past Australian forager behaviour on the data provided by very few analysed sites. At this still early stage in the history of Australian archaeology, regional research projects have an integral role to play in providing a more reliable and comprehensive database on which to base theories about the past (Bird and Frankel 1991b). This project has a regional focus that adds to the database presently accumulating through current research in the Northern Territory and elsewhere in Australia (Hiscock and Walters 1994; Veth and Hiscock

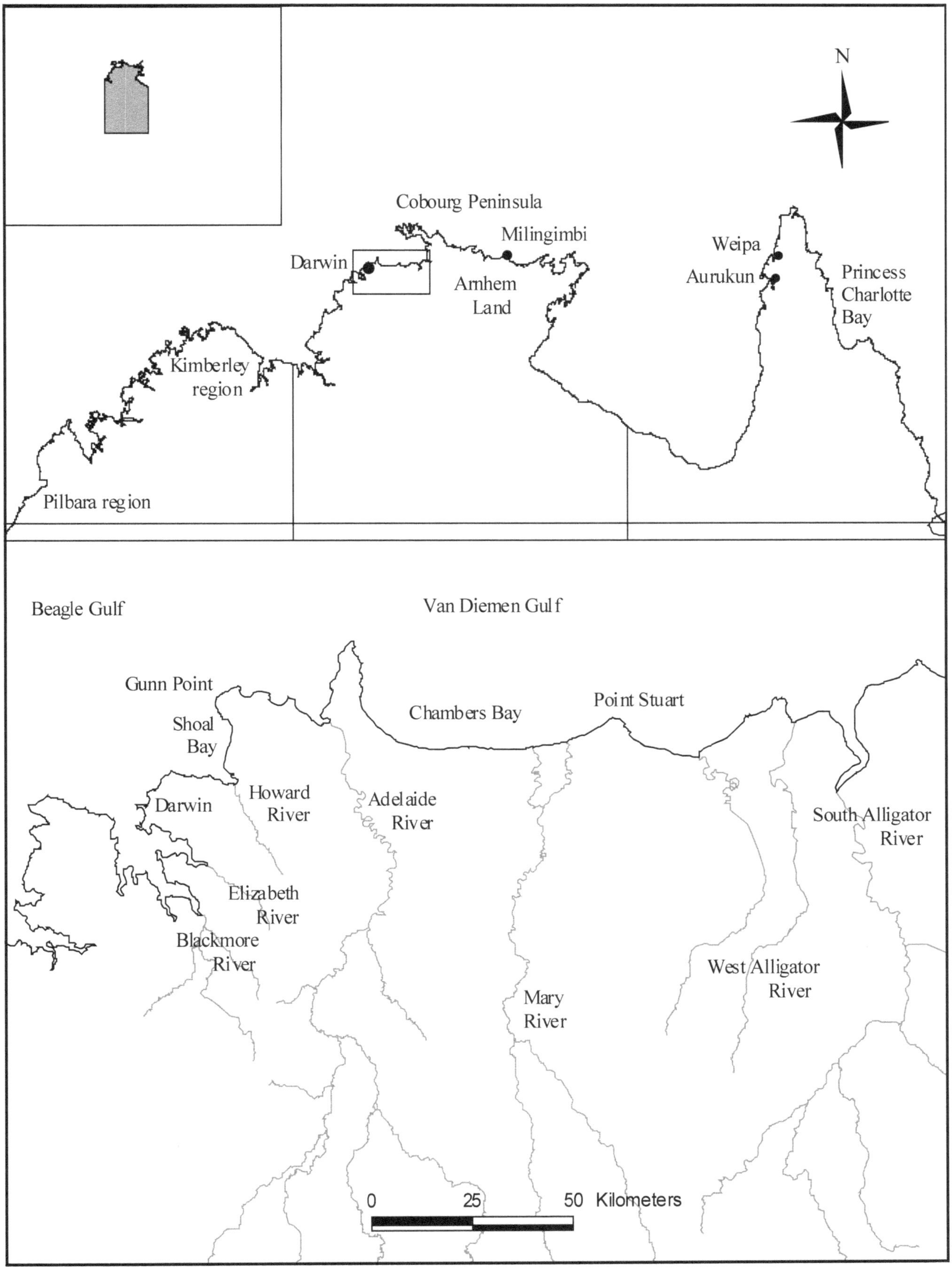

Figure 1.1 Location of the Darwin region

1996). It is one of the small regional studies considered within its historical and ecological context (Bailey and Parkington 1988:5), which are necessary to fit in to the larger whole and provide a base on which broad models may be built (McBryde 1974). It takes a micro-regional approach (cf. Flood 1980) to attempt to reconstruct groups of people's lifestyle on a local scale but covers a macro-time scale in terms of the life-spans of generations over several thousand years in the late Holocene. The time period under discussion covers the late Holocene period to 1869 AD when the European settlement of Darwin was established, with the main focus being between 2000 BP and 500 BP.

Thus the archaeological evidence examined from the Darwin region spans some fifteen hundred years in the late Holocene, a relatively short time scale in archaeological terms but a long time in human terms, representing a period in which about 75 generations of people may be born, live their lives and die (Horton 1993:12). A relatively fine-grained analysis of individual sites accumulated over decades or centuries, in conjunction with a more coarse-grained examination of the cluster of sites spanning millennia, provide the opportunity to explore temporal and spatial problems of scale identified by Frankel (1993) in constructing regional sequences.

Previous archaeological work in the Darwin region

Historically, the Darwin coastal region has been a rather neglected area in terms of archaeological research. Investigations in the north of the Territory have largely focussed on the more remote and therefore more exotic Arnhem Land region (Allen and Barton nd.; Brockwell 1989, 1996a; Hiscock 1996; Hiscock and Mowat 1993; Hiscock *et al.* 1992; Jones 1985a; McCarthy and Setzler 1960; Mowat 1995; Roberts 1991, 1994; Schrire 1982; White 1967). Studies closest to Darwin have been to the east in the Adelaide River and Mary River regions (Brockwell 1996b, 1996c; Crassweller 1996; Guse 1992; Smith 1981; Smith and Brockwell 1994), and north-east of Darwin, at Chambers Bay (Baker 1981), Cobourg Peninsula (Mitchell 1994), Groote Eylandt (Clarke 1994) and the historical settlement of Port Essington (Allen 1969). Many of these studies were undertaken as part of archaeological research for graduate and postgraduate degrees and generally explore chronological patterns of occupation of these regions.

When I began this research previous archaeological fieldwork within the Darwin coastal area was limited to minor briefs for the now defunct Museum and Art Gallery of the Northern Territory's (MAGNT) sites register and a growing number of cultural resource management (CRM) surveys undertaken for development projects (eg. Caldwell Engineers 1983; Guse 1995; Guse and Hiscock 1993; Heritage Surveys 1995; 1997; Hiscock 1992; 1993b; 1995b; Hiscock and Mitchell 1991; Mulvaney 1993; Smith 1981; Stone 1989a). These studies generally record and describe individual sites, determine their relative significance and the likely impact of the proposed development, and make recommendations regarding management strategies or mitigation procedures as appropriate under the *Northern Territory of Australia Heritage Conservation Act 1991*. A bibliography of environmental and archaeological reports relevant to the Darwin region provided in Richardson (1996a), shows that knowledge of the number and distribution of pre-contact sites in the Darwin region had been obtained in a somewhat piecemeal fashion.

Nonetheless, what could be ascertained from these reports is that the most numerous site type recorded on the register is shell middens, followed by stone artefact scatters, some in association with the middens. Other coastal site types reported include stone quarries, burials and fish traps. Most known stone artefact scatters in the Darwin region were located through CRM surveys, usually conducted inland of recorded shell middens in bounded areas selected for development, which were small enough for relatively intensive surveys (eg. Hiscock 1993b; Hiscock and Mitchell 1992; Stone 1989a).

In 1994 one small section of the Darwin Harbour coast was intensively surveyed as part of research for an Honours thesis (Burns 1994). As a result of this small project, and research arising from consultancy work (Hiscock and Hughes in press), 28 sites, mostly shell middens, were described on Middle Arm Peninsula, part of the Darwin Harbour shoreline (see Chapter Five, Figure 5.2). Hiscock and Hughes (in press) estimated a minimum density of 22/km^2 for shell mounds at Haycock Reach on the peninsula. Situated at the junction of the hinterland high ground and mangrove zone, these discrete, generally circular shell deposits are composed predominantly of the marine bivalve *Anadara granosa*. Radiocarbon dating of *Anadara* shell from some of the middens indicated that they were formed in the late Holocene (Hiscock 1997; Hiscock and Hughes in press). These sites appeared similar in age, size and shell content to some middens described on the coastal plains of Chambers Bay (Baker 1981:62-82) and in the South Alligator River region (Woodroffe *et al.* 1988:96).

Initial excavations in this area were undertaken on shell middens at Bayview Haven (Guse and Mowat 1993; Hiscock 1992; 1995a; Wedd 1993), and on Middle Arm peninsula (Burns 1994) (see Chapter Five, Figure 5.2). Excavations at Bayview Haven were undertaken as a salvage operation prior to destruction of the site with development of a residential estate, to determine if the shell material had an anthropogenic origin and to obtain samples for radiocarbon dating. The operation was used as a training exercise in excavation techniques for undergraduate students at Northern Territory University (Guse and Mowat 1993). Excavation of MA1, a mounded shell deposit on Middle Arm peninsula, was undertaken for comparative analysis with that of mounds built from shell and sand by the scrubfowl *Megapodius reinwardt* on nearby Channel Island. That analysis demonstrated distinct differences in the composition of these two types of shell mound, consistent with archaeological interpretation of MA1 as a cultural deposit, *contra* Stone's (1989b) hypothesis of a megapode origin for both types of mound (Burns 1994).

Excavations of these middens found they were dominated by the bivalve mollusc *Anadara granosa*, with lesser numbers of other taxa such as the bivalves *Marcia hiantina*, the oyster *Saccostrea echinata* and the gastropods *Nerita* sp., *Terebralia semistriata*, *Telescopium telescopium*, *Chicoreus capucinus*, *Cassidula angulata*, and *Volema cochlidium*. Other faunal material such as animal bone, fish otoliths and macropod teeth, as well as stone artefacts, charcoal and ochre were also identified.

The presence of a number of shell middens, as revealed through the surveys on Middle Arm Peninsula, suggested that there might be many more sites along sections of the Darwin coastline where archaeological investigations had not yet been carried out. Consequently, more extensive surveys were undertaken in the region as part of the research for this study, in order to determine the number, distribution and type of sites in the region, expected to be predominantly shell middens.

Historic context of the Darwin region archaeological record

In 1873, the report of Thomas Reynolds, the Commissioner of Crown Lands and Immigration, of his visit to the Northern Territory, stated:

> Plenty of good building stone at Palmerston, but there is an absence of limestone: there are however, large accumulations of shells just outside the township - some of these fifty to one hundred feet long, several feet high, and ten to twelve feet wide. *These accumulations should be preserved, as a large proportion of them will be required for public buildings* (my emphasis).

The burning of shells to make lime for building or agriculture was common practice in the early days of European settlement in Australia (see Allen 1969 for description of the practice in Port Essington). This evidence of destruction (see also Douglas 1871; Knight 1874; 1876:182; NTRS 828), during the process of European settlement, of what were probably Aboriginal shell middens, suggests that these sites were once much more prevalent around the old Palmerston settlement we now call Darwin. In the short history since European settlement - a mere 130 years - Darwin city and surrounds have been transformed into an urban landscape which would be unrecognisable to past Indigenous inhabitants if they were alive today.

During surveys I observed remnants of Aboriginal shell middens, one on the side of the only major road out of Darwin, the Stuart Highway, and a number of dirt tracks paved with shells from nearby middens, as well as recently mined shell mounds, probably for gardening purposes. Surveys also revealed extensive destruction of shell mounds around Darwin during World War Two, which were quarried for construction of military camps and roads (and see Caldwell 1983; Hiscock and Hughes in press). One site on the banks of the Elizabeth River was so disturbed by World War Two activities that the original dimensions are now indeterminable. The midden material was utilised for cement foundations and a rectangular arrangement of seventy 44-gallon drums filled with shells, resembling a garage of some sort (Burns 1996a:29). Through such activities the process of European settlement irrevocably altered the physical remains of past lifestyles. It is probably fair to say that the pattern of distribution of shell middens around Darwin is in part the result of the level of development of the area.

Physical removal of archaeological remains is only part of the story of the neglect of anthropological and archaeological study in this region. The traditional owners of the Darwin coastal region had disappeared from ethno-historical and anthropological records by the early 20th century, into the realm of administration records, as a "problem" for the European colonisers (Wells 1995:17). This notional disappearance of Indigenous groups commonly occurs within decades in areas that were the focus of European settlement (Read 1991), as such groups did not fit the stereotype of more isolated Aboriginal people, considered less changed from more "traditional" socio-cultural organisation and so worthy of anthropological study (Cowlishaw 1992:24 in Wells 1995). The corollary, when the anthropological history of a people is not recorded, has been little or no formal recognition of the existence and significance of material remains of a pre-contact past in the Darwin region for well over a century.

Relevance of the research

Like their counterparts overseas, archaeologists in Australia have debated the role of molluscs in prehistoric economies (Bailey 1975a, 1975b; Erlandson 1994:274; Waselkov 1987:118-123). Views range from the idea that molluscs are a minor component of a broad-based economy (Bailey 1975a, 1975b; Cribb 1986b), and/or as a secondary resource fallen back on in times of scarcity, and/or or as a seasonal staple (Erlandson 1988; Meehan 1982a), or are the mainstay of past coastal economies (Beaton 1985, 1986). Some overseas researchers suggest that intensive mollusc exploitation made possible by the high marine productivity due to upwelling conditions, in places like coastal California and Peru, provided the impetus for the appearance of complex maritime economies (Erlandson 1994:277; Moseley 1975).

Inexorably linked to the question of the role of molluscs and middens is the timing of the appearance of large shell mounds on the coast and the questions of continuity and change in resource procurement strategies (Bailey 1993:15). Cultural shell deposits have been described as "information traps" (Cribb 1986b:171) in the archaeological literature. They not only illustrate prehistoric cultural practices, but also act as an indicator of past marine and lacustrine environmental conditions (Cribb 1986b:172), allowing an investigation of the interaction of Holocene landscape change and human land use (Sullivan and O'Connor 1993). Issues examined include the degree of dependence on marine and terrestrial resources in coastal areas, and the extent to which shell deposits are representative of past coastal economies (eg. Beaton 1985; Gaughwin and Fullagar 1995; Hallam 1987). Explanations for the appearance of the north Australian *Anadara* mounds, predominant in the last 3000 to 500 years, either favour environmental causes (O'Connor and Sullivan 1994b; Sullivan and O'Connor 1993; O'Connor 1999a; White and O'Connell 1982:156-7) or see the onset of mounds as a reflection of a wider social phenomenon (eg. Veitch 1999).

The archaeological record and intensification

At present the pre-contact archaeological record in Australia predominantly dates to the mid to late Holocene period, which has led some researchers, following Lourandos (1983) to interpret this more abundant material and some qualitative changes in site characteristics as indicative of a process of intensification. Intensification, defined as increased productivity and production involving socio-economic change, is linked to changes in settlement, subsistence and demographic patterns and associated with increasing complexity of social networks in the mid-late Holocene in Australia (Lourandos 1983, 1985a, 1985b). Although Lourandos' theory was based primarily on research in southwestern Victoria, he claimed studies in other parts of Australia (eg. Attenbrow 1982; Hughes and Lampert 1982) indicated "increasingly complex and competitive social networks and their related ceremonial institutions" concurrently developed throughout the continent (Lourandos 1983:92). There are a variety of ways in which this purported process of intensification is said to be expressed archaeologically. The original archaeological criteria defined by Lourandos (1983:82) as indicators of intensification are:

- More intensive use of individual sites
- Increased establishment of new sites
- Increased use of marginal environments
- Increased complexity of site economy (ie. resource management strategy)

Freslæv and Frankel (1999:248) build on these indicators in their challenge to intensification theories in the more recent archaeological literature, with regard in particular to the coastal archaeological record. They propose three outcomes that may be expected from a process of intensification involving increased populations, increased sedentism and regional elaboration of social networks between groups:

- Greater diversification in the use of resources, so that the region became more densely occupied and the total resources of the region more effectively utilised (the broad spectrum revolution);
- Greater technological investment in the effective exploitation of single resources;
- More intensive use of existing resources (new sites, more deposit, greater predation) (FreslÆv and Frankel 1999:248)

Lourandos' (1983:82) criteria have been employed by some archaeologists around Australia in their interpretation of site characteristics and site distribution (eg. Barker 1991, 1999; Morwood 1987; Ross 1985; Williams 1988) in support of the theory of intensification. In particular, mounding of deposits, both earth mounds, in southeastern Australia, and shell mounds such as those at Weipa, have been seen by some as archaeological expressions of a pan-Australian change, in behaviour and increasingly complex social organisation in the late Holocene, associated with decreased residential mobility and increased population (eg. Flood 1990:125; Lourandos 1983, 1993; Williams 1988). Regionalised patterns in north Australian rock art are also argued, in support of the intensification hypothesis, to represent demographic changes in the Holocene, with population growth and fissioning into local groups (David 1991; David and Chant 1995; David and Lourandos 1998).

Some researchers have dismissed these theories (eg. Beaton 1983; Bird and Frankel 1991a, 1991b; Head 1986; Hiscock 1986; Rowland 1989). Bird and Frankel (1991a:1) point out the social evolutionary paradigm underlying such theories, which imply that earlier Holocene and Pleistocene societies were less intensive in production and less complex in social organisation. They argue that changes seen in the archaeological record are not necessarily cumulative and directional as implied by the intensification theory, but can be interpreted as " a continuous series of short-term responses to local conditions". Others point out the lack of regional correlation, between the timing of changes in site numbers and rates of cultural discard across Australia (eg. Attenbrow 1987:352-8; Hiscock 1986) that are taken as evidence of a package of widespread social changes in the mid-late Holocene (eg. Bowdler and O'Connor 1991; Lourandos 1985b).

As the late Holocene sites in this coastal study appear to fit some of the above criteria, my interpretations bear on the general "Great Intensification Debate" (Lourandos and Ross 1994), on postulated human population changes, and/or changes in social organisation during the mid-late Holocene (eg. Beaton 1983, 1985, 1990; Cribb 1986b; Hughes and Lampert 1982; Jones 1985b; Lourandos 1984, 1985a; Rowland 1989). I therefore employ FreslÆv and Frankel's (1999:248) expected outcomes of intensification, which is specific to the archaeology of coastal regions, in my interpretation of the archaeological record in the Darwin coastal region (see Chapter Eleven). Since Lourandos (1983, 1985a, 1985b) put forward his model of intensification in the early 1980s, the debate has largely revolved around the relative importance or roles, of social versus environmental factors, in explaining perceived temporal and spatial patterns in the mid-late Holocene in the archaeological record. Two main camps have emerged in this debate: those who favour primacy of socio-demographic explanations (eg. Barker 1991, 1999; Lourandos 1983, 1985a, 1993; Morwood 1987; Ross 1985; Williams 1988; Veitch 1999) and those who favour ecological, technological or environmental/taphonomic explanations (eg. Bird and Frankel 1991a, 1991b; Head 1986; FreslÆv and Frankel 1999; Jones 1985b; O'Connor 1999a; Richman 1999) for observed changes in the archaeological record. As will unfold in the following chapters, the data from this study provides more support for views in the latter camp.

Contributions to the debate to be examined include propositions that sites represent adaptations to specific regional and local environments (O'Connor 1990, 1999a, 1999b), and that population increases are inherent in the nature of human populations and growth, requiring new resource procurement strategies, such as an economy focussed on molluscs (Beaton 1985). Some researchers (Godfrey 1989; Head 1986; Rowland 1989) argue that perceived increases in site numbers, which underlie interpretations of population increase and/or intensification, are the result of post-depositional factors that affect the survival of earlier sites. The dynamic nature of coastal environments and factors in coastal site preservation are well-discussed in the archaeological literature (Rowland 1999:144-5). Factors recognised as contributing to site survival and archaeological visibility include burial of early sites in depositional environments or removal by coastal erosion, especially through storm surges or cyclones, and possible sea level oscillations after stabilisation around 6000 BP (Flood and Frankel 1989; Godfrey 1989; Head 1986; Rowland 1989, 1999:27; Woodroffe *et al.* 1988).

With respect to shell mounds, research shows that although by dint of their solid alkaline bulk these are more protected than other site types, such as shallow shell deposits or scatters, this is no guarantee of their survival, as mounds too are subject to innumerable climatic and other physical and chemical post-depositional factors (see Bird 1992; Head 1983; O'Connor and Sullivan 1994b:27). On the other hand, some researchers also note that the high archaeological visibility of shell mounds that are relatively well-preserved may lead to bias in both the focus of research and interpretations on what proportion of a past economy these large volume deposits represent (eg. Bailey 1994; 1999; Bailey *et al.* 1994; Bird and Frankel 1991a; Cribb 1986b; Rowland 1999; cf. Beaton 1985, 1986). Environmental considerations also include discussion on changing ecosystems and the availability of certain mollusc resources at particular localities and during different time periods in the past (eg. Allen 1996; Bailey 1999; Beaton 1985; Chappell and Grindrod 1984; Hiscock 1999; O'Connor 1999a; Sullivan 1982; Veitch 1999).

Current hypotheses on economic systems in the mid to late Holocene in northern Australia

Dramatic environmental changes during the Holocene have figured predominantly in interpretations of archaeological sites in north Australia, particularly since a palaeoenvironmental record of post-transgressive landscape changes in the Alligator Rivers region (Woodroffe *et al.* 1985, 1987, 1988) has become available (Hiscock 1999; Hiscock and Kershaw 1992). Models of economic systems in north Australian coastal areas in the mid to late Holocene, recently summarised by Hiscock (1999) have been related to environmental change and the establishment of freshwater conditions, particularly in the Alligator Rivers region, and have focussed on the implications of these changes for settlement and subsistence patterns and demography.

In syntheses of archaeological work in Arnhem Land, Allen (1987, 1989, 1996, Allen and Barton n.d.) propose that after 3000 BP people abandoned occupation of the East Alligator River rockshelter sites, Nawamoyn, Malangangerr, Paribari, Malakunanja II and Ngarradj Warde Djobkeng, situated on the inland margin of the floodplains, in favour of occupation of open sites on the floodplains. This change in settlement and subsistence strategies is said to correlate with an environmental transition, from mangrove forests through saline flats and grasslands to developed freshwater wetlands. The model of abandonment is based on radiocarbon estimates of around 3000 and 4000 BP for the upper levels of midden deposits from two of the rockshelters, Ngarradj Warde Djobkeng and Malakunanja II. Each of the rockshelter deposits is dominated by the *Potamididae* and *Geloinidae* Family mangrove gastropods *Cerithidea* and *Geloina.* Allen (1987:5) argues that the ecological requirements of these molluscs, and the small crab *Sesarma* sp., indicate that people were collecting from the landward fringe of the estuarine mangrove swamps that existed at that time close to the shelters.

Theories on abandonment of rockshelter sites around 3000 BP (Allen 1987, 1989, 1996, Allen and Barton n.d.; Schrire 1982) have been challenged by Mowat (1995), who argues that the dating of midden deposits is not comprehensive enough to preclude their use after 3000 BP. She provides evidence for occupation spanning 5000 years, of shell mounds on the West and South Alligator Rivers coastal plains, concurrent with occupation of the rockshelter sites. Mowat's (1995:123,163) argument for continuity of occupation of the coastal plains is only based on six dates for four of seven shell mounds on the west bank of the West Alligator River, and seven dates obtained by Woodroffe *et al.* (1988) for the South Alligator River plains sites. However, it seems likely that, just as Mowat's work has filled in some of the hitherto unrepresented periods in terms of the archaeological record, further research in the area would build up a more complete picture of continuous occupation in the region. Similarly, the apparent hiatus in occupation of rockshelter sites after 3000 BP possibly represents cessation of shell deposition, with less visible activities continuing until the appearance once again of visible remains in the form of freshwater shell in the last 1000 years (Allen 1989:95).

Of particular relevance for this study is the composition of the midden deposits in these sites, dominated by molluscs primarily from the mangroves. The *Potamididae* Family gastropods *Cerithidea* sp. and *Telescopium telescopium*, and *Geloinidae* Family bivalves *Geloina* (*Polymesoda*) *coaxans* from the mangroves dominate in both the East Alligator and Magela Creek rockshelter midden deposits and in the open midden sites on the South Alligator plains dated to about 4500 to 3000 years BP (Allen 1987, 1989, 1996, Allen and Barton n.d.; Schrire 1982; Woodroffe *et al.* 1988). Within that same period, open midden sites dominated by open mudflat bivalves *Anadara* and *Marcia*, of the *Arcidae* and *Veneridae* Families respectively, accumulated on the coastal plains of the West Alligator River (Mowat 1995:98-123). On a map of the area showing the Western Arnhem Land sites (Figure 1.2) it can been seen that the East and South Alligator sites are situated near the middle reaches of their respective rivers, tens of kilometres further inland than the West Alligator sites adjacent to the lower reaches of the river near the coast. As Allen (1987, 1989, 1996; Allen and Barton n.d.) suggest, the most parsimonious explanation for the mollusc assemblages is that they represent exploitation of the local environment.

Increased site numbers in Arnhem Land over the last 1000-1500 years is interpreted as a dramatic increase in population and changes in economic organisation to systems similar to that documented ethnographically, in response to formation of the highly productive freshwater wetlands (Allen 1989:113-5; Jones 1985b:291-3). This is based partly on radiocarbon estimates and data from rockshelter excavations and partly on increased numbers of large open sites on the South Alligator sub-coastal floodplains margins, also dated on geomorphic grounds to the late Holocene. For example, at Anbangbang I rockshelter near the South Alligator River, large quantities of shell, bone, plant material and stone artefacts in the upper levels are translated into greater site use in the last 800-1000 years, relative to previous periods, represented by much less quantity of cultural material in the lower levels (Jones and Johnson 1985a:64). Decay of organic materials in the lower levels is dismissed as a relevant factor in this analysis.

The large open sites, mainly stone artefact scatters, of the South Alligator sub-coastal floodplains are interpreted by Meehan *et al.* (1985:135) on the basis of ethnography, site size and high artefact density, as base camps for several hundred people, occupied seasonally for several months every year. More recent interpretations, following further surveys in the area, which found a continuum of site sizes from very small to extremely large (Hiscock 1991:27; Hiscock *et al.* 1992), suggest that large high density stone artefact scatters are just as likely to represent palimpsests of a number of recurring occupational events by small groups (Foley 1981).

Rather than seeing increased population as an explanation for large sites, some researchers propose a model of relative sedentism in settlement strategies on the South Alligator floodplain. In this model, people occupy the large floodplain margin sites for several months in the dry, with large gatherings linked to ceremonies and trade/exchange, then move to smaller sites in the nearby woodlands during the wet (Brockwell 1989, 1996a; Guse 1992). Hiscock (1996) provides support for this hypothesis through an examination of assemblage differences in frequency of bipolar cores between the two groups of wetland sites. He suggests that more intensive stone reduction, that is more bipolar working, indicates more conservation of raw material, and therefore lower residential mobility. Thus the large sites on the floodplain margin, which have a high frequency of bipolar cores, are taken to indicate relatively low residential mobility, while the woodland sites, with a low frequency of bipolar cores, are taken to indicate relatively high residential mobility.

Arguments for population increase based on more productive environments have been qualified by Allen and Barton (n.d:107), who argue that the posited marked increase in productivity of freshwater swamps over the "sterile hypersaline tidal flats" of the transitional period, in terms of resources available (Jones 1985b:294) has not been demonstrated. In addition, the postulated regional nature of population increase across north Australia (Jones 1985b:291-3) is challenged by Hiscock (1999:100), who points to the apparent decrease in site numbers with the cessation of accumulation of *Anadara* mounds about 700-1000 years ago around Darwin Harbour. Hiscock (1997) also links the cessation of *Anadara* mound accumulation with environmental change and disappearance of productive shell beds and suggests that the decline in productivity may have resulted in decreased population in this locality (Hiscock 1999:100). He puts forward the possibility that site increases in one area, such as in the South Alligator wetlands in the last 1000 years may represent a "reorganisation of population distribution and resource use" rather than a regional population increase.

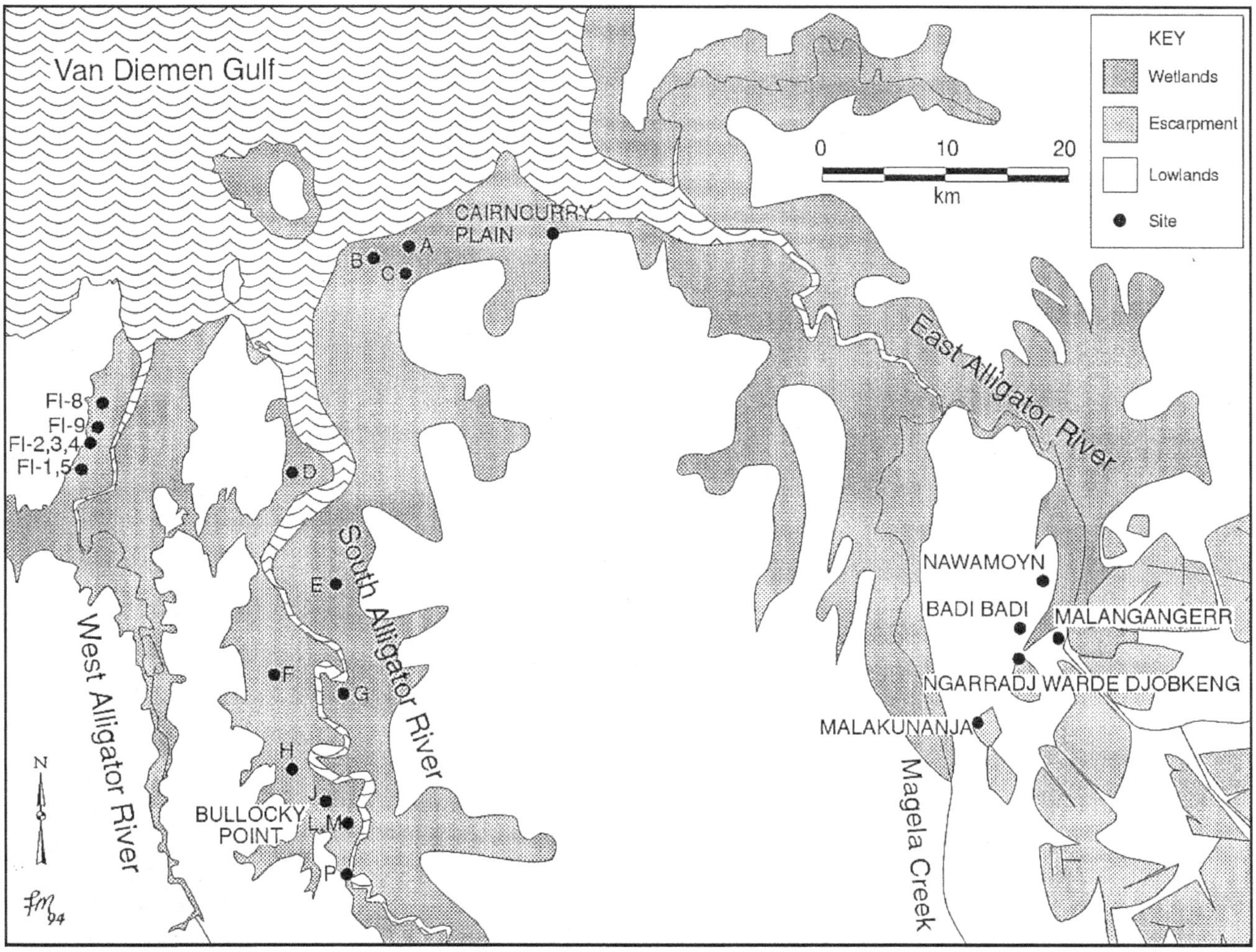

Figure 1.2 Location of Western Arnhem Land sites mentioned in the text [from Mowat (1995:44)]

Reorganisation of population distribution, with changes to patterns of settlement and resource use, have also been posited for north Australian coastal groups within the last few hundred years, following Macassan contact. Both archaeological (Mitchell 1994a, 1994b, 1995) and ethnographic (Berndt 1951a; Thomson 1949) evidence points to systems of regional exchange in Arnhem Land intensifying after Macassan contact began. On the basis of chronological differences, in extent of imported stone and faunal remains such as sea turtle and dugong, and in size and structure between pre and post-contact middens, Mitchell (1994a) hypothesises that major economic changes were stimulated by Macassan contact. These include a change in Aboriginal settlement patterns to larger groups resident for longer periods on the coast, close to foreign camps, with a decline in importance of molluscs, adoption of dugout canoes and subsequent increase in capture of large marine animals and an expansion of the ceremonial exchange system, as well as possible population decrease due to introduced disease (Mitchell 1994b:197).

Similar patterns of dynamic coastal resource procurement systems in dynamic coastal landscapes as seen on north Australian coasts are shown in the growing database of archaeological sites around the Australian coast (Hall and McNiven 1999). This provides further data with which to explore the significance of early and long-term human use of the coast and the proliferation of shell mounds and other site types in the late Holocene. Current schools of thought include the idea that ethnographically observed economic strategies of low mobility, marine-focussed resource procurement and extra-regional ceremonial exchange have only been in place for 500 to 1000 years, and earlier periods with vastly different environmental conditions, have no ethnographic parallel (Allen 1997; Gregory 1998:283-5; Hiscock 1999; Mitchell 1994a:412-7). These propositions are examined in the light of data from research for this study.

Structure of the study

In Chapter One the social, historic, and archaeological context in which this research was undertaken has been outlined. Chapter Two describes the environmental context, and Chapter Three the ethnography and historic accounts of Indigenous economies in northern Australia and resource use in the study area. Chapter Four details the methods used in this research project. The results of archaeological surveys are presented in Chapter Five. Chapter Six discusses the factors that require consideration in midden analysis, when theories are being constructed on past human behaviour, based on data from archaeological sites. In Chapter Seven, I describe the results of excavations of a few of the sites located around Darwin Harbour and explore the factors contributing to variability in these midden sites. In Chapter Eight sites at Hope Inlet are investigated through excavation, and the results analysed, in view of considerations explored in Chapter Six.

Chapters Nine and Ten consider the consonance between the archaeological data, the ethno-historic data, the environment and interpretations of human/environment interaction built on the basis of the data. In Chapter Eleven some of the broader issues that this research addresses are discussed, including the role of molluscs and mounds in past Indigenous economies, and the implications of the findings of this research for popular paradigms such as intensification. My conclusions follow in Chapter Twelve.

2 Changing environments, seasons and resources

Introduction

Environmental variability and the extent of its effects on human history, is a long-standing issue in interpretations of culture change in archaeology. Recently, the issue has been comprehensively revisited by Jones *et al.* (1999) and Rowland (1999), in the light of the vast bulk of data that has emerged from climatic and environmental research over the last decade. Both these works identify correlations, between environmental oscillations that impact on ecological systems, and archaeological sequences indicating cultural change (Rowland 1999:34). Whilst it is clear that correlation does not equal causation, I take the view that since humans, along with all organisms, are participants in a reciprocal relationship with the whole, called the environment (Levins and Lewontin 1985; Walters 1992:167), shedding light on one aspect of the system lights up the other. Or as Bird and Frankel (1991a:10) put it, "Whether one regards the environment as a primary cause of culture change, or a context within which change occurs, it is still necessary to place sites in their setting". In order to obtain an understanding of late Holocene human/environment interaction in this case, based upon interpretation of archaeological remains, it is useful to examine the environmental context in which archaeological sites were formed. To provide as comprehensive an understanding of the palaeoenvironment as possible, both regional trends and local conditions, as seen through data on global climate change, geomorphology, and present day landform types and environments, including faunal and floral resources, are discussed in this Chapter.

Two landscape types

Unlike the coasts of Arnhem Land and the Kimberleys, which are backed by upland plateaus and ranges, the Darwin hinterland landscape is one of low relief. Like the rest of the north Australian coastline however, this section of the coastal zone is a depositional, low-energy shoreline that is still prograding, through both deposition from seaward and terrestrial sedimentation (Davies 1972; Sullivan 1996:3). Despite their broad similarities, two distinct types of coastal landscape, in terms of their geomorphic units, occur within the Darwin study area (Figure 2.1). In the south undulating to rugged low hills, and in the north extensive tidal flats and sandy plains, dominate the topography (Pietsch and Stuart-Smith 1987). Around Darwin Harbour the landscape comprises a relatively steep shoreline around a large shallow embayment and along the banks of the associated riverine system. In contrast, Hope Inlet has a landscape of low relief, wedged between the large peninsulas of Port Darwin and Gunn Point, and comprising small coastal plains covering an area less than 100 square kilometres. The Northern Territory coastline is macrotidal, with a tidal range up to 8m in Darwin Harbour, and seawater inundates the supratidal flats of Hope Inlet periodically.

Despite their differences, each of these areas is an estuary, defined by Barnes (1974:2) as "a region containing a volume of water of mixed origin derived partly from a discharging river system and partly from the adjacent sea; the region usually being partially enclosed by a land mass". In the earth's geological history each individual estuarine system is a transient phenomenon and many on Australia's coast have come and gone during the Holocene, with implications for their plant and animal communities, including humans (Barnes 1974; Hodgkin and Hesp 1998).

Current climate

Currently, the Darwin coastal region has a monsoonal tropical climate with two recognised seasons, a dry season extending between April and November, and a wet season between December and March. The contrast between the wet season and the "annual drought" of the dry season is stark (McDonald and Alpine 1991:20), with most of the average annual rainfall of 1624mm falling within the four months of the wet. Although this seasonal pattern is relatively predictable, variability in rainfall patterns both spatially (Whitehead *et al.* 1990) and temporally (Taylor and Tulloch 1985) is a feature of current climates in the wider Darwin region (Bowman 1992:26). Temperatures generally are high, ranging from 34°C to 24°C in the wet, when the heat is exacerbated by the humidity, and from 32°C to 17°C in the dry. The area is subjected to periodic cyclones during the wet, which may cause considerable coastal landform changes and destruction of vegetation. Also seasonal is wind direction, with east to northwest winds prevailing during the wet season and the easterlies during the dry (McAlpine 1969).

From a contemporary Aboriginal perspective, varying slightly in timing between different groups along the north Australian coast, six major seasons are recognised and differentiated on the basis of changes in weather conditions and available resources (Davis 1984; Russel-Smith *et al.* 1997; Thomson 1939:214-5). Climate is likely to have had a major influence on past human/landscape interaction, but research suggests the current climatic patterns have a relatively recent history in the region (Lees 1992b; McGlone *et al.* 1992; Rollins *et al.* 1986; Shulmeister and Lees 1995). Accordingly, the climatic and environmental

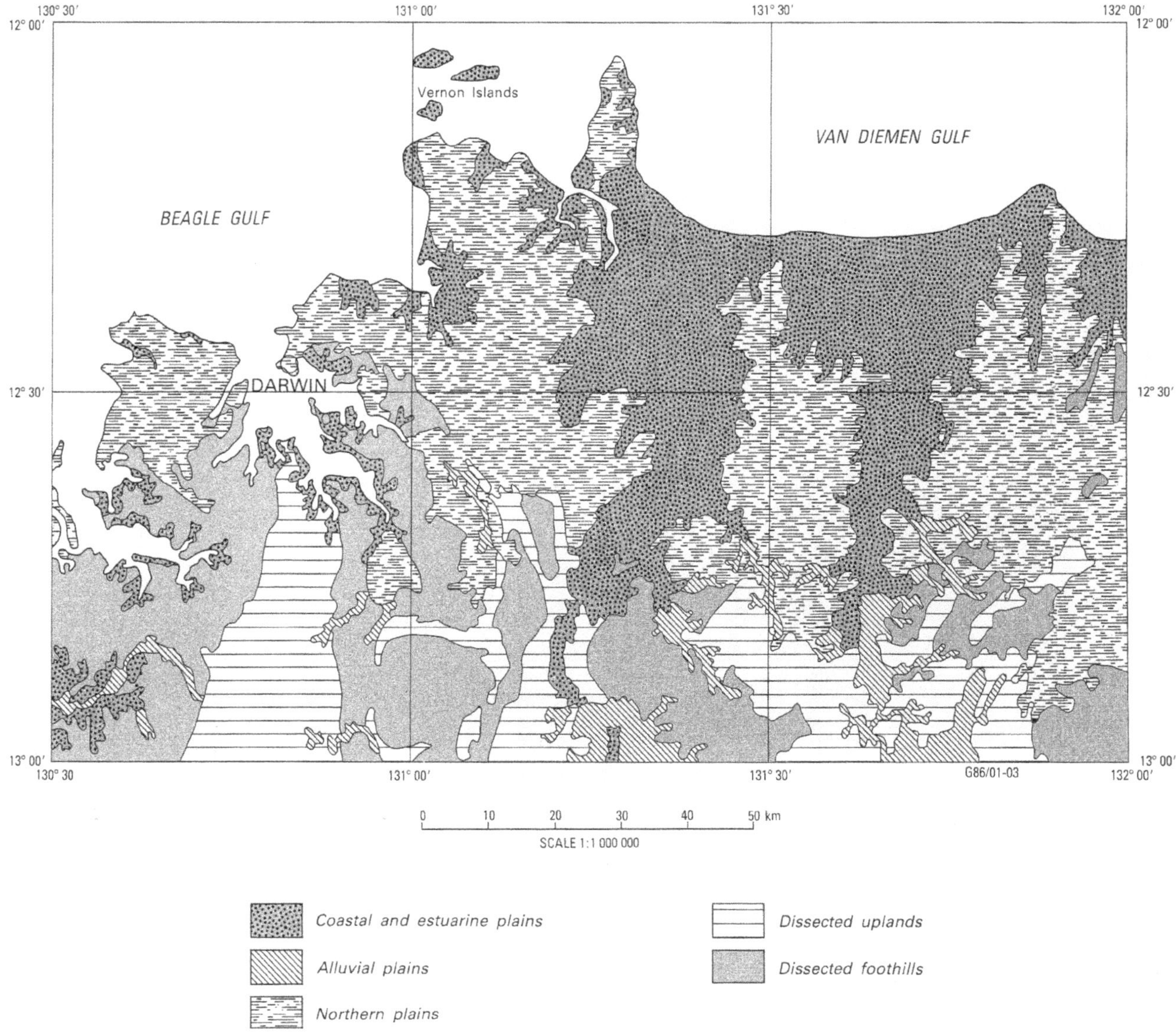

Figure 2.1 Map showing different geomorphic units of Hope Inlet and Darwin Harbour.

changes of the mid-late Holocene are addressed to explore the implications for chronological change in the past Indigenous economies.

Geomorphology

Coastal landscapes, sea levels and climate change

Coastal landscapes in this part of northern Australia are mostly of Holocene age (Semeniuk 1985). The available environmental evidence suggests that the Holocene was a period of dynamic climatic and subsequent environmental and ecological variability, both on a global and local scale (Meese *et al.* 1994; Stahl 1996), including tropical areas such as the Darwin coast (Woodroffe 1993). It is within the constraints of this environmental context that human resource procurement strategies must operate, so an understanding of the processes and effects of change helps to illustrate the possibilities of human/landscape interaction in the late Holocene. Environmental variability in the Holocene includes long-term cyclical climatic trends and sea level fluctuations or oscillations, short-term high magnitude/low frequency events such as the El Nino/ Southern Oscillation (ENSO), droughts, floods and cyclones, and the more regular and therefore relatively predictable annual change between the wet and dry seasons (Rowland 1999:15-16).

Studies on the Northern Territory coast provide evidence that sea levels stabilised following the last post-glacial marine transgression, at their present levels around 6000 years BP (Grindrod *et al.* 1999:465; Woodroffe *et al.* 1987:200; Woodroffe *et al.* 1988). Woodroffe *et al.* (1993:262) state that sea level rose a final 5 to 7m during the “big swamp phase” of widespread mangroves from 6800 to 5300 years B.P. It is possible to infer from geomorphic evidence on Cobourg peninsula that sea levels may have been up to one metre above present levels around 5000 years BP, though local variation is often attributed to hydro-isostatic loading of the continental shelf (Chappell *et al.* 1982; Lees 1992b:6; Woodroffe *et al.* 1992:107).

Elsewhere, recent global and Australian evidence suggests sea level oscillations at around 3000-3500 BP (Flood and Frankel 1989; Rowland 1999:27), and a rise around 700 BP, coinciding with a warming period called the Little Climatic Optimum and a rapid fall in the subsequent transition to the Little Ice Age (Nunn 1998). Apart from these indicators that sea level may have been higher since 6000 BP however, discerning sea level changes, especially through such methods as pollen records, is difficult in this region due to the large tidal range (Woodroffe and Mulrennan 1993:34; Woodroffe *et al.* 1987:199) and remains speculative at present.

Estuarine evolution

The model of estuarine evolution and Holocene deposition demonstrated for the Alligator Rivers is said to apply broadly, with local variations, to other systems in the Top End (Woodroffe 1995:80). This includes the relatively minor systems of Darwin Harbour and Hope Inlet, also macrotidal estuaries (Michie 1988). The post-glacial process of marine flooding and sedimentary infill formed the present day Darwin Harbour, a drowned river valley system with associated rivers, tributaries and peninsulas (Semeniuk 1985). Since the Holocene transgression extensive tidal sandy mudflats have built up in the embayment and riverine zones of Darwin Harbour, and in the more completely infilled chenier plains of Hope Inlet (Michie 1988:38; Woodroffe 1995).

A number of palynological studies demonstrate that marked landscape change occurred on coasts east of Hope Inlet over the last few thousand years, and show that as estuarine systems developed, seaward mangrove communities were succeeded by more landward ones, and then by freshwater grasses and sedges (Russel-Smith 1985; Woodroffe 1995; Woodroffe *et al.* 1985). In a model of estuarine evolution and Holocene deposition for the Alligator rivers region, Clark and Guppy (1988:679-81) propose a process for the transition phase from mangrove to freshwater environments, in which, as sedimentation continued after sea level stabilisation, forests of *Rhizophora* decreased and *Bruguiera*/*Ceriops* and *Avicennia* increased. Diversity of mangroves and the tidal salt swamp vegetation also increased, creating a mosaic of environments further influenced by rainfall and vegetation. Channels formed on the tidally inundated flats as the surface built upwards and outwards to a level where high tide flooding became rare, and brackish and freshwater swamps formed in residual depressions on the landward edge of the coastal plains (Clark and Guppy 1988; Bird 1964:114).

Freshwater wetlands on blacksoil plains were established by different times around the region, around 1400 BP in the South Alligator area (Hope *et al.* 1985; Russel-Smith 1985), 1700 BP in the Magela floodplains (Clark *et al.* 1992a:48) and closer to home on the Mary River paleoestuarine and coastal plains by 2000 BP (Woodroffe and Mulrennan 1993:61). Similar processes can be posited for the Hope Inlet area, though only minor black-soil floodplains developed in this area, around the mid-reaches of the Howard River (Pietsch 1985). Chenier plains dominate the Hope Inlet estuary today, consistent with a westward increase in the relative areas of coastal plains, through marine mud accretion, for rivers along the van Dieman Gulf, associated with decreased fluvial input (Woodroffe *et al.* 1993:271-2).

Recently, a stratigraphic profile of Holocene sedimentation has been obtained for Shoal Bay, based on augur holes and sections through the chenier plain at Camerons Beach (Woodroffe and Grime 1999). Woodroffe and Grime (1999:305-6) determined the surface topography of this plain by "levelling along selected transects which were tied to benchmarks and related to Australian Height Datum (AHD), which approximates mean sea level". Mangrove wood found in a lower intertidal sand unit around 1.0-1.5m AHD, approximately two metres below the surface of the plain, was dated to 6130±100 years BP, corresponding with the time of sea level stabilisation at its present level in northern Australia.

At Camerons Beach shell samples obtained from both within the most landward chenier ridges and from a sandy mud stratigraphic unit revealed in a backhoe pit, approximately 1.0-1.3m AHD (above mean sea level) and 450 metres back from the current shoreline, dated to around 2300 years BP (Woodroffe and Grime 1999:306,313,318). Shell from the pit, the bivalve *Austriella sordida* (syn. *A. corrugata*) lives in mangrove muds (Richard Willan pers. comm.). *Insitu* beds of this bivalve occur at the base of a mud unit between 1.0m and 2.7m AHD (above mean sea level). The present day mudflats at Camerons Beach reach an elevation of approximately 3.0m and chenier ridges around 4.0m AHD (above mean sea level) (Woodroffe and Grime 1999:306,309). The stratigraphic sequence is interpreted as indicating that the Shoal Bay chenier plains and ridges formed more recently, and over a shorter time period than elsewhere in the region, with the first apparent major phase of progradation at around 2300 years BP (Woodroffe and Grime 1999:318-9).

Woodroffe *et al.* (1993; Woodroffe and Mulrennan 1993) provide evidence that progradation of the coastal plains of the South Alligator, Adelaide and Mary Rivers to the east was most pronounced between 5000 and 3000 years BP. They (Woodroffe *et al.* 1993:272) noted increased coastal sedimentation and later progradation continuing in a westerly direction from the Alligator Rivers system and episodic chenier ridge formation as the coastal plains continued to prograde (Woodroffe 1995:80).

The association between anthropological shell middens and chenier ridges

Availability of shell from source areas (shell beds), as required for the formation of shell middens, has also been identified as an environmental factor necessary for the construction of chenier ridges (Chappell and Grindrod 1984:222). As O'Connor and Sullivan (1994b:25) point out, this explanation indicates an environmentally controlled association between anthropological shell middens and cheniers. Chappell and Grindrod (1984:221-222) suggest that when the amount of mud deposition in intertidal and subtidal areas is high, productivity of shell beds is reduced. Subsequently, Clark *et al.* (1992b:100) argue that chenier ridges are built during times when the amount of sediment deposited from rivers is low, during periods of relatively aridity.

Chenier ridges, elongated deposits of sand and shell developed on mudflats or saline tidal flats, known as chenier plains, occur in tropical coastal locations (O'Connor and Sullivan 1994b:16), where there are alternating conditions of an abundant supply of mud and longshore transport of sand-sized sediments (Ostvos and Price 1979). These and a stable or recessive sea level are prerequisite conditions for chenier formation (Cook and Polach 1973; Todd 1968). Research in the wet-dry tropics of northern Australia suggests that cheniers and mudflats are formed by a complex interaction of storm impact, the mangrove fringe, embayment geometry, availability of shell, and mud deposition (Chappell and Grindrod 1984; Lees 1987; Sullivan and O'Connor 1993).

The recent study of the Camerons Beach chenier plain at Shoal Bay found that a major storm, Cyclone Tracy, was not a dominant factor in chenier ridge formation (Woodroffe and Grime 1999). The authors found evidence in support of Chappell and Grindrod (1984) that evolution of the chenier plain and ridges was a moderate process, driven by an internal dynamic. They argue that processes incorporating concurrent mudflat deposition and foreshore erosion occur at different points along the coast, with steady, ongoing landward migration of sandy shelly ridges (cf. Lee *et al.* 1994).

Woodroffe and Grime (1999) call for caution in the dating of chenier ridges using shell from within the ridge. They observed current winnowing of 670-1060 year-old shells from fossil beds and suggest possible incorporation of older shell into more recently formed chenier ridges. Lees (1991:12) discusses discrepancies in dates on shell in chenier ridges at Point Stuart, sometimes of thousands of years, which indicate the shell predates the ridge formation, and states that the youngest shell age is often assumed to be the best estimate of when the chenier ridge formed. But even though the age of shells within cheniers is known to predate deposition events, this difference in age is thought to be small, as Chappell and Grindrod (1984:208) argue, based on the evidence of comparable ages from intercalated middle and chenier layers in the Princess Charlotte Bay chenier plains.

Despite the ambiguities inherent in dating cheniers, which necessitate care in their interpretation, they can provide a useful benchmark for archaeological interpretation of use of a continually changing landscape, when combined with dating of shell from middens found in spatial association. Radiocarbon dating of chenier ridges at Camerons Beach, Shoal Bay, suggests two major episodes of chenier ridge formation. The first is thought to have occurred after progradation of the chenier plain began around 2300 years ago, when at least three ridges formed, and since around 1000 years ago, with at least four ridges being formed (Hickey 1981; Woodroffe and Grime 1999).

A narrow discontinuous chain of chenier ridges lies on the Hope Inlet plains, about five kilometres east of the chenier ridges at Camerons Beach and one kilometre east of the Howard River mouth. They run parallel to the mangrove fringed shoreline, approximately 50-100m inland and rise less than 50cm above the plains surface (Plate 2.1). The lower part of these ridges appears to be partially buried by muddy sedimentation subsequent to their formation (Sullivan and O'Connor 1993) (see below). Cultural shell mound deposits occur at intervals along this chenier ridge chain. The seaward proximity and low crest height of this ridge (see Chappell and Grindrod 1984:199, 206) suggest that this ridge formed within the most recent ridge building period, after 1000 years BP.

About 60km northeast of Hope Inlet at Point Stuart, Chambers Bay, dating of shell from chenier ridges by Clark *et al.* (1979), and subsequently Lees (1987), indicate that five chenier ridges formed in about the last 1270 years. On the lower Mary River coastal plains between Point Stuart and Hope Inlet, Woodroffe and Mulrennan (1993:42,55,66) dated two chenier ridges at around 2000 years and 1000 years and conclude that the chenier building phase began around 3500 years ago. South west of Darwin, in the Victoria River Delta, five chenier ridges formed between around 2000 years to 1200 years ago (Lees 1992a).

Current geomorphological evidence indicates widespread development of mudflats in Shoal Bay around 2300 years ago (Woodroffe and Grime 1999), which coincides with an infilling phase of a prominent paleochannel on the eastern plains of the Adelaide River, and a phase of infill on the Mary River to the east (Woodroffe *et al.* 1993; Woodroffe and Mulrennan 1993:63-4). Woodroffe and Mulrennan (1993:63-5, 98-9) hypothesise that the infilling occurred through a positive feedback process related to asymmetric flood and ebb tidal velocities and wet season fluvial flood discharge. In this process macrotidal estuaries infill with deposition of tidal sediment from seaward, but in channels where the catchment is too small, wet season flooding may not be large enough to reopen the channel and move sediment downstream again. The evidence in support of a marine source for estuarine infill and progradation of chenier plains in north Australian estuaries includes the distribution of foraminifera, and lack of correspondence between fluvial sediment volumes and catchment area (Kench 1999; Woodroffe *et al.* 1993). Material from the cheniers at Camerons Beach, Shoal Bay has been sourced to marine sediments such as the estuarine tidal flats at Hope Inlet and the coral reefs and sandbanks off Gunn Point (Michie 1984).

Links between cheniers and climate

Studies elsewhere have found that the interactive process between sediment-bearing longshore currents and tidal flow from estuary mouths is important to the growth of chenier plains, which are often found adjacent to estuaries that receive little or no sediment from the channels passing through them (Todd 1968). A marine rather than fluvial source for infill and muddy progradation does not suggest direct local climatic control, but is consistent with a long-term arid period in the sense that the estuary developed though predominantly marine deposition within broad-scale low fluvial input (Kench 1999:375; Lees and Clements 1987:312; cf. Lees 1992b; Chappell and Grindrod 1984; Woodroffe and Grime 1999).

Woodroffe and Grime (1999) assert that there is poor correlation between chenier-building phases in the North Australian region. Others (Lees and Clements 1987; Lees *at al.* 1990; Shulmeister and Lees 1992) argue that there is a statistically significant correlation between the timing of formation of ridges across the region, and link regional chenier ridge building across northern Australia with drier climatic conditions in the late Holocene. Lees (1987; 1992b) and Lees and Clements (1987) posit an overall arid trend after 4000 BP, interrupted by wetter periods between 3500 and 2800 years BP, 2100 and 1600 BP and a number of times in the last 1000 years.

Evidence in support of an overall decrease in late Holocene precipitation in northern Australia is provided by an increasing number of studies. This includes the geomorphic evidence based on dating of cheniers noted above (Clarke *et al.* 1979; Hickey 1981; Lees 1985; 1987), and sediment and pollen records from dunefields activated between 2600 and 1800 BP at Cobourg Peninsula, Cape Flattery and Cape Arnhem (Lees 1992b:6; Lees *et al.* 1990), and on Groote Eylandt (Shulmeister 1992; Shulmeister and Lees 1992). Shulmeister (1992:113) argues that the widespread occurrence of dune emplacement around this period, together with pollen data from the dunefield lake deposits on Groote Eylandt, suggests a climatically driven process across northern Australia, characterised by increased aridity and increasing climatic variability after 4000 BP, with an increase in precipitation again in the last 1000 years. Evidence also exists for a pan-Australian general trend toward aridity or arid periods during the late Holocene (eg. Ahmad 1996; Bowler *et al.* 1976; Singh 1983) and a northwards migration of aridity (Semeniuk 1996; and see O'Connor 1999a:46).

Global climate change

A growing number of researchers argue that increasingly variable climatic conditions seen in the late Holocene are explained by the beginnings of climate dominated globally by the modern ocean-atmosphere system, ENSO and Walker Circulation systems (Lees 1992b; McGlone *et al.* 1992; Rollins *et al.* 1986; Shulmeister and Lees 1995). Meteorologists use an index called the Southern Oscillation Index (SOI), a measure of relative atmospheric pressure over the central equatorial Pacific, and the western Pacific/eastern Indian Ocean, and related to climatic events all over the earth, to predict weather (Clark *et al.* 1992b). Extreme negative value of the SOI usually coincides with an anomalous warming of the east equatorial Pacific Ocean, called El Nino (Rasmusson and Carpenter 1982). In years of very low pressure off Darwin, the SOI is positive, and rainfall is high over northern and eastern Australia, India and northeastern Africa, while in years of prolonged and unusually high pressure off Darwin the SOI is negative, and droughts prevail throughout the region (Ridpath *et al.* 1991:210). Other global correlations occur with the ENSO phenomenon. For example, during El Nino western coastal South America experiences heavy rainfall, while eastern and northern Australia is drier than average and during warming periods a high number of severe floods in coastal Peru correspond with severe droughts in Australia (Clark *et al.* 1992b:105).

McGlone *et al.* (1992) examine records from eastern Australia, South America and New Zealand and conclude there is little evidence for ENSO pre-7000 BP and that the modern strongly ENSO-affected climate didn't begin until 3000 BP. Similarly, Shulmeister and Lees (1995) provide evidence from pollen records from Groote Eylandt, of environmental change from continuously increasing rainfall (effective precipitation) during the early Holocene, to reduced rainfall and increased climatic variability after 4000 BP, and argue that this change is due to the onset of ENSO. Under this system in tropical Australia, climate is more variable, and the monsoons less reliable than in the mid-Holocene. Instead of effective monsoons every year, there are cycles of strong and weak years. In strong years although the rainfall may be as high as during the period prior to 4000 BP, reliability of rain is reduced. In weak years there may be failure of the monsoon rains and drought (Shulmeister and Lees 1995).

Geographic, palaeoecological and archaeological research suggests some global correlation of climatic change expressed as extremes of floods and droughts in the late Holocene. Global climate change has been suggested around 2650 BP, caused by a temporary weakening of the monsoon circulation and resulting in a drier period in tropical areas (Van Geel *et al.* 1996). Geological records from north

America suggest that flood magnitudes increased after 1000 years ago, and especially between 700 to 500 years ago, in the Mississippi region, in response to modest climate change during the transition from the Medieval Warm Interval to the cooler Little Ice Age (Knox 1993:431). Large floods are recorded about 400 years ago in California around the same time as floods or droughts elsewhere around the world (Schimmelmann *et al.* 1998).

In Australia sporadic major floods have been dated at irregular intervals back to 4000 years in the Kimberleys (Gillieson *et al.* 1991; Wohl *et al.* 1994), and seven large floods in the last 1200 years in northeastern Queensland (Wohl 1992), including some around 400 years ago. Sedimentary flood records indicate nine major floods on the Katherine River in the Northern Territory back to 450 years BP (Baker and Pickup 1987) and eight on the Finke River in arid central Australia back to 900 BP (Baker *et al.* 1987; Pickup *et al.* 1988). This evidence is consistent with the proposition of a more arid period of reduced wet season precipitation in northern Australia increasingly interspersed by sudden episodes of high rainfall and extreme floods (Lees 1992b; Lees *et al.* 1990). Significant climatic variability on time scales of centuries, related to the ENSO phenomenon, is argued to have occurred in the Alligator Rivers region of the north Australian coast during the last 2000 years, based on the stratigraphic record from the Magela plain (Clark *et al.* 1992b:108-9).

For the purposes of this study it is the implications of environmental change for local conditions experienced by foraging peoples that is important (Jones *et al.* 1999). Episodic, unpredictable environmental phenomena, such as torrential rains, flooding and droughts, associated with ENSO events, disrupt both marine and terrestrial habitats and therefore resource availability and predicability (Moore 1991:30; Sandweiss 1996). Of greater impact may be changes over more intermediate time spans of decades to centuries, in estuarine morphology through sedimentation resulting in expansion and decline of habitats, particularly for mangrove and mudflat floral and faunal communities utilised by past human communities (Kench 1999:372; Woodroffe *et al.* 1985).

Present day environment: landforms, flora and fauna

Darwin Harbour

Present day Darwin Harbour is a broad shallow embayment, indented with many small promontories, the largest being Middle Arm Peninsula between the major arms of the harbour, East Arm and Middle Arm (Figure 2.2). The topography of the Darwin area is generally described as dominated by undulating surfaces cut across steeply dipping Proterozoic and flat-lying Cretaceous rocks less than 30-40 metres above sea level. Darwin city was built on this surface which has been eroded at parts of the coastal edge to form low cliffs up to 15 metres high with intermittent sandy beaches (Nott 1994; Woodroffe and Grime 1999). Outcrops of rock and cobbles within Quaternary sediments that would have been suitable for stone artefact manufacture by past inhabitants occur locally in the hinterland around Darwin Harbour. Quartz and quartzite are the most common materials. Outcrops of other raw material, such as sandstone and chert are occasionally observed (Pietsch 1983; Pietsch and Stuart-Smith 1987:21).

Darwin Harbour acts as a large estuary during the wet season, receiving quantities of freshwater from the main rivers, the Elizabeth, Darwin and Blackmore Rivers, and the many small tidal creeks which empty into it (Semeniuk 1985:85). Sources of freshwater in the Darwin area include numerous ephemeral freshwater soaks, swamps, creeks and shallow drainage lines that dissect the hinterland, and some more permanent waterholes and watercourses that persist through the dry (Wood *et al.* 1985). These include Berry Springs and the large Marrara, and Noogoo swamps and Knuckeys, Ironstone and Marlows lagoons, all situated inland within a few kilometres of the estuarine wetlands of the harbour coast. Heavy rainfall during the wet season results in fresh groundwater being available at shallow depth in most areas throughout the year (Story *et al.* 1969:17).

The coastline around most of the harbour consists of a wide fringe of low closed mangrove forests, which merge into extensive tidal mudflats formed from intertidal marine alluvium, mud, clay and silt (Pietsch 1986). Mangroves in the Indo-Pacific generally follow a zonation pattern of seaward *Avicennia marina*, a *Rhizophora stylosa* zone, a *Bruguiera/Ceriops* zone and a landward mixed mangrove zone dominated by *Avicennia marina* (Macnae 1968). Studies of the 20,400 hectares of mangroves around Darwin Harbour, the largest single stand of mangroves in the Northern Territory (Woodroffe 1995) show more complex variations of this pattern for both the Darwin Harbour and Hope Inlet areas. At Darwin Harbour a *Sonneratia alba* zone occurs at the most seaward edge, and *Rhizophora stylosa* occurs at the seaward edges of the main tidal flat and along tidal creeks, sometimes found with *Avicennia marina, Bruguiera sp., Camostemon schultzii* and *Aegiceras corniculatum. Bruguiera/Ceriops* occurs on the main tidal flat and mixed species low closed forests of dominant *Ceriops tagal*, with other species such as *Lumnitzera racemosa* and *Avicennia marina* (Woodroffe and Bardsley 1988; Brocklehurst and Edmeades 1996).

At the junction between mangroves and hinterland occasional sandy, shelly chenier ridges and small areas of saltflats occur on several of the harbour promontories. The hinterland landforms that rise above the harbour basin generally comprise dissected upland terrain, low ridges and hills vegetated with eucalypt woodland and patches of monsoon vine thicket typical of the northern monsoonal coast, and intervening alluvial flats, with shrubs and grassland (Wood *et al.* 1985).

Vegetation in the upland areas is mostly open and mixed forest dominated by eucalypt species such as *E. bleeseri* (Smooth-stemmed Bloodwood) or *E. tectifica* (Northern Box), with *E. tetradonta* (Stringybark), *E. miniata* (Darwin Woolly Butt) *E. foelscheana* (Fan-leaved Bloodwood) and *Erythophleum chlorostachys* (Ironwood), and an understory of shrubs and low trees such as *Terminalia ferdinandia* (Billy-goat plum) and *Cycas armstrongii.* (Zamia or cycad palm). Patches of monsoon vine forest occur in wet well-drained areas, with species such as *Dioscorea transversa* (long yam) and *Sterculia quadrifida* (Bush peanut). On the low swampy alluvial flats is woodland with *E. polycarpa* (Swamp Bloodwood), *E. papuana* (Ghost Gum) and *Livistona humilis* (Fan or cabbage palm) and stands of *Melaleuca* (paperbark) and *Pandanus* (Wood *et al.* 1985). Grassland is extensive in low lying areas subject to seasonal inundation, dominated by *Eriachne* spp. and *Themeda australis* in better drained sites and perennial sedges such as *Scleria poaeformis* in poorly drained heavy clay soil areas (Fogarty *et al.* 1979; 1984). Expanses of waterlilies species, *Nymphaea* spp. and small patches of the water chestnut *Eleocharis* spp., exploited by both magpie geese and Aboriginal people, are reported on semi-permanent lagoons that last well into the dry, such as Knuckeys Lagoon (Hodgson 1997:28; Parks and Wildlife Commission 1999), but do not appear to have been as extensive as at Hope Inlet.

Hope Inlet

Today Hope Inlet is a short prograding coastal inlet at the southern end of the open sandy shoreline of Shoal Bay (Figure 2.2). At the southwest extremity of Shoal Bay are tidally inundated marine swamps, including the large Leanyer swamp, and east of these Kings Creek and the small Howard River flow northwards into the shallow seas of the inlet on either side of Howards Peninsula. The discontinuous chenier ridge on the Hope Inlet plain, discussed above, lies just over a kilometre east of the Howard River mouth (Plate 2.1).

In this area water tables remain high even in the dry, and the lower horizons of soils on the coastal plains are perennially waterlogged (Williams 1969:75). In some coastal areas fresh water can be found in depressions between sand dunes behind the mangroves (Christian and Stewart 1953:117). Numerous minor waterholes and lagoons occur inland, including the larger Fisher, Lyons, Korebum and Limul Limul lagoons with associated freshwater swamplands, and seasonal creeks and shallow drainage lines that dissect the hinterland (Wilson and Bowman 1987; Wood *et al.* 1985). Rainforest communities occur at Howard Springs and Holmes Jungle, associated with a perennial water supply (Fogarty *et al.* 1979).

Dense mangroves hundreds of metres wide fringe the seaward edge of the coastal plains, which consist of seasonally inundated low level muddy sandy saltflats extending about 12km north-south and two to five kilometres east-west. At the seaward edge *Bruguiera parviflora*, *Rhizophora stylosa* and *Avicennia marina* dominate the low closed mangrove forests. Further from the sea the forests become lower, with *Ceriops tagal*, *Lumnitzera acemosa* and *A. marina* dominant. Mangroves also line the tidal channels that dissect these supratidal muddy saltflats.

Toward the landward edge of the saltflats are grey cracking clays, seasonally inundated by freshwater for months. Mixed vegetation communities occur in these wetland areas, in complex patterns associated with micro-topography and water depth. These include grassland/sedgeland communities dominated by *Schoenoplectus litoralis* and *Cyperaceae* spp. reeds in transitional saline/freshwater areas, *Eleocharis* spp. (water chestnut) and *Oryza* spp. (wild rice) in flooded areas and *Ischaemum rugosum*, *Eriachne burkitti* and *Sorghum* spp. grasses in more elevated areas. Perennial freshwater swamps with *Typha orientalis* rushes or *Scleria poaeformis* sedges occur in broad drainage channels between the hinterland projections. Paperbark forests of *Melaleuca cajuputi* and *M. leucadendra* and of *Pandanus spiralis* fringe the grasslands and drainage lines (Wilson and Bowman 1986, 1987). The adjoining hinterland uplands, a laterized plateau of Cretaceous sediments, slopes down gently at angles less than 10° toward the coast and juts out as a succession of small promontories onto the saltflats. There is a very limited range of locally occurring outcrops of rock in the Hope Inlet area. The only sources of raw materials suitable for stone artefact manufacture observed were porcellanite, a fine-grained sedimentary rock (Pietsch 1985:14) on the lower slopes of the hinterland, and one grey/blue quartzite outcrop on the saltflats at the southern end of Shoal Bay.

Baker (1981:16) identified the transitional zone of marginal woodland at Chambers Bay as one of high floral and animal diversity and density, and thus of considerable economic importance to Aboriginal people, and this applies to the similar landscape at Hope Inlet. Vegetation on the transitional slopes includes large stands of the cycad palm *Cycas armstrongii*, patches of monsoon vine forest, semi-deciduous closed forest of *Melaleuca cajuputi, M. viridiflora* (Paperbark), *Bombax ceiba* (Kapok tree) and *Acacia auriculiformis* (Black wattle), low open forest with *Lophostemon lactifluus* (Swamp box) and *E. papuana* (Ghost Gum), and open forests with *Eucalyptus miniata* (Darwin Woolly Butt) and *E. tetradonta* (Stringybark). A comprehensive list of plant species in the area is given in Wilson and Bowman (1986; 1987:169). There is now documented evidence on contemporary Aboriginal use of most of the plant resources from all habitat types such as those listed above, in monsoonal northern Australia (eg. Russel-Smith 1997).

Faunal resources

Despite a dearth of scientific research on flora and fauna in the Darwin region wetlands, a few superficial surveys

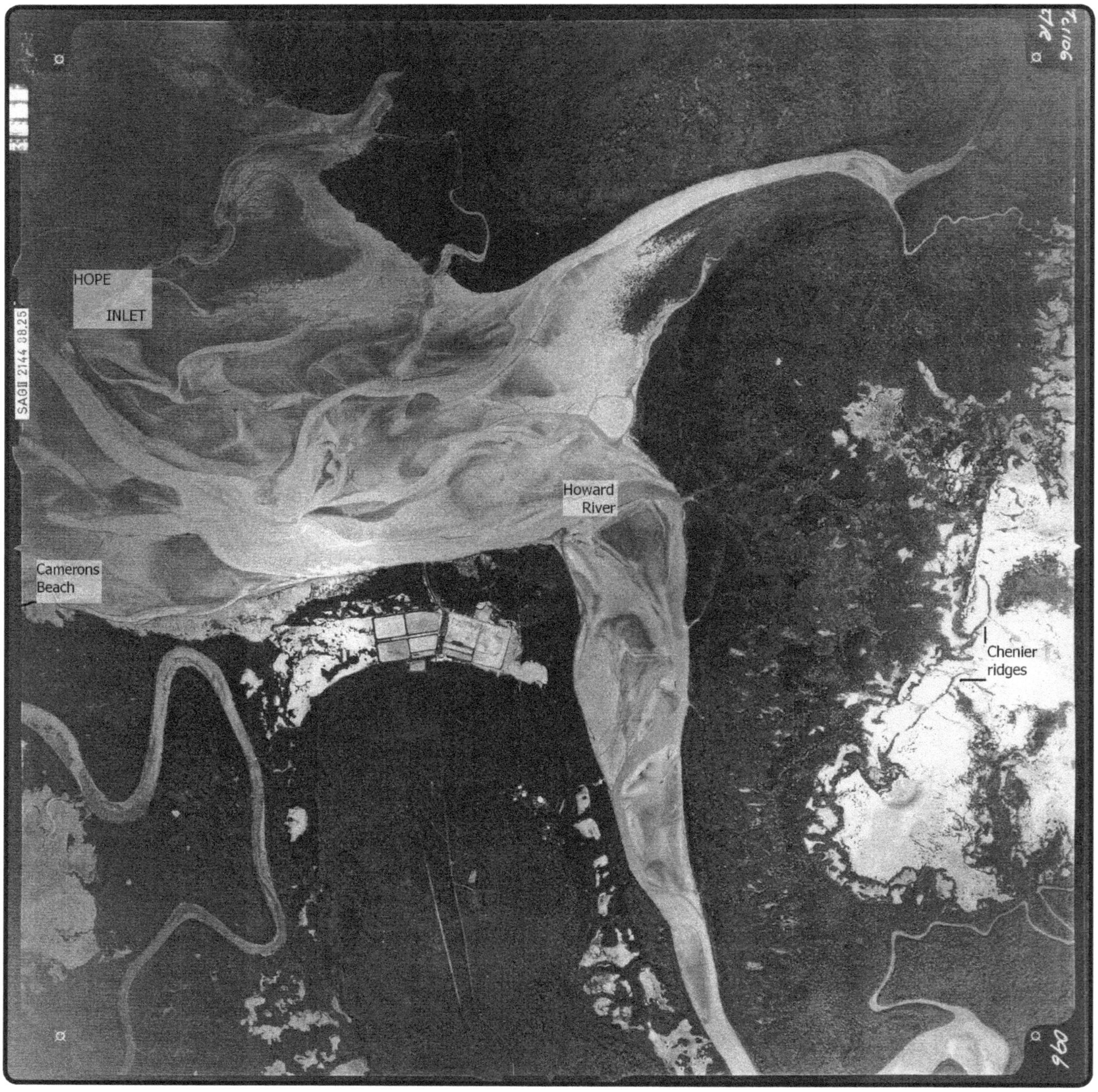

Plate 2.1 Map showing location of Hope Inlet chenier ridge

undertaken give some indication of the numbers and diversity of faunal resources, particularly in the mangroves, swamps, monsoon forest and woodlands of Howards and Shoal Bay peninsulas (Woinarski *et al.* 1988). Surveys in habitats such as the Leanyer swamp area and Howard River Nature Park recorded hundreds of bird species and a range of mammals, reptiles, fish, insects and invertebrates, as well as feral cats, dogs, buffaloes and pigs (Cities Commission 1974; Conservation Commission of the Northern Territory 1990a; Gow 1978). Buffaloes have largely been eradicated through Government programs, but pigs remain numerous in the Hope Inlet area. Weeds are widespread, including the horehound *Hyptis*, presumably spread by the buffalo previously stocked by Koolpinyah Station (Conservation Commission of the Northern Territory 1990b; personal observation). The common agile wallaby (*Macropus agilis*) is found on the sub-coastal lowlands and floodplains, and the less abundant antilopine kangaroo (*Macropus antilopine*) is found in coastal woodlands (Williams and Newsome 1991).

During fieldwork at Hope Inlet we observed live crabs, fish, molluscs, snakes, frill-necked lizards, skinks, goannas and wallabies, and numerous birds including magpie geese, brolgas, kingfishers, sea eagles and an emu, as well as large centipedes in the excavation pit of a saltflats mound! Numerous and diverse fauna are recorded as frequenting or living in tropical mangroves (Hutchings and Saenger 1987), including a variety of bird species and mammals such as native rodents, bandicoots, possums and flying foxes (Frith 1973; Williams and Newsome 1991; Vardon and Tidemann 1999). Reptiles recorded include crocodiles, snakes and goannas such as the mangrove monitor (*Varanus indicus*), for which mangroves is a primary habitat. Crustaceans and

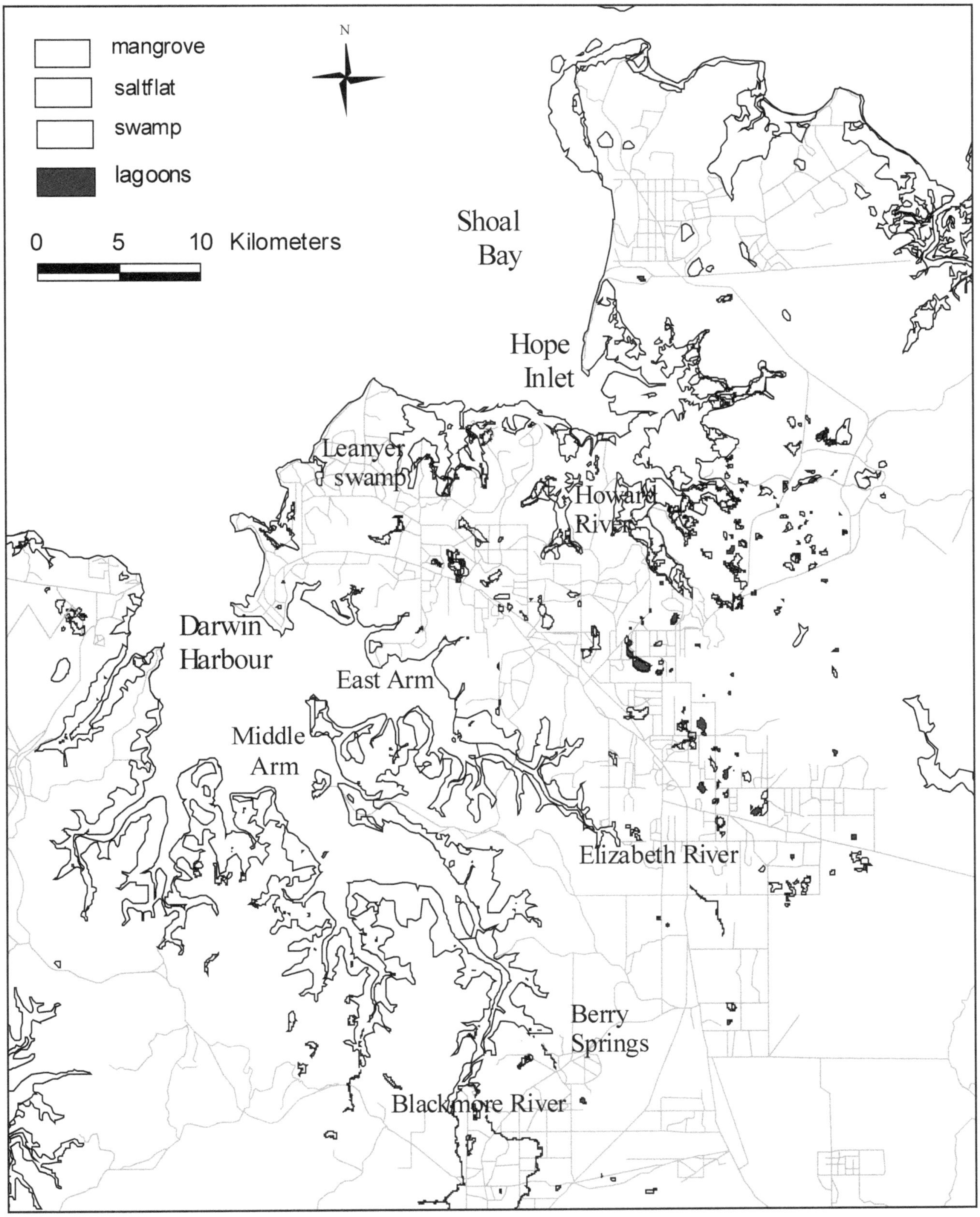

Figure 2.2 Map of the Darwin region

molluscs living in the mangroves include the well-known Mud crab (*Scylla serrata*), the mud lobster (*Thalassina anomola*) and a variety of other crabs, and molluscs. Today the most common macro-molluscs recorded are the mangrove gastropods *Telescopium*, and different taxa of *Terebralia* and *Nerita* (Caldwell 1983:76; Crowe 1997; Smith *et al.* 1997), as well as *Littoraria* and *Chicoreus* (Hanley 1993; Hutchings and Recher 1983; Knuckey 1999). A list of extant macro-mollusc taxa recorded in the Darwin region is given in Table 2.1 (overleaf).

Fish are numerous in the Darwin Harbour and Hope Inlet estuaries, with over 400 species recorded (Larson 1988; Larson and Martin 1989; Larson and Williams

1997). In addition to resources from the mainly estuarine environment, many small nearshore rocky and coral reefs in Darwin Harbour (eg. East Point, Weed Reef, Channel Island) and outside the harbour near Shoal Bay (eg. Lee Point, Gunn Point) support a large number of shrimp, fish, crab and sponge species (Bruce 1988; Caldwell 1983:78; Hooper 1988; Larson 1988; Larson and Williams 1997; Martin 1978; Wolstenholme *et al.* 1997). Studies on reef foraminifera in some of the older Shoal Bay cheniers indicate that reefs had become established in the area around 2400 years ago (Michie 1984; 1988).

Resources and seasonality

Seasonality of climate has a marked influence on the ecology of plants and animals in these monsoonal wetlands (Haynes *et al.* 1991:76). This is clearly illustrated by Nix and Kalma's (1972:74) "growth indices" for vegetation in Darwin, which shows the dramatic difference in growth between the dry and wet seasons, with little plant growth between the months of May and October (and see Jones and Bowler 1980:17).

Seasonal availability of resources is seen in the seasonal habitats of the sedgelands and transitional margins, where snakes, including the olive python (*Bothrochilus olivaceus*) and Water python (*B. fuscus*), are common in the mid-late dry, hunting frogs and rats. Other species such as Northern snake-necked turtle (*Chelodina rugosa*), the Freshwater snake (*Styporhynchus mairii*) and Mitchell's water monitor (*Varanus mitchelli*) are found in the flooded paperbarks during the wet (Braithwaite *et al.* 1991). Seasonal migration of fish is common in the estuarine systems, inhabited by many small fish species (Bishop and Forbes 1991). Large numbers of fish enter mangrove forests at high tide and move back to deeper waters as the tide falls. Barramundi, *Lates calcarifer*, the best researched fish in the Territory, and Threadfin salmon (*Polynemus sheridani*), migrate from coastal waters and rivers to supra-tidal swamps, used as nursery areas for spawning at the start of the wet. Juveniles inhabit the swamps and floodplains in the wet and move upstream as the dry approaches (Davis 1985; Storrs and Finlayson 1997). Fork-tailed catfish, the dominant fish biomass in lower reach habitats, spawn annually at the start of the wet and may migrate into estuarine creeks at this time (Rimmer and Merrick 1983).

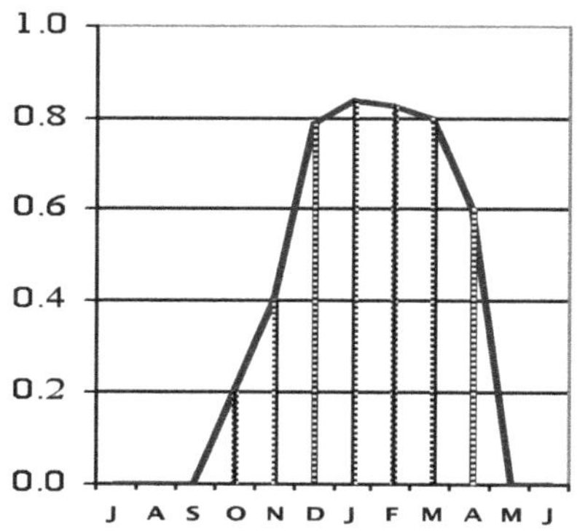

Figure 2.3 Graph showing seasonal vegetation "growth indices" for Darwin (after Nix and Kalma 1972:74)

The Hope Inlet wetlands have been identified as an important wet season breeding habitat for waterfowl, and dry season refuge and feeding area for magpie geese and other geese, ducks and birds (personal observation; Shurcliff 1993; Conservation Commission of the Northern Territory 1990b). Large flocks of migratory birds including the most abundant and important in contemporary and past Aboriginal waterfowl hunting, magpie geese (*Anseranas semipalmata*) or *Gakingga* as the Larrakia call them, are attracted to the exposed intertidal mudflats at low tide and the swampy marshes (personal observation; Hodgson 1997:11; Russel-Smith *et al.* 1997:176; Shurcliff 1993; Storrs and Finlayson 1997; Whitehead *et al.* 1988).

The magpie goose seasonally and annually moves around geographically dispersed wetlands, flying to one area in the wet for breeding, building floating nests in swamps of mixed sedge and wild rice, feeding on the rice and other seeds. In the dry they move to persistent swamps and waterholes, feeding on sedge tubers and seeds and other grasses and water lilies (Conservation Commission of the Northern Territory 1990a; Morton *et al.* 1990; Tulloch 1985; Whitehead and Tschirner 1991; Whitehead *et al.* 1988). The Mary and Adelaide Rivers have been identified as an important nesting areas and the Alligator Rivers as feeding areas (Bayliss and Yeomans 1990; Storrs and Finlayson 1997:20). At Hope Inlet the dry-season refuge and feeding area incorporates the well-used Howard Swamps hunting reserve, some 16km^2 encompassing freshwater sedge swamp, saline mudflats, *Melaleuca* swamp forests and dry open woodlands (Whitehead and Tschirner 1991:101). Today some 3000 shooting permits are issued annually for the hunting season from August to November, mainly taking magpie geese concentrated on small areas of contracting perennial swamps, as Indigenous hunters were observed to do without the benefit of guns at the time of first European contact (see below).

Effects of extremes in rainfall (high or low) on resources

The truism: "There is no such thing as a normal season" (Bauer 1984:3) sums up the variable nature of the north Australian monsoonal climate. Variability in intensity and duration of wet season rainfall in the Darwin region is known to have significant short and long term effects on both plant communities, and animal populations (Taylor and Tulloch 1985:291; Wilson and Bowman 1987). Seasonal resources may diminish with failure of the monsoon rains and drought in response to the ENSO phenomenon (Shulmeister and Lees 1995). Taylor and Tulloch (1985:290) list some of the studies that have recorded the dramatic effects of extremes in rainfall in

Family	Taxon / Common name	Habitat	Reference for recorded taxa in Darwin area
Arcidae	*Anadara granosa* roughbacked cockle	Mudflats/sandy mud	Blackburn 1982
Veneridae	*Marcia (Tapes) hiantina*	Sand, sand/mudflats	Hodgson 1997
Melongenidae	*Volema cochlidium (syn. Volegalea wardiana)*	Sand/mud, mangroves	Blackburn 1982; Knudson 1997
	Syrinx aruanus false trumpet shell	Sand/mud	Blackburn 1982
Volutidae	*Melo amphora* baler shell	Sand/mud	Blackburn 1982
Cymbiinae	*Cymbiola flavicans*	Sand/mud eggs attached to mangrove roots	McMichael 1960
Placunidae	*Placuna placenta* windowpane oyster	Sand/mud	Blackburn 1982
Potamididae	*Terebralia semistriata* mud whelk	Mangroves (*Avicennia, Brugiera* and *Ceriops*)	Crowe 1997; Smith et al. 1997
	Telescopium telescopium long bum	Mangroves (*Rhizophora*)	Blackburn 1982; Caldwell 1983; Crowe 1997; Smith et al. 1997
	Cerithidea sp.	Mangroves roots and mud (*Avicennia, Brugiera*)	Caldwell 1983
Ellobiidea	*Ellobium aurisjudae* and *Cassidula angulata*	Mangroves mud (*Rhizophora, Brugiera*)	Blackburn 1982; Wells 1997
Neritidae	*Nerita sp.* periwinkle	Mangroves roots/rocks	Blackburn 1982; Caldwell 1983; Knudson 1997
Muricidae	*Chicoreus capucinus* (syn. *Naquetia permaestus*)	Mangroves roots and mud	Blackburn 1982
Corbiculidae	*Geloina (Polymesoda) coaxans* mud mussel	Mangroves mud	Hodgson 1997
Ostreoidea	*Saccostrea echinata* oyster	Mangroves roots/rocks	Caldwell 1983

Table 2.1 A list of recorded macro-mollusc taxa in the Darwin region

the Darwin-Katherine region. Very low rainfall in some years reduces the growth of rice and native grasses, and result in failure to breed or high mortalities in magpie geese or other waterbirds. Extremes of high rainfall may cause rapid flooding of floodplains and high mortalities of native rodents, and variation in rainfall is also associated with abundance of prawns and fish (Taylor and Tulloch (1985:291). And of course it is well known that record monsoonal rains temporarily destroy shell beds, the source of molluscan resources, as Meehan (1982a:163) reported on the coast of the Blyth River region in the 1970s.

Yet Meehan (1982a:163) also noted that the gap in mollusc resources was simply filled by the local Anbarra group through increased harvesting of yams and other vegetable foods from the jungle patches, which matured earlier in response to the prolonged rains. They also hunted wallabies, freshwater turtles and goannas, while fish continued to form a substantial part of the diet. The implication of this is that although the more visible remains of mollusc gathering may cease, that does not mean that people no longer exploit other resources that are more favourably affected by increased rains. Yet as Jones (1980b:111-2) describes for the coastal Gidjingali in Arnhem Land in the 1970s, coping with the drought side of environmental variability may have been much more difficult for past foragers, and "inherent in the Aboriginal economic system, and a key determinate in population size and the structure of seasonal territorial movements" (Jones 1990:41).

> The flow of seasons dominates Gidjingali life: it governs the foods eaten,...plans for the hunt,... location of camps, and...timing of ceremonies. The flow is, however, not symmetrical, for Barra (the northwest monsoon) dominates the seasons. With a heavy wet season - although uncomfortable and sometimes, if cyclones occur, dangerous - there is an abundant early dry season when the waters recede. Then, even in the late hot dry season, fruit can be found, wells if dug deep enough still flow, and swamp bottom muds are damp and easy to work while digging for roots. A small Wet means a harsh dry season, with a tendency for even the best waterholes to become brackish or salty, for the plains to become brick hard, and for hunger to stalk in the anxious Barparanga (build-up) months. The end of the Dry was formerly the stress period in Gidjingali society (Jones 1980b:111-2).

Such ethnographic studies indicate that forager resource procurement strategies must operate within the parameters of seasonal resource availability of an area. This study therefore discusses the relationship between environmental resources and the human population in the light of ethnographic and historic observations in the period following European contact.

3 Ethnographic and historic records of Indigenous coastal economies in the tropical Top End

Introduction

Problems with using ethnography or ethnohistory to interpret archaeological material are well recognised in archaeology (eg. Davidson 1988; Gregory 1998:283-5; McBryde 1979; Murray 1988). One of the problems lies in assumptions of continuity between people's past and present behaviour, over a scale of time which may span thousands of years. Another is the bias brought to written accounts by the observer, depending on their level of knowledge and experience of what they were observing. Biases also enter the ethnohistorical record through the particular focus determined by the observer's own personal interests, or also in the case of anthropologists, in the theoretical paradigm of the day (Baker 1981:48; Gregory 1998:55). Gregory (1998:53-6) has summarised a number of approaches taken by archaeologists in using ethnographic and historic data.

Ethno-historical information is discussed in this study partly to offset the notional disappearance of Indigenous groups here who were dispersed from the late 19th century with the white focus of settlement, and accentuate the fact that pre-contact Indigenous economies did indeed exist in the Darwin region (Read 1991; Wells 1995:17 and see Chapter One). In addition, the approach taken in this study uses analogies derived from ethnographic and historic observation of contemporary practices as ideas on possible past human behaviour to be used for comparison against the types of archaeological material present (see Chapter Nine). Differences as well as similarities are considered to try to avoid "uncritical projection of ethnographic patterns identified as traditional onto the distant past" (Gould 1980; Huchet 1991; Wylie 1985; Stahl 1993:252). It is also useful to remind us that the archaeological record tells an incomplete story, since not all activities observed in ethnographic studies leave readily visible archaeological remains.

Observations on the practices of Indigenous groups in the Darwin area date from the late 1800s and are limited in number and fairly general, though the major sources, from Herbert Basedow, Paul Foelsche, Thomas Parkhouse and Baldwin Spencer, can be assumed to be relatively reliable. Basedow, a physician, and Spencer, a biologist, were the first two Chief Protectors of Aborigines in the Northern Territory after settlement (Austin 1997:20). Paul Foelsche was the Inspector of Police and lived in Port Darwin for ten years from 1870 (Curr 1886:250). Parkhouse was the Accountant and Paymaster of South Australian railways at Port Darwin for some years. Such ethno-historical sources describe the Indigenous inhabitants of the Darwin area as the Larrakia (Foelsche 1882,1886; Tindale 1974), one of a group of non-Pama-Nyungan language speakers of the coastal and riverine Top End (McConvell 1997), who were noted by Parkhouse (1895) to be closely allied and intermarried with the Wulnar people occupying the territory to the east and west of Adelaide River.

Cultural contact

According to historic records, from the 16th century Dutch, Portuguese and Spanish ships navigated the northern coast of Australia, but actual contact with the Indigenous inhabitants was negligible. From at least 1720 AD, and possibly as early as 1100 AD (Macknight 1976:153-4,158; McIntyre 1984:51) Aboriginal people along the coasts and islands north of Darwin were in regular contact with the Macassans (Macknight 1986; Mitchell 1994a). The Macassan trepang fishers however, did not frequent the Darwin coast, as Port Darwin was not a suitable area for harvesting trepang. When the government officials wanted to impose a licensing system in 1882 for Macassans harvesting trepang, the fishermen complained it was a hardship to have to visit Port Darwin to procure licences (Macknight 1976:49; 108). The only archaeological evidence of Macassan visitors to the coastal area around Darwin is a purported Macassan well at Gunn Point (Conservation Commission n.d.). Thus it is likely that contact between the Indigenous groups in the Darwin area and the Macassans would have been of an indirect nature.

The Larrakia would have known of the Europeans long before settlement was set up in their traditional country, due to the advance guard diffusion of ideas, disease and material culture, well documented in the literature (Urry 1979:3). Some Larrakia were part of a Wulna group of up to 200 who in 1866 camped close to the Escape Cliffs settlement at the mouth of the Adelaide River for the wet season, and exchanged fresh food and their labour and sexual favours for sought after European items (Ritchie 1998:5,6; Manton 1866; Young 1866). Escape Cliffs was one of four previous, failed European settlement attempts on the northern coast, following in the steps of Fort Dundas, Raffles Bay, and Port Essington that had gone before. It was abandoned in 1867, two years before the settlement of Palmerston, or Darwin as it is now known, was established in 1869. The extent of influence of previous contact was shown by events such as the Larrakia in the early days of the settlement of Palmerston singing "My Old Kentuckey Home", which "they had picked up from the whites at Escape Cliffs" (Reid 1990:35).

Earlier British records of Indigenous inhabitants of the Darwin area were made during a surveying expedition in H.M.S. Beagle in September 1839, at Hope Inlet and Shoal Bay, named thus by John Lort Stokes. The records show that members of Stokes' party who were collecting shells from "the extremity of a long flat", met an Indigenous group of families totalling 27 people at Shoal Bay. This group showed one of the officers of the crew who said he was thirsty a small well near the beach. The crew found the water too brackish, but reported that the Aboriginal people drank from it (Stokes 1846:12). Fifteen days later, the same group, with an additional five men, appeared where the crew were camped at Emery Point, Darwin Harbour. They expressed great delight at a deep well dug by the Europeans, and drank deeply from it. Such reports hint at scarcity of water at this time of the year –the late dry season- and indicate the importance of water resources to the local inhabitants.

Stokes noted that there was influenza of some type raging through the group (Stokes 1846:18-21). Foelsche estimated the population of the Larrakia "tribe" to be around 500 persons in the late 19th century (Curr 1886:251). Reports of a smallpox epidemic prior to the settlement at Escape Cliffs in 1864, which killed so many people that they could not bury them all (Foelsche 1882), suggests that the population in earlier times may have been much higher than this. Populations of up to 1500 people have been estimated at contact for the coastal plains of the Adelaide and Mary Rivers to the east (Ritchie 1998:4).

Most information on the Indigenous inhabitants of the Darwin region comes from historical records and ethnographic accounts dating from the late 1800s, well after the effects of the previous failed settlements and Macassan contacts would have been felt. Such records indicate that the sphere of knowledge on Larrakia practices was confined to what was seen in small groups of people who, like the Wulna at Escape Cliffs, set up camp close to the European settlement. The main Larrakia camp, at the corner of what are now central Darwin city streets, comprised several circles of huts or "wurleys" described as egg-shaped bark constructions, each circle formed with three to seven huts. Another camp consisting of three rings of huts containing Wulna related to the Larrakia, was situated 150 yards from the main camp (Parkhouse 1895). Outside of this sphere was the unfamiliar bushland (to Europeans), and even in the 1930s Larrakia said to be "detribalised" were recorded as still determinedly and periodically "going bush" for unknown activities (Barclay 1937). But the camping of local Aboriginal people on the fringe of the white Palmerston settlement, along with other Indigenous groups moving into the area, and their unequal relations with whites, reflects the already extensive disruption to their existing socio-economic system, exacerbated by sought after tobacco, alcohol and opium and unintended disease (Bauer 1964; Eylmann 1994; Rowley 1970).

Clearly, previous settlements and the concentration of European settlement in the Darwin area from the late 1800s had a substantial impact on the cultural and physical landscape within which the Indigenous people had lived their lives in the past. Consequently, the ethnohistoric literature from the early years of European settlement provides only limited information on traditional activities in the Darwin region at the time (eg. Basedow 1907; 1925; Foelsche 1882; Parkhouse 1895). Nonetheless, this information is useful as an aid to interpreting what remains of the archaeological record and understanding the nature of coastal occupation in the past, so is considered in this study.

Indigenous knowledge and use of resources in the Top End as recorded historically

Resource procurement

Early records suggest that at the time of first European settlement faunal resources were much more abundant in the Darwin area than in the present day (Grey *et al.* 1988:165; Knight 1876:179). The surveyor George Goyder reported that Shoal Bay was an area rich in molluscs, fish and birds (Kerr 1971:108). In the early days of European settlement the ethno-historic literature describes fishing as a principal occupation in the Darwin area, with molluscs and crabs observed to contribute daily to meals (Basedow 1907:23,26; Foelsche 1882:12-14). Larrakia women are reported fishing in shallow water using hand nets made from spinifex shoots and fibre twine or tree bark (Basedow 1907:24, 1925:130; Foelsche 1882:13). A trident spear with a shaft of reed or bamboo and three barbed prongs of ironwood attached with resin or paperbark was said to be favoured by the Larrakia. These were used to spear fish and stingrays in shallow water on the beach or off reefs and rocks and to spear fish in deeper water from the bow of a canoe.

Canoes were also used for fishing and hunting of dugong and turtles. (Basedow 1907:22-25, 1925:131,162-4), to transport items such as turtles and molluscs (King 1969:89; Reid 1990:38) and to travel from one section of the coast to another. Eylmann (1994:33) reported Larrakia travelling 30km by bark canoe from Palmerston to Southport in a day, and that they went out to the open sea for days in these boats. Reports describe the use of two types of canoe, the sewn bark canoe and dugout canoe, around Darwin Harbour (Basedow 1925:164; Holland 1976; Kerr 1971:68) and on the Adelaide River (Hordern 1989:164).

Ethnohistoric reports indicate that the Indigenous people employed extensive knowledge of the environment, animal behaviour and migration patterns and a variety of technologies in hunting, fishing and gathering. Kangaroos were ambushed along well-used paths to water, and magpie geese were captured from hides built in preparation of their annual arrival (Basedow 1907:19-27; Foelsche 1882:12-14). To ambush emu, Larrakia hid themselves in trees they knew bore seeds that attract emu. Hunters would also entice birds within reach of their throwing stick by imitating their calls. Magpie geese, whistling duck, cockatoos, plovers and many other birds were captured this way (Basedow 1907:21-2, 1925:138-9; Foelsche 1882:13). Fish and crabs were procured from reef pools or from constructed fish traps using the tides, or from rivers creeks and waterholes by spearing, netting or using certain poisonous barks or leaves to stupefy the fish (Basedow 1925; Foelsche 1882). Traps were built of either stone such as the fish trap at Nightcliff (Hodgson 1997:38), or of wood, (Basedow 1907:23, 1925; Foelsche 1882, 1886:255). Fish traps built of wood and wire mesh said to be in traditional Aboriginal style were recorded at Mindil Beach and other beaches in the early 1900s (Hodgson 1997:33-4).

Utilisation of floral resources is another aspect of Aboriginal economy that unfortunately does not survive well in the archaeological record. Yet ethnographic evidence shows that in the past, plants contributed substantially to food and material culture necessary for the maintenance of Larrakia lifestyles (Basedow 1907, 1925; Hodgson 1997), as for other Aboriginal groups (eg. Meehan 1982a:150-51). Continued historical use of many of the natural resources of the Darwin region by Aboriginal people have been documented by Hodgson (1997) and Povinelli (1992). A similar, if more extensive variety of edible plant and animal resources and their seasonal availability in the Darwin region is listed by Baker (1981:49-52) and by Russel-Smith *et al.* (1997) for the Western Arnhem Land region.

Fire

There is evidence of past use of fire by the Larrakia and neighbouring groups in their resource procurement strategies, in the manner of traditional practices of burning the landscape in places like Arnhem Land (Haynes 1985), referred to by Jones (1969) long ago as "fire-stick farming". At Gunn Point, the northern end of the Shoal Bay, are relict stands of coniferous forests of the fire-sensitive tree *Callitris intratropica.* Vegetation studies here and elsewhere suggest that past Aboriginal burning regimes existed in this area, which may have preserved these and other biogeographically restricted species, now being destroyed by recent changes in fire patterns (Bowman 1988; Bowman and Panton 1993; Bowman and Wightman 1985; Price and Bowman 1994).

Ethnographic and historic records show Aboriginal use of fires for hunting game, for example, to drive wallabies into ambush using the topography and dry season winds, and lighting fires beneath paperbark trees to confuse, suffocate and bring to the ground magpie geese roosting in the trees at dusk. Records describe fires being used for signalling and to ease movement across the country after the wet, by burning the tall dead spear grass which covers much of the landscape (Stocker 1966:225). In Arnhem Land fires were not lit until the crop from food plants such as yams (eg. *Dioscera* spp. and *Ipomoea* spp.) was collected, as burning destroyed the vines of these plants (Thompson 1949 cited in Stocker 1966:225).

Fire also has the effect of increasing nut yields of some plants such as cycads, which are well known ethnographically as a feasting food for funerals and other ceremonies (Basedow 1907:7; Beaton 1982; Meehan 1982b:106). Fires have varying effects on yields of the cycad nuts or seeds, which develop over a long period from March through December, depending on the stage of seed development, height of the female plant and on the intensity of fire (Parks and Wildlife Commission of the NT 1997). Therefore the use of fire would need to follow a particular regime to increase cycad nut yields (see Beaton 1977). In the light of this information and the presence of large stands of the cycad *Cycas armstrongii* on the hinterland margins at Hope Inlet, it is interesting to note Harney's (1951:245) statement that along the Northern Territory coast ritual grounds are almost always located near cycad forests.

Trade and exchange, totemic beliefs and ceremonial life

Although ethno-historical information is scant, what there is indicates that the people of the Darwin region were very much part of ceremonial exchange and trade routes at the time of first European contact. Petrie (1950a:52, cited in Micha 1970:290) recorded the trade routes, said to be travelled by totemic culture heroes of the Dreamtime: "Djamara came out of the earth at the northern point of Dampier-land. According to one version, he travelled to the east, and today still lives in the region of Darwin". According to Petrie (1950c:39, cited in Micha 1970:290-1):

> "the Djamara path corresponds with an east-west [trade] route connecting Darwin and Arnhem Land with a trading centre, such as Kingana on Vansittard Bay...On the path between Darwin and Kingana, there are six stations or trade centres known in the Northern Territory. Barter took place at all localities. The route is described as a well-defined trade route, as a "big road from long way".

Ethno-historic accounts also reveal the rich material culture and ceremonial life practised by the Larrakia and neighbouring groups (Basedow 1907, 1925:248-382; Foelsche 1882:4-7, 1886:255; Parkhouse 1895), and comment on the large gatherings at Port Darwin during the wet season (Parkhouse 1895:639). Items obtained from the Aboriginal locals at Port Darwin in the late 1800s demonstrates extensive use of natural resources for material culture, associated with the less tangible symbolic and ritual aspects of societal life (Koepping 1979:1). These include a range of body ornaments such as armlets and necklaces made of reed beads or shells, headdresses of emu or cockatoo feathers, or with bills of the spoonbill or other birds or kangaroo teeth attached, bamboo nose sticks, bark belts and fur headbands, or implements such as yam bags, wooden digging sticks, spears and ceremonial bullroarers (Basedow 1907; Foelsche 1882; Kerr 1971:111).

A variety of ceremonies were held to celebrate gatherings and battles with neighbouring groups, and initiation of the young, and funerals (Foelsche 1882:4-7; Spencer 1912:19). The anthropologist Ronald Berndt (1951) describes the cyclical seasonal ritual and ceremonies such as the *Kunapipi* which were performed by Northern coastal groups including the Larrakia, in order to ensure continuation of the human species and a constant supply of food (Berndt 1951a:234; 1951b). Large quantities of food were required to feed people gathered for ceremonies, and cycad seeds were collected and processed for mourners' feasts at funeral ceremonies (Basedow 1907:7). The Larrakia differed from surrounding groups at the time of European contact in not practising circumcision for initiation ceremonies (Spencer 1912:20) and according to Elkin (1950), possessed no sections. An underlying ideology that Spencer (1912:33-4; 1914:204) reported as shared in common with other Australian Aboriginal groups however, was their totemic belief system. Larrakia country is part of an area mapped by McConvell (1985:31) as an area of matriclans and matrilineal social totemic groups. Parkhouse (1895:640) also recounts a conversation with a Larrakia boy that indicates a belief in totems and patrilineal descent of totems: "My boy being told to remove a green jumping frog, and misunderstanding my remark, besought me not to kill it, alleging, 'That one my father'".

Australian Aboriginal Totemism has been well documented ethnographically amongst many groups across Australia (eg. Elkin 1978[1933]; Frazer 1910; Hale and Tindale 1934; Hart *et al.* 1988; Sharp 1939). Totemism is defined as a religious and social belief system in which humans identify themselves as having an intimate and special relationship with certain animals, plants, and natural phenomena. Members of the same totemic group regard each other as kinsmen or brothers and sisters, call themselves by the name of their totem and believe they are descended from it (Frazer 1910:5,53). Totemism and its meaning in terms of philosophical views of the world and systems of ordering the universe, have been extensively analysed (see Elkin 1969; Kolig 1988). A foray into the intricacies of this belief system is beyond the scope of this archaeological work. However, the relevance of totemism is the way in which this belief system is intricately linked to Aboriginal subsistence and economic systems by intimately relating people to landscape and resources (Jones 1990:25). This is explored in Chapter Ten, with reference to the link between shell mounds and ritual practices associated with this belief system.

These accounts reveal some of the extensive religious and ecological knowledge employed in resource procurement strategies by these coastal peoples at the time of European contact. This includes fishing and mollusc gathering, which constituted an important part of the subsistence strategies as they did elsewhere in northern Australia, where large shell mound sites have also been recorded. In the next chapter I discuss the methods used in this study to investigate the archaeology of the Darwin region, which materialises predominantly in the form of shell middens and mounds.

4 Methods

Survey methods

Archaeological surveys of the coastal region, outside the built up residential areas of Darwin and outlying town of Palmerston, were undertaken in 1995 and 1996. A combination of two survey strategies was employed: purposive (non-random) and stratified random sampling. Purposive surveying was undertaken in areas considered to be of high archaeological potential. These areas were selected on the basis of information from the Museum and Art Gallery of the Northern Territory (MAGNT) sites register, previous archaeological experience in the Darwin region, and the identification of large mounds on satellite imagery by the then Conservation Commission of the Northern Territory (CCNT) now the National Parks and Wildlife Commission. They included the hinterland margins around Darwin Harbour, along the banks of tidal rivers, creeks and lagoons, and the hinterland margins and saltflats of Hope Inlet.

Purposive surveys around Darwin Harbour covered all that area of the margins of the mangroves along the coastline between the Bayview Haven site, along both sides of the Elizabeth River banks and around the coastline of Middle Arm Peninsula and along the eastern banks of the Blackmore River (see Chapter 5, Figure 5.2). Pedestrian forays were made at intervals into clearings in the mangrove zone, and the hinterland coverage extended from the margins inland for up to three kilometres. During the course of surveying the random sample localities, some areas outside of these localities, which seemed likely places to contain sites according to my fieldwork experience as an archaeologist (Burns 1996a:6, 27) were also purposely inspected. These included parts of the east and south coastline of West Arm and a small area on the western side of the Blackmore River where a quarry was located (see Chapter 5, Figure 5.2).

Purposive surveys at Hope Inlet included coverage of 90% of the saltflats areas and margins of the mangrove zone and intensive inspection of a network of mangrove-lined tidal channels and perennial ponds on the saltflats midway between the hinterland and seaward mangrove fringe (see Chapter 5, Figure 5.3). The margins of each of the promontories extending into the saltflats were intensively inspected as well as inland for up to three kilometres, concentrating on places where a localised change in environment was observed, including areas surrounding all major swamps and hinterland lagoons and their associated shallow drainage lines (see Figure 2.2 and Figure 5.3). Howards Peninsula was surveyed on its northern and eastern margins adjacent to small saltflats areas and mangroves leading on the eastern side to the Howard River. Studies indicate that this purposive surveying strategy of areas where site locations are predicted, is much more effective than random sampling for cost-effective location of sites that exhibit highly clustered distributions (Redman 1987:251; Schiffer *et al.* 1978:5), as suggested by the results of previous surveys (see Chapter One).

In general, previous surveys carried out in the region have employed purposive sampling strategies to fulfil the aims of consultancy reports for development projects (eg. Guse 1995; Heritage Surveys 1995; Hiscock and Mitchell 1992). Thus the known distribution of sites in the Darwin region was in the main obtained through purposive sampling. Purposive sampling is an inherently biased method, since it follows the self-fulfilling prophecy of only surveying those areas where sites are expected to be found (Sundstrom 1993:93). To reduce this bias whilst working within the limited resource constraints of a PhD project, which preclude completely surveying such a large area, additional surveys were undertaken following the well known procedure of stratified random sampling (see Hughes and Hiscock 1981:5,27; Read 1975; Schiffer *et al.* 1978). In this method separate random samples are selected from different spatial units, or strata (in this case land units) within the whole region, which gives an increased level of accuracy over simple random sampling (Drennan 1996:87; Read 1975:58). These random surveys were carried out as a pilot project for a Territory wide survey of archaeological sites, funded by the Australian Heritage Commission through the National Estate Grants Program (Burns 1996a). As a pilot project constrained by time and resources, around 8km^2, less than 1% of the 2,195km^2 of the Darwin Catchment Area shown in Figure 4.1, was able to be surveyed for the random sample, so these results must be regarded as preliminary.

Geomorphologists have classified four broad land systems in the Darwin coastal region: the Dissected Foothills, Koolpinyah Surface, Alluvial Plains and Coastal Plains environmental zones. Each of these land systems is composed of a series of land units, which are relatively homogenous in soil, landform and vegetation features (Wood *et al.* 1985:1). Land is grouped into units on the basis of nine landform types as follows (after Fogarty *et al.* 1984:9):

Land Unit Groups	Landform Types
1	Rugged hills and slopes
2	Low rises and gravelly sideslopes
3	Flat to gently undulating upland surface
4	Gentle lower slopes
5	Upland drainage lines

6	Broad lowland plains
7	Major channels and fringes
8	Depressions and billabongs
9	Estuarine fringe

These landform types are further subdivided into some 30 land units (eg. 1a, 1b, 1c) according to criteria such as soils, drainage and vegetation. The site catalogue records of the MAGNT, recorded to the land system level, indicate that archaeological sites in this region are concentrated in the coastal plains and dissected foothills land systems. Random surveys were carried out at the more detailed level of the landform types that make up the land systems. Information on the distribution of archaeological material was obtained through surveying of randomly selected sample areas of each land unit within the study region, following the land unit classification scheme in Fogarty *et al.* (1984).). Land units 8b and 9b, consisting of swamp depressions and soft saline muds respectively, were excluded from the survey due to their inaccessibility. The portions of randomly selected study areas surveyed are termed "localities", and are numbered and prefixed according to the mapsheets on which they are located (eg. The first sample locality on the Noonamah 1:50,000 mapsheet (5172) is N1. The inclusion of this random sampling in the overall survey strategy allows a quantitative analysis of areas of presumed low artefact density, a fundamental requirement to ensure the adequacy of a regional investigation (Dunnel and Dancey 1983:272; Sundstrom 1993; Thomas 1975).

A selection of survey areas was ensured by randomly choosing a section from each of the land units. Two or more areas were selected for each land unit, so that if one was inaccessible or extensively disturbed, the next area could be examined. The land units vary greatly in the areal extent to which they occur in the study region. To allow for this difference, two portions that were spatially separate were selected in land units that covered much larger areas. Following this method, pedestrian surveys were conducted in 30 transects, each approximately 250 metres by one kilometre in randomly selected sections of the different landform types throughout the study region. This figure varied locally depending on conditions on the ground, such as accessibility, extent of vegetation and visibility. The advantage of transects lies in their large edge effect, useful in the detection of archaeological materials, and in estimating relative density and degree of clustering (Schiffer *et al.* 1978:12). Non-random sampling outside the sample areas included inspection of boundary areas. This allowed an examination of variation within different parts of the same landform type, and of the effects of transitional boundaries between different land units.

Environmental factors such as climate, geology, landforms and associated soils and vegetation are known to influence the availability of water, plant and animal resources, raw material for stone artefacts, and suitable campsites, and therefore the nature and distribution of prehistoric sites in the landscape (Sullivan and Bowdler 1984). These environmental factors also affect the survival and visibility of sites in ground surface surveys (Hughes and Sullivan 1984:35). Such archaeological studies suggest that a preferred environmental setting for prehistoric Aboriginal activity was one which provided close proximity to as many landscape elements, and landform types, as possible, in order to exploit resources.

There are well-documented factors which affect the success of an archaeological survey (see Guse 1995:13; Hiscock 1995b:4). Limiting factors to reconstructing the patterns of Aboriginal exploitation of an area from archaeological surveys are (after Anderson 1984:11) those:

1. that determine what is left at a site, such as cultural preference and curation practices.

2. that modify, disturb or degrade what was originally discarded, either cultural or natural.

3. that affect the ability of the archaeologist to locate remains, including the intensity of the survey, accessibility and visibility.

1. Cultural preference and curation practices:

Include collection, discard and mounding behaviour and caching of artefacts.

2. Cultural and natural factors:

Cultural factors include the substantial alterations to the Darwin landscape by European activity in the last 150 years. For example, around Darwin Harbour watercourses have been altered, areas extensively cleared for residential development and agriculture, mangroves destroyed and creeks and parts of the coastline infilled. The Hope Inlet area was previously under a pastoral lease held by Koolpinyah station which ran cattle and buffalo, and much of the sandy coast in this area has been subject to sandmining activities for many years. Material from many prehistoric sites such as shell middens have been utilised historically for construction purposes as discussed in Chapter One (see Hiscock and Hughes in press), and formal stone artefact types extensively removed by collectors, some of which have been lodged in MAGNT.

Natural factors include the burial of sites, as the Darwin coastal landscape is a depositional one, which is still prograding from seaward deposition and terrestrial sedimentation (Sullivan 1996). The annual tropical rounds of heavy rains and dry season fires, and chemical or biological action contribute to weathering and also affect the survival of archaeological remains.

3. Survey intensity, accessibility and visibility

The intensity of the survey has been closely correlated with the number of sites found, and the ability to identify artefacts (cf. Richardson 1988). Survey intensity is determined by the interval spacing between surveyors and the speed of surveying. The survey methods chosen by the archaeologist take into account environmental conditions, such as the extent to which vegetation reduces the ground surface visibility. Access to much of the study area is generally restricted to four to six months during the dry due to wet season flooding, and this is also the period of highest ground surface visibility, so fieldwork was continued over the two dry seasons of 1995 and 1996. Some areas are accessible only by boat, but four wheel drive access was gained to most areas in the dry season via dirt tracks and fencelines.

For the purposive surveys, in order to maximise visibility of the ground surface, areas were surveyed where possible after the annual dry season fires had burnt off vegetation. In areas of low or no visibility the survey was slowed. Ground surface visibility varied across the study area, from high on the bare open saltflats of Hope Inlet, to low in some densely vegetated areas such as monsoon vine thickets, or around creeks and waterholes, which were not burnt. Unfortunately, when the random survey was undertaken for the Heritage project in November 1995, as dictated by time constraints, ground surface visibility was reduced in many areas by increased vegetation growth due to early wet season rains. This may have biased the results, in favour of more obtrusive sites such as shell mounds, as opposed to surface scatters of stone artefacts. Survey areas were intensively inspected on foot by the author with the assistance of archaeology students, with interval spacings between surveyors of 40-50m, giving a medium intensity survey. A hand held Global Positioning System (GPS) was used to obtain grid references and locate the sites on maps.

For the purposes of this survey, sites were defined according to common archaeological convention of distinguishing between relatively dense, discrete concentrations of archaeological material or features and the sparsely distributed materials, or background scatter, that surround them (eg. Binford 1982:5). Variability in density of archaeological material is a result of not only the nature and extent of past human activities, but also of geomorphic processes that affect the visibility of the material in the contemporary landscape, as discussed above (Hiscock 1995b). Notwithstanding the debate on the site concept (Thomas 1975), these sites fit within generally accepted definitions of what constitutes an archaeological site or "archaeological place" as it is termed in the Northern Territory Heritage Conservation Act 1991 (Hiscock 1994a:56). When concentrations of archaeological material such as shell and earth mounds coalesce they are classed as one "composite" site (after Cribb 1996:160), and clusters of archaeological material in close proximity are given the same site number but distinguished with different suffixes, for example HI36, HI36a, HI36b.

Definitions of the terms used in this study for types of sites that occur in the Darwin coastal region generally follow those used by other archaeologists (eg. Bowdler 1983:135; Gregory 1998:222; Hiscock 1995:3):

Shell middens, which are deposits containing more than (an estimated) 50% by weight of shells, occurring somewhere in the open, near a beach or estuary or rocky shoreline, or an inland lake or river. These shells have been deposited by humans exploiting marine, estuarine and riverine resources. Middens may take the form of a thin layer of shell over, or just below (sub-surface), the land surface, or a thick mound of shell. Shell middens are classed as mounds or mounded deposits in this study if they have an (estimated) depth of more than 30cm. Shell middens are usually circular, but may also be elongated or irregular in shape, or doughnut-shaped (shell rings) around a central area containing little or no shell. Unstratified surface scatters of shell also occur.

Stone artefact scatters, may contain flaked or ground stone artefacts and/or hearth stones. Artefact scatters may occur as surface scatters of material or as stratified deposits where there have been repeated occupations. These scatters do not necessarily imply that prehistoric people actually camped on the site; rather they indicate only that some type of activity was performed there.

Earth mounds, which are mounded, usually circular, deposits predominantly composed of earth, which may contain shells, bones and stone artefacts.

Quarries, which are localities at which stone has been procured.

Engraving sites or rock art, where motifs are abraded in rock.

Stone arrangements, which are sites containing arrangements of stone not attributed to natural processes, such as fish traps and ceremonial arrangements.

Each of these site types or their components may be found in association with each other.

After locating large numbers of sites, quick methods of measurement (following Bailey 1975a:V:9-10) were employed in order to obtain field data on dimensions for as many of the sites as possible. Mounds were measured across key cross-sections of short axis (width) and long axis (length) with a tape measure or by pacing, and height estimated relative to eye level. Dimensions of excavated sites were mapped using a dumpy level. Various methods

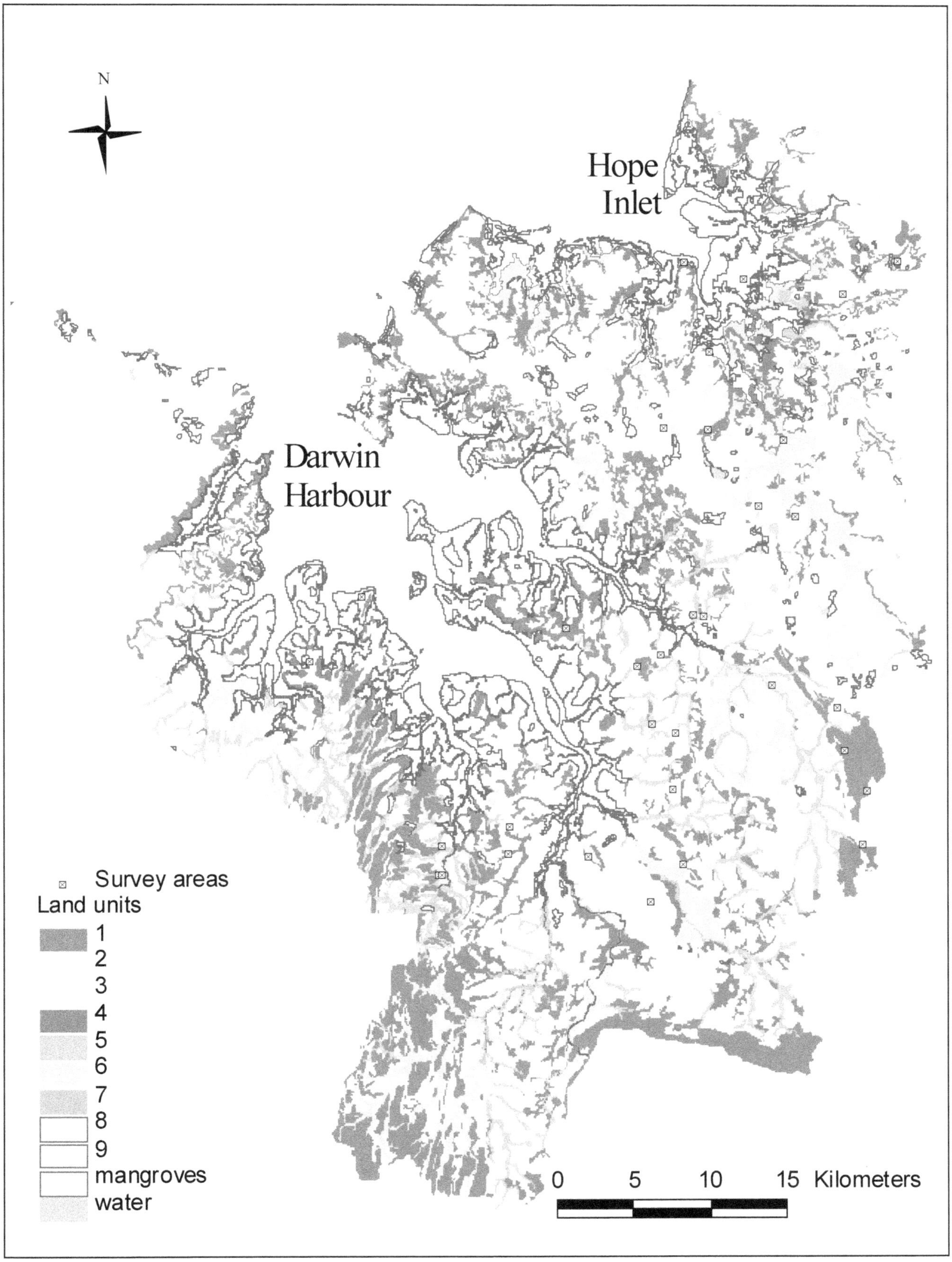

Figure 4.1 Map showing land units and random survey areas in the Darwin coastal region

have been used in the archaeological literature to estimate mound volumes (eg. Bailey 1975a:V:9-10; Sorant and Shenkel 1984). For middens that are circular in plan, Bailey (1975a:V:9) used the formula for a segment of sphere $V=\pi h/6[h^2+3R^2]$, where V = volume, R = radius of base area and h = height. The method used in this study is a generalisation of the formula for a segment of sphere extended to an elliptical plan. Thus volumes of mounds were calculated following the formula $V= L_2/L_1 \times \pi(L_1^2h/8 + h^3/6)$ where L_1 = the short axis, L_2 = the long axis, and h = the height, for most of the mounds, derived from the formula for the volume of a spheroid. This assumes the mounds are elliptical in plan and circular in vertical section and a flat base. For the large mounds with steeper, less rounded sides the formula is adjusted to L_2 = the short axis and L_1 = the long axis (Matthew Fegan pers. comm.). Using this formula allows a rough approximation of the total quantities of shell deposited over the total site area, which is adequate given that many of the mounds are partially buried and the calculated volume must be considered an underestimate (see Chapter Five and cf. Sorant and Shenkel 1984).

Characteristics including environmental context, artefact types, dominant shell taxa and the presence of any other cultural material on the surface such as human or animal bones, were recorded for each midden. For the scatters of shell and stone artefacts, types of raw materials and artefact types were recorded. To gain a rough idea of average density the number of artefacts in one or two one-metre square grids were counted. These site types are worthy of further investigation not possible within the time constraints of this project, which concentrated on excavations and analysis of the mounds.

Stone artefacts were classed as either flaked or ground stone. Included in the ground stone category are artefacts exhibiting signs of pounding, pecking or hammer-dressing, termed crumbling techniques by Hiscock and Mitchell (1993), and/or exhibiting signs of grinding, sawing, engraving, drilling scouring and polishing, termed abraded (Hiscock and Mitchell 1993). Types of flaked stone artefacts were identified according to characteristics as defined by Hiscock (1984:128-9); as flakes if they exhibited one or more of the following characteristics:

1. A ring crack where the percussor struck the core.

2. A distinct positive bulb of percussion.

3. A definite eraillure scar beneath a striking platform,

and as cores if they exhibited one or more negative flake scars but no positive flake scars, and as retouched flakes if they exhibited flake scars extending onto the ventral surface and/or deriving from the ventral surface. Flakes retouched on one side are termed *unifacial points*, and on both sides *bifacial points*. Artefact types from the Middle Arm mound sites were identified by Peter Hiscock.

Excavation methods

Methods differed slightly for excavations at Hope Inlet undertaken for this project and previous excavations at Darwin Harbour, which were originally carried out to answer different questions. At Darwin Harbour excavations were undertaken in 1995 on two middens, which appeared more weathered and therefore possibly older, than another shell deposit excavated in 1994 as part of an Honours thesis (Burns 1994). In each of these middens small test pits 0.5m x 0.5m were dug in order to test the hypothesis derived from observation of more weathered shell on some middens, that these were older than middens on which the shell appeared less weathered. At Hope Inlet excavations were undertaken in 1996 on three shell mounds, an earth mound containing shell, and one subsurface shell midden. These sites were selected for excavation for comparative analysis of the different site types, and across the different environmental zones; on the saltflats, the hinterland margin overlooking the saltflats, and inland in the eucalypt woodland. Larger one square metre pits were excavated in each of these deposits to obtain comparative information on their contents, stratigraphy, and chronology.

Sampling is necessary in shell midden studies not only because sorting and analysis of the bulk of even a small midden is prohibitively time-consuming and thus expensive, but also for conservation purposes. In this study the friable nature of these deposits made fewer holes the preferred option in terms of conservation. So as to undertake a comparative inter-site analysis within the limited resources of this study, bulk samples comprising half of the one metre excavation square, designated Section A, from a number of deposits were analysed, rather than a number of smaller column samples from different areas of one deposit (cf. Bowdler 1983:138; O'Neil 1993:527). The excavated samples represent less than 5% of each of the deposits. However, the size of the samples analysed (between 320,000cm^3 and 800,000cm^3) in the present study is equal to or larger than, the column sample sizes of 2000cm^3 and 6250cm^3 recommended as a minimum by Bowdler (1983) and Barz (1977:38) respectively, and those used by other Australian researchers (eg. Mitchell 1993:185; O'Connor and Sullivan 1994:20; Sullivan 1984:2).

Many researchers have discussed the best way to adequately sample middens (eg. Ambrose 1967; Barz 1977; Claassen 1991b; Greenwood 1961; O'Neil 1993; Meighan *et al* 1958; Waselkov 1987) and Treganza and Cook (1948) excavated a whole midden to test its homogeneity (see also Ranson 1980). Although it has been demonstrated that shell middens cannot be presumed to be homogenous deposits (O'Neil 1993; Waselkov 1987:143), observation of other sections of these mounds exposed through disturbance by goannas suggested that the bulk samples excavated broadly represent overall mound content. My own excavations and other midden studies suggest that the main concern regarding the heterogeneous nature of middens, is that rare or clustered features or artefacts may not be recovered. However, it has been shown that larger or more sampling sections do not necessarily resolve these inherent problems (O'Neil 1993:524, 527-8). Much information can still be gained from the gross content obtained through excavation, with the qualification of uncertainty as to how representative the samples are of the total deposit (Brown 1975) and recognition that methods of analysis affect the results and therefore interpretations of archaeological sites (Spiess 1988:174).

Field methods generally followed guidelines provided by Johnson (1980). The excavation squares were marked out using string, and at Hope Inlet each 1m x 0.5m half of the excavation squares, designated Sections A and B, was excavated so it could be analysed separately. All the material from Section A of each site was sorted and examined in detail in the laboratory. Keeping as much as possible within any natural strata observed, the deposit was removed in arbitrary excavation units, or spits of average thickness 3cm, basically consisting of two to three 10-litre buckets per half square. Thus a relatively constant spit volume was ensured, which reduces the effects of variance caused by the arbitrariness of the spit boundaries, of these artificially constructed, relatively small analytical units (Frankel 1988:42; Walters *et al.* 1987:89). Levels were checked by recording depths at the corners and centre of the bottom of each excavation unit, using a dumpy level.

Depending on the nature of the excavated deposit, it was either removed in bulk, or weighed with hand-held spring balances and sieved on site through 6.4mm and 3.2mm sieves and all sieve residue material >3.2mm taken to the laboratory for analysis. Numerous studies have demonstrated that 3.2mm is the appropriate minimum mesh size for the recovery of small faunal taxa, which is important for reconstructions of past environment and resource procurement strategies as undertaken for this project (Gordon 1993; James 1997; Jerardino and Yates 1997: Shaffer and Sanchez 1994:528). During the excavations shell and charcoal samples were collected for radiocarbon dating, and pH levels taken at regular intervals. Prior to leaving the sites, the excavation pits were backfilled to prevent further disturbance.

Dating

Radiocarbon determinations for this research were obtained for most sites through standard radiometric analysis and accelerator mass spectrometry technique for one site. Paired samples of the marine shell *Anadara granosa* and charcoal were obtained from some sites, where they were considered to be in secure stratigraphic association. This was undertaken for two purposes. The first is as a means to aid assessment of the extent of mixing of deposits (Head *et al.* 1983:110). The second is to add to a growing database of dates in the region, with which it is hoped to address questions that have been raised on the applicability of the standard Australian correction factor, for radiocarbon determinations on marine shell from the Northern Territory coast (Woodroffe and Mulrennan 1993:40-1; Woodroffe *et al.* 1988:98; Scott Mitchell pers. comm.). While a factor of 450±35yrs has been determined by previous studies as necessary to correct the marine reservoir effect for coastal Australia (eg. Gillespie and Polach 1979; Head *et al.* 1983; Bowman and Harvey 1983; Bowman 1985), some shells from middens in the South Alligator river region gave ages younger than 450 years, implying a smaller correction factor may be appropriate (Woodroffe *et al.* 1988:98). Initial interpretations of dates for sites in the Darwin region suggest they will facilitate refinement of this value, when sufficient dates for a statistical analysis are obtained. For the purposes of this study, the present correction value is applied during calibration of the radiocarbon estimates, using the University of Washington's CALIB 3.03 computer program marine calibration model with a standard Delta R correction value of -5±35 (Stuiver and Braziunas 1993; Stuiver and Reimer 1993). Thus the corrected and calibrated dates on marine shell in this study are considered minimum estimates for the ages of the deposits.

Laboratory methods

Excavated material was weighed, and where necessary for identification, wet sieved through 6.4mm and 3.2mm mesh sieves, and dried. Components from both the 6.4mm and 3.2mm mesh sieves were sorted by hand into the different categories of shell taxa, bone and crab, fish otoliths, charcoal, stone artefacts, non-artefactual stone, ochre, vegetation and other. Stone artefacts were counted and weighed and sorted into the raw material categories quartz, quartzite, dolerite, chert, tuff and other. This allowed an examination of the relative extent of local and imported stone raw material between different sites and between sub-surface assemblages and surface scatters, as well as between the different landscapes. Raw numbers of stone artefacts in each sample were standardised by calculating artefact density per cubic metre, to allow comparison between sites and between the two areas of Darwin Harbour and Hope Inlet.

Only macroscopic charcoal, vegetation and bone were measured (by weighing) in this analysis. Most of the bone (upwards of 80% in each of the deposits) was recovered from the 3.2mm sieve. Much of it was burnt, very fragmented and difficult to identify to specific taxa. The extent of fragmentation meant it was not feasible to estimate the dietary contribution in numbers or size of fish or other animals using NISP or MNI counts (cf Casteel 1976; Erlandson 1994:57; Horton 1984). As Ringrose (1993:151) argues, too much meaning can be assigned to faunal data through highly detailed quantitative analysis, which may not be appropriate given the uncertain representative nature of the assemblage due to pre and post-depositional processes. Identifiable non-molluscan faunal remains are simply tabled as presence/absence to the level of taxa possible. Fish bone and otoliths were identified by Helen Larson of the Museums and Art Gallery of the Northern Territory (MAGNT), and through comparison with a collection of fish skeletons and otoliths held at MAGNT. Other bone and teeth was identified with the help of Dirk Megirian and Fiona Mowat also of MAGNT, and Ian Walters, Sally Brockwell and Chris Crassweller of the Northern Territory University and through comparison with drawings and photographs in Goodfellow (1993) and Triggs (1996). In this analysis the focus is on gross patterns (Amorosi *et al.* 1996:151), and all components of the deposit are weighed to allow quantitative comparison within and between sites, of non-molluscan faunal material, molluscan fauna and other organic material such as charcoal, as well as stone artefacts.

Shell taxa and their habitats were identified with the help of Dr. Richard Willan of MAGNT, and the use of photographs from Blackburn (1982), Meehan (1982a) and Wells and Bryce (1988). Procedures for the analyses of small samples, as developed by previous researchers were generally followed (Bowdler 1983:139). Each shell taxon was weighed (Wt) and the minimum numbers of individuals (MNI) counted. The diagnostic features used for calculating MNI varied with the prevalence of that feature, so that the most prevalent feature for each taxon was used to estimate MNI. Thus MNI was calculated on spires for most gastropods, such as *Telescopium* and *Terebralia,* on flanges (apertures) for *Cassidula, Ellobium*, and *Nerita*, and on the larger number of left or right umbo and hinge for the bivalves *Anadara* and *Marcia*. For the oyster (*Saccostrea echinata*) the opercular shaped upper valves, or lids, with more than half the adductor muscle scar visible (Slack-Smith 1998:272; Thomson 1954:133) were used as diagnostic features for calculating MNI.

Despite the extra time involved, the combination of both weight and MNI methods of analysis provides a more accurate description of the proportions of shell taxa and allows comparison with non-shell components (Coleman 1966:37, cited in Bowdler 1983:140; cf Nichol and Williams 1981; Rowland 1983; Mowat 1995:81; Walters *et al.* 1987:85). Use of shell weight alone underestimates the actual flesh content provided by the lighter shells, and may therefore bias interpretations of the contribution to subsistence provided by the relative proportions of different shell taxa in the deposit (Coleman 1966:37, cited in Bowdler 1983:140). For example, Meehan (1982a:141-2) demonstrated that the thin, fragile bivalve shell *Marcia hiantina* provides a higher flesh/shell ratio and protein content than the much heavier gastropod *Telescopium telescopium*.

Shell weight loss may also occur due to fragmentation and chemical leaching processes, and the extent to which this occurs varies between different shell taxa. MNI counts are less affected by this as the most durable shell parts, such as hinges and flanges, are counted. The discrete nature of MNI counts as a categorical variable is also more appropriate to research questions on environmental change and the nature of resource exploitation that deal with inter-site comparison of relative frequency of mollusc taxa (Mason *et al.* 1998:309, 319), as addressed in this project. Examples provided by Mason *et al.* (1998) demonstrate that using weight or MNI methods may provide results that present very different interpretations, and that MNI is the much more reliable measure to answer questions about the past on the environment and human exploitation of resources.

However, the use of MNI counts alone as a method of analysis disguises significant differences in size, both of the shell taxa, and individual shell specimens. It also provides no measure of the extent of shell fragmentation, information that may give an indication of discard practices of mound accumulation, including rate of discard, as well as post-depositional processes. A measure of shell fragmentation also allows quantification of the variable extent of shell destruction rate between the different taxa in the mounds. This information facilitates a more accurate reconstruction of the original quantities of the different molluscs deposited. In order to obtain this information, weight of broken shells for each taxon was measured during sorting and the proportion of breakage calculated. All shell pieces, retained by both 6.4mm sieves and by 3.2mm sieves, which are less than half the whole shell, are classified as fragmented.

Weight measures allow a more meaningful, if still problematic comparison of the relative proportions of total quantity of shell and non-shell components in a sample than MNI counts (see Grayson 1984; Mason *et al.* 1998: 319). Employing both methods of analysis (MNI and Wt) minimised the deficiencies inherent in either type (Bowdler 1983:140) and enabled calculation of Wt/MNI for each taxon, which gives an indication of relative average shell weight, and therefore of relative shell size. For each of the excavated samples, identified shell taxa are listed in descending order of frequency of MNI, with Weight and Wt/MNI. This figure is only a rough approximation as MNI count is affected by the extent of fragmentation (in that identification of individuals becomes more difficult with increased fragmentation). This can be seen in the analysis below, where Wt/MNI for *Anadara* shell increases in sites as fragmentation rate increases, in particular between the Middle Arm sites with a much higher fragmentation rate than the Hope Inlet sites. However, the figure is still useful in that it provides at a quick glance an indication of large differences in size of different taxa and of the same shell in different sites, taking fragmentation into account.

Shell size

Measurement of any changes in shell size within and between the excavated deposits is another feature examined in this study, as an aid to reconstructing the palaeoenvironment, the past mollusc habitats exploited and the relationship between these and cultural selectivity (Claassen 1998:131). This follows previous research elsewhere on cultural shell deposits, which use shell size as an indicator to examine extent of human exploitation of molluscan resources and/ or environmental change over time (eg. Jerardino 1997; Spennemann 1987; Swadling 1976).

In order to take account of the many problems associated with identifying the natural and/or cultural factors actually determining average shell size in a shell deposit (see Claassen 1998:47-9,133-7; Spennemann 1987 and Chapters Six and Ten), and other archaeological factors which may affect the interpretation, such as sampling technique and sample size, this analysis takes a very broad brush. Shell size of the dominant mollusc is considered in conjunction with a range of other data, such as shell breakage rates and proportions of other mollusc taxa and other fauna. In addition the data are considered on different temporal and spatial scales, at both intra-site and inter-site levels (Frankel 1988:43). I would argue that the large sample sizes used from a number of sites are adequate for the broad chronological trends considered here (cf. Ford 1989; Spennemann 1987).

For this analysis, during sorting of the excavated material, I calculated average weight per *Anadara* shell (indicating shell size), for each spit in the three shell mounds. To do this I divided the weight by the minimum number of individuals (Wt/MNI), of whole, unbroken *Anadara* shells from each spit. The accuracy of this time efficient method of determining average shell size for 121,393 shells through three deposits was confirmed by measuring with a ruler the maximum valve length to the nearest millimetre (following Spennemann 1987:85), of whole *Anadara* shell from three spits, at the base, middle and top, for each of the shell mounds.

Distribution of cultural material

In order to adequately illustrate the distribution patterns of cultural material seen in analysis of the deposits, both raw data and density values (or rates of discard) of the material recovered are given. Thus quantitative data, of the components recovered per excavation unit and totals for the whole sample, and relative percentages of total shell taxa for the sample are presented for each site, for comparison. Rates of shell fragmentation are shown as percentages of total shell in the sample. The distribution of organic material shell, charcoal and bone, and of stone artefacts is presented as density g/kg and no/kg of each component per total weight of the excavation unit.

Density values (number or mass per unit volume or total mass) are used to estimate relative contribution of different food remains in a site in reconstructions of prehistoric diets (eg. Bailey 1975b; Buchanan 1988) and as an index of occupation intensity at a site (see Schiffer 1987:282; Jones 1985b:291; Jones and Johnson 1985b:180-3). Using density values does not have the problems of relative percentages in that the proportion of one shell taxon is relative to all the other taxa. So for example, changes in a dominant taxon will change the relative frequency of another lesser taxon, even though the actual numbers for the lesser taxon remain fairly constant (Grayson 1984).

Problems with using density values include possible compaction of some lenses of deposit, and the need to ensure, rather than assume, that changes in density of artefacts is independent of changes in rate of deposition of the dominant component of the matrix (Jerardino 1995:21). In shell middens usually the major control of bulk density is shell, and as Hughes (1980:13) has argued there is a roughly inverse linear relationship between bulk density and other organic matter in sandstone rockshelter deposits.

In order to provide a check against compaction, shell dissolution or winnowing processes which affect the vertical stratigraphic development of the deposits, I have used O'Connor *et al.*'s (1999:66) suggestion of calculating simple replicated bulk density for the spits (average 3cm vertical increments). Using graphs showing bulk density (total spit weight per total spit volume) for each of the deposits (see Chapter Eight) as a benchmark elucidates whether the patterns exhibited by the shell, charcoal, bone and stone artefact components are not simply an artefact of the relative densities of materials that make up the total volume. For example in Site HI66, because the proportion of bulk matrix to charcoal is so large: 200/1 or more, and charcoal exhibits peaks of four times its mean value, there would have to be a large removal of bulk matrix to account for the change in their relative proportions. That is, for a quadrupling of charcoal there would have to be a quartering of the bulk matrix. The ratios are so extreme that it is unlikely that there could have been removal of this much bulk matrix, which demonstrates that the patterns seen in this case are unlikely to be the result of winnowing.

Density values in this analysis are considered in context with radiocarbon dating of deposits and rate of shell fragmentation, from which an assessment of the deposition rate can be made. Thus a number of independent measures of the nature and intensity of occupation are provided (cf. Hiscock 1986; Jerardino 1995). The results of these archaeological surveys and excavations using the methods outlined in this chapter are given in Chapters Five, Seven and Eight.

5 Archaeological sites on the Darwin coast

Introduction

Chapters Two, Three and Four described the environmental and ethnographic context of the archaeological record in this region, and methods used to locate archaeological remains. Discussed was the number of factors that determine recovery of archaeological material in this region. This includes the nature of the two different landscapes, comprising the drowned river valley of Darwin Harbour and prograding coastal plains of Hope Inlet, formed through geomorphic processes. It also includes some of the resources available in the region and ethnographic and historic records of use of these resources by contemporary Indigenous peoples of the north Australian coast that may produce the types of discarded material seen in the landscape. In this Chapter the results of both random and purposive surveys are given and similarities and differences in site types and distribution patterns recorded in the two landscape types are examined. Since the predominant site type on the Darwin coast is stratified shell deposits, from a selection of which excavated material provides the primary data for this study (see Chapters Six and Eight), more detailed descriptions are provided of these sites and of the next most frequent - stone artefact scatters, than of less common sites such as quarries.

Results of surveys

Random surveys located only five pre-contact Aboriginal sites in the Darwin coastal region and two areas of relatively high background scatter. Chronological classification of sites not radiocarbon dated as pre-contact are made on the basis of site characteristics, including the presence of flaked stone artefacts and absence of historic material such as flaked glass or metal. In addition, the dominant shell taxa in most sites, *A granosa*, no longer occurs in any significant quantity in the local Darwin environment of extensive mangrove flats, considered to have formed within the last 700-500 years (Hiscock 1997, and see Chapter Ten).

The results of the random surveys is a low density of archaeological materials when compared with the results of the purposive surveying, which located over 300 new sites unevenly distributed in clusters across the landscape; 63 around Darwin Harbour, and 239 in the Hope Inlet area (Figures 5.1, 5.2 and 5.3). However, considering that less than 1% of the total study area of 2,195km^2 could be surveyed in the small pilot project for which the random surveys were undertaken, these results are not surprising. Given the small number of sites located through the stratified random sampling method, the predictive power based only on the results of the random survey is limited.

Notwithstanding this, these explorative results correspond with those of purposive surveys for this study and the MAGNT register, in that the most common coastal site type are shell middens, followed by stone artefact scatters (see Chapter One), and that a large proportion of archaeological sites are situated on or within 300m of land unit boundaries (Burns 1996a:37). These results also corroborate previous studies, which indicate that highly clustered distributions are more effectively and cost-efficiently located through predictive purposive surveying strategies than through random sampling when the total area being sampled is large (Redman 1987:251; Schiffer *et al.* 1978:5). This is reinforced by the fact that another fifteen archaeological sites were recorded through non-systematic purposive surveys of some areas observed outside the random sample localities during the course of surveying, that seemed likely places to contain sites according to my fieldwork experience as an archaeologist (Burns 1996a:6, 27).

Random surveys

Of the five sites found through random surveys, four are shell middens located in the coastal plains environmental zone and one is a surface scatter of stone artefacts and shell, located in the dissected foothills environmental zone. Ground surface visibility varied greatly for these surveys, generally in relation to extent of vegetation, and the poor visibility in some sections may have resulted in identification of more obtrusive sites such as middens, as opposed to the less visible stone artefact scatters. All five sites are situated on land unit boundaries. The four middens are all located in a similar environmental context in landform type 9, on beach ridges or tidal flat close to the present day mangroves. The stone artefact scatter is located on the crest of a low hill in open eucalypt woodland at the junction of landform types 2 and 3, less than 600 metres from the mangrove zone (Burns 1996a).

A relatively high density background scatter was observed in the rugged hills and slopes (landform type 1) around Darwin Harbour, in areas with numerous outcrops of the raw materials quartz and quartzite. This suggests that stone artefact scatters may possibly be present in this landform type, outside the surveyed area. Details on the random sections surveyed and the location and results are mapped in Figure 4.1.

Purposive surveys

Purposive surveys revealed the presence of large numbers of previously unrecorded archaeological sites in the Darwin coastal region. Site types recorded during these surveys occur in similar proportions to that observed in the (MAGNT) sites register, being mostly shell middens and mounds composed predominantly of *Anadara granosa.* Other site types were recorded less frequently, sometimes in association with the shell deposits. Stone artefact scatters are the next most prevalent site types around Darwin Harbour, and scatters of shell and stone artefacts, and earth mounds at Hope Inlet. The combination of strategies, of purposive and random surveys, provides a regional context on site distribution patterns within which particular sites may be selected for more detailed investigations though excavations (following McBryde 1962:12).

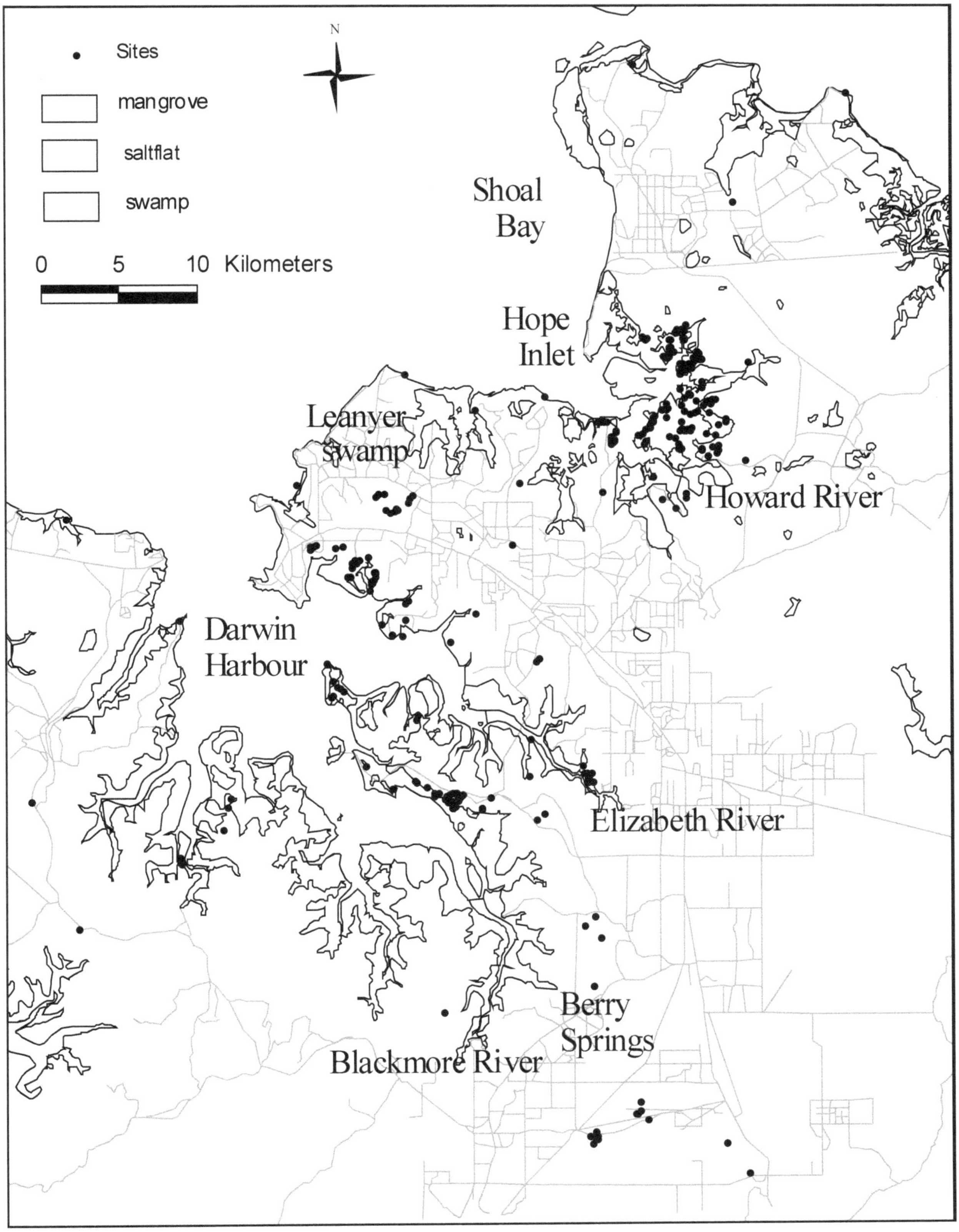

Figure 5.1 Location of sites in the Darwin region

The number and type of archaeological sites recorded by the author during these surveys at Hope Inlet and around Darwin Harbour (the Darwin Harbour list includes those recorded by other researchers) is given in Table 5.1 (overleaf). Categories of shell middens and mounds includes unstratified shell scatters, shell rings and composite mound sites (see below and Chapter Four for definitions). Stone artefact scatters containing shell and those with stone only are classed separately to show one of the differences in the characteristics of this site type between the two areas.

Archaeological sites around Darwin Harbour

Much of the Darwin Harbour shoreline, in particular the southern and western aspects, is still relatively inaccessible, and in all likelihood the record of site distribution obtained through surveys for this project is incomplete. A range of sites, including *Anadara* shell middens and stone artefact scatters, continues to be recorded in the Darwin area through consultancy surveys (eg. Alford 2000; Burns 1999; Heritage Surveys 2000). However, investigations for this project are the most comprehensive undertaken to date in the area. Examination of the landscape around Darwin Harbour generally confirmed the distribution patterns noted previously by Burns (1994) and Hiscock and Hughes (in press), with the predominant coastal site types being shell middens followed by surface stone artefact scatters.

Shell middens

Anadara shell middens typically occur in clusters at the hinterland-mangrove junction on the many promontories of

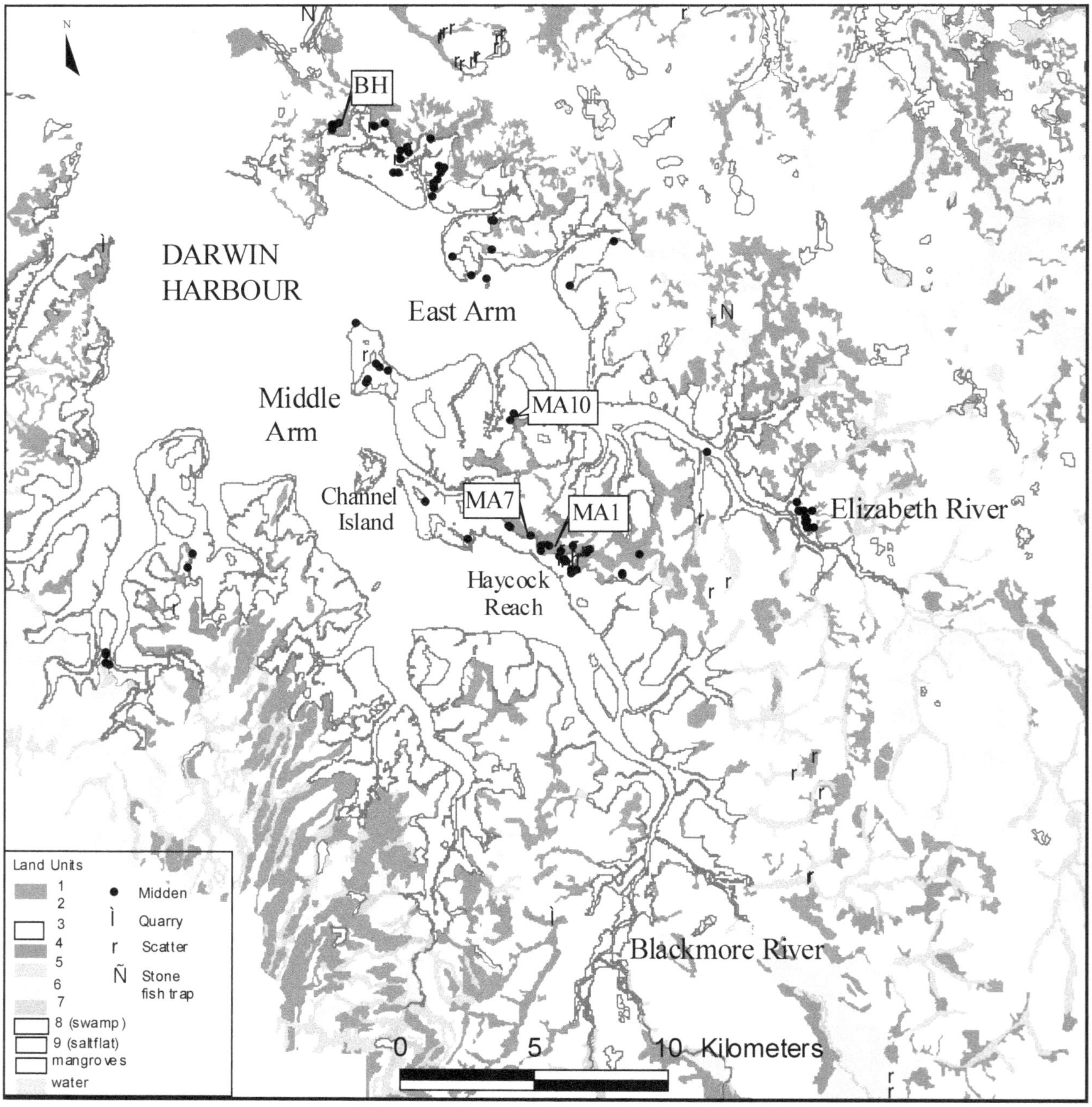

Figure 5.2 Location of sites around Darwin Harbour

Site Type	Darwin Harbour	Hope Inlet
Shell middens and mounds	96	197
Sub-surface middens	0	2
Earth mounds	0	16
Stone artefact and shell scatters	2	12
Stone artefact scatters (stone only)	39	1
Quarries	4	1
Engraving site	1	0
Stone arrangement	2	0
Total	144	239

Table 5.1 Site types recorded at Darwin Harbour and Hope Inlet

the indented harbour shoreline (Figure 5.2). A few middens are situated on rock outcropping from the intertidal sandy mudflats amongst the mangroves. Most are situated on the hinterland margin at the foot of the raised hillslopes, or on the crests of discrete low hills and rocky knolls overlooking the mangroves, some with panoramic views of the harbour and surrounding countryside (Plates 5.1 and 5.2).

The shell middens are generally circular or elongated in shape, and range in size from shallow deposits seven metres in diameter and 20cm in depth, to mounded deposits up to 14m in diameter and two metres in depth. Surface inspection indicated a fairly consistent similarity in midden composition. In addition to the dominant *Anadara granosa*, other shell such as the bivalves *Marcia hiantina* and *Saccostrea echinata* (oyster), and the mangrove gastropods *Chicoreus capucinus, Nerita sp., Telescopium telescopium, Terebralia semistriata* and *Volema cochlidium* were observed in lesser numbers. Large, broken single specimens of *Syrinx aruanus* (false trumpet shell) were observed on some sites as well as the typical suite of shells. A few stone artefacts, mostly consisting of flakes manufactured from local quartz, are common on the surface of the shell middens.

Exceptions to the archetypal *Anadara* dominated midden were found in one small cluster of mounded shell deposits along a hinterland ridge overlooking the mangrove-fringed banks of the Elizabeth River (Figure 5.2). These deposits are composed predominantly of the mangrove tree oyster *Melina ephippium* and lesser numbers of other taxa found in the mangroves such as *Nerita sp., Telescopium telescopium, Geloina coaxans* and *Volema cochlidium* in a matrix of black soil. Careful inspection found no *Anadara* shell on the surface of these deposits, or in any of the material revealed by goanna burrowing. Mounds composed predominantly of *Melina* have not been recorded previously in the Northern Territory, so these sites are significant in terms of their rarity.

Quarries

Recorded quarries within the Darwin area are relatively rare, even though outcropping quartz occurs commonly on hill crests or low laterite ridges. This is possibly a result of difficulty with identification of this site type, which may be partly due to the low visibility of quartz artefacts against the dense background quartz stone ground cover surrounding outcrops (Plates 5.3 and 5.4) (Burns 1996a:28; Hiscock and Mitchell 1992:13). Quarries are associated with stone artefact scatters and sometimes found in association with middens (Burns 1997). Results of surveys also demonstrate raw material rationing in sites, with a decrease in density of artefacts manufactured from a particular raw material extracted from a quarry, and increased reworking of the stone artefacts, as distance from the quarry increases (Burns 1996b:20; Guse 1998:17; and see Roddom 1997).

Stone artefact scatters

Stone artefact scatters made up only a small proportion of sites recorded during surveys for this project. Most known stone artefact scatters in the Darwin region have been recorded through consultancy surveys (eg. Guse 1995; Hiscock 1993b; Hiscock and Mitchell 1992; Navin and Officer 1995; Stone 1989). Scatters are commonly found inland of the shell middens, in close proximity to outcropping stone, usually at the junction of different ecological zones. They are often located on low laterite ridges overlooking alluvial flats (Plate 5.5), where seasonal fresh water would be available, and are associated with the numerous coastal swamps, rivers and creeks in the area. Artefact types recorded in surface scatters are mostly flaked stone, including retouched flakes, bifacial points and cores, and the occasional ground stone such as hammerstones and axes (Guse 1995).

Quartz is the dominant raw material component on all sites (Plate 5.6), with some quartzite, sources of which are locally available in numerous outcrops within a few kilometres of the coast. Other raw materials, such as sandstone, dolerite, chert and Gerowrie tuff, are occasionally observed. Small outcrops of sandstone are found in the Berry Springs area (Pietsch and Stuart-Smith 1987:21). Dolerite and tuff are not available close to the coast, and must have been transported from sources much further inland. Raw

Plate 5.1 Typical *Anadara* mound on Middle Arm Peninsula at foot of hillslope overlooking sandy strip and mangroves in background

Plate 5.2 Shallow *Anadara* midden on hillcrest with views across harbour and mangroves to Darwin city skyline

material sources for some of these artefacts may have been outcrops of Gerowrie tuff and dolerite up to fifty kilometres south of Darwin in the Adelaide River region. Conclusions from consultancy surveys are that locally available rocks such as quartz were predominantly exploited as sources of raw material for stone artefact manufacture, and that scatters are concentrated in areas with suitable sources of stone raw materials and water (Guse 1995; 1996; Guse and Gregory 1994; Hiscock and Mitchell 1992).

Archaeological sites at Hope Inlet

Investigations at Hope Inlet showed this area to be an archaeologically rich area, with large numbers of *Anadara* shell mounds, comparable in both type and distribution within the landscape, to *Anadara* mound sites described elsewhere in northern Australia (Bailey 1977, 1994; Beaton 1985; Cribb 1986; McCarthy and Setzler 1960; Meehan 1982a, 1983; Mulvaney 1975:247-8; Peterson 1973; Roberts 1991, 1994). Commonly found in clusters of two or more, sites are spread across the environmental zones, predominantly on landform type 9, which includes the saltflats, chenier ridges and beach ridges on the hinterland margins. A few sites are located on the boundaries of all other landform types except type 7 and type 8, as would be expected in these seasonally inundated environments.

Shell middens

Anadara shell middens and mounds at Hope Inlet are found on each of five hinterland promontories and adjacent saltflats (Figure 5.3) (21% of all sites are on the hinterland margins). Tidal channels, mangroves and swamps separate each of the promontories and an associated saltflats area from each other. Middens located on the saltflats (61% of sites) extending between the hinterland promontories and the mangrove fringe, are usually close to the mangrove lined tidal channels. Some appear to be compacted and salt encrusted through periodic inundation by seawater, partially buried under mud, or situated on partially buried chenier ridges. On two of the promontories, large "composite mounds" comprising a series of partly coalescent shell and earth mounds, similar to those described by Cribb (1996:160) at Aurukun, are situated on lateritic ridges which border the seaward edge of the hinterland promontories (Plates 5.7 and 5.8). In the main, the deposits exhibit in three forms, either circular, elongated or occasionally doughnut-shaped with a central area devoid of shell. Smaller shell mounds, middens, shell rings and earth mounds are located a few hundred metres inland from the large mound complexes, and on the other promontories.

These are situated approximately 50-100m apart just within view of one another, in mixed and open eucalypt woodland (18% of sites). Two sub-surface middens were also located in flat cleared areas in the woodland, revealed only by the presence of midden material brought to the surface by goanna activity.

Shell deposits are found in a continuum of sizes (Plates 5.13 and 5.14), from scatters and shallow deposits of shell one metre in diameter and 20cm in depth, to large mounds and composite mounds extending for up to 300m along lateritic ridges, the largest mound rising seven metres above the surrounding landscape. Other researchers (Bailey 1994:114; Beaton 1985:4) have noted a similar continuum in the size distribution of *Anadara* shell mounds on Cape York Peninsula. The largest mound has steep sides on the short axis and both a moderate slope and long gentle slope with easy access to the top comprising the long axis (Plate 5.7), which Bailey (1975a:VII:14) describes as a consequence of the accumulation process of the mounds. Like the Weipa mounds of Cape York (Bailey 1975a:VII:13), measurements of the dimensions of over 90% of the middens indicate that the majority are less than one metre in height (85%) and 100m^3 in volume (62%) (Table 5.2). Middens with an estimated height >2m make up only 7% of sites, and of these sites a large proportion (65%) are located on the hinterland margin.

On the surface of the saltflats sites, small numbers of stone artefacts were observed, mostly flaked quartz, along with an occasional edge ground axe or pestle (Plate 5.11). Laterite rocks are also a feature at many sites on the saltflats, and were presumably transported there for use as hearth material. Also observed on some sites were animal bones, crab exoskeleton, ochre, and otoliths, as well as large, broken shells such as *Melo amphora* (baler shell) and *Syrinx aruanus* (false trumpet shell), mentioned earlier to be present on some Darwin Harbour middens (Plate 5.12). Although not recovered in any excavated sites, human bone, which had been brought to the surface by the burrowing activity of goannas, was observed on a few shell mounds at Hope Inlet. There are a number of references to the Larrakia practice of secondary burial of bones of the deceased (Basedow 1907:5, 1925:208; Foelsche 1886:256), and the small quantities of human bone seen in a few mounds is consistent with this practice. The only obviously historic items seen were a few World War II aircraft ammunition shells. Non-shell material recorded on the surface of the more inland sites comprised mostly pre-contact material such as stone artefacts, quartz stone and ochre. European post-contact material such as buffalo and pig bones and pieces of tin and other metal, were seen on only a few hinterland sites.

Shell and stone artefact scatters

While quartz and quartzite flakes commonly occur in small numbers on the surface of middens, greater concentrations this type of small, sharp, flaked stone artefact were found in scatters on the saltflats containing both stone artefacts and shell. These scatters range in size from approximately 20m^2 to 800m^2. More obtrusive components of many of these shell and stone artefact scatters than the typical small quartz flakes are the larger stone artefacts. These include many rounded or irregularly shaped pestles or pounders made from quartz, quartzite and sandstone, and

Plate 5.3 Quarry site MA20, showing quartz outcrop and surrounding quartz ground cover

Plate 5.4 Close-up of white quartz stone artefacts at base of outcrop, site MA20

Plate 5.5 Stone artefact scatter site MA13, showing a few white quartz stone artefacts in foreground

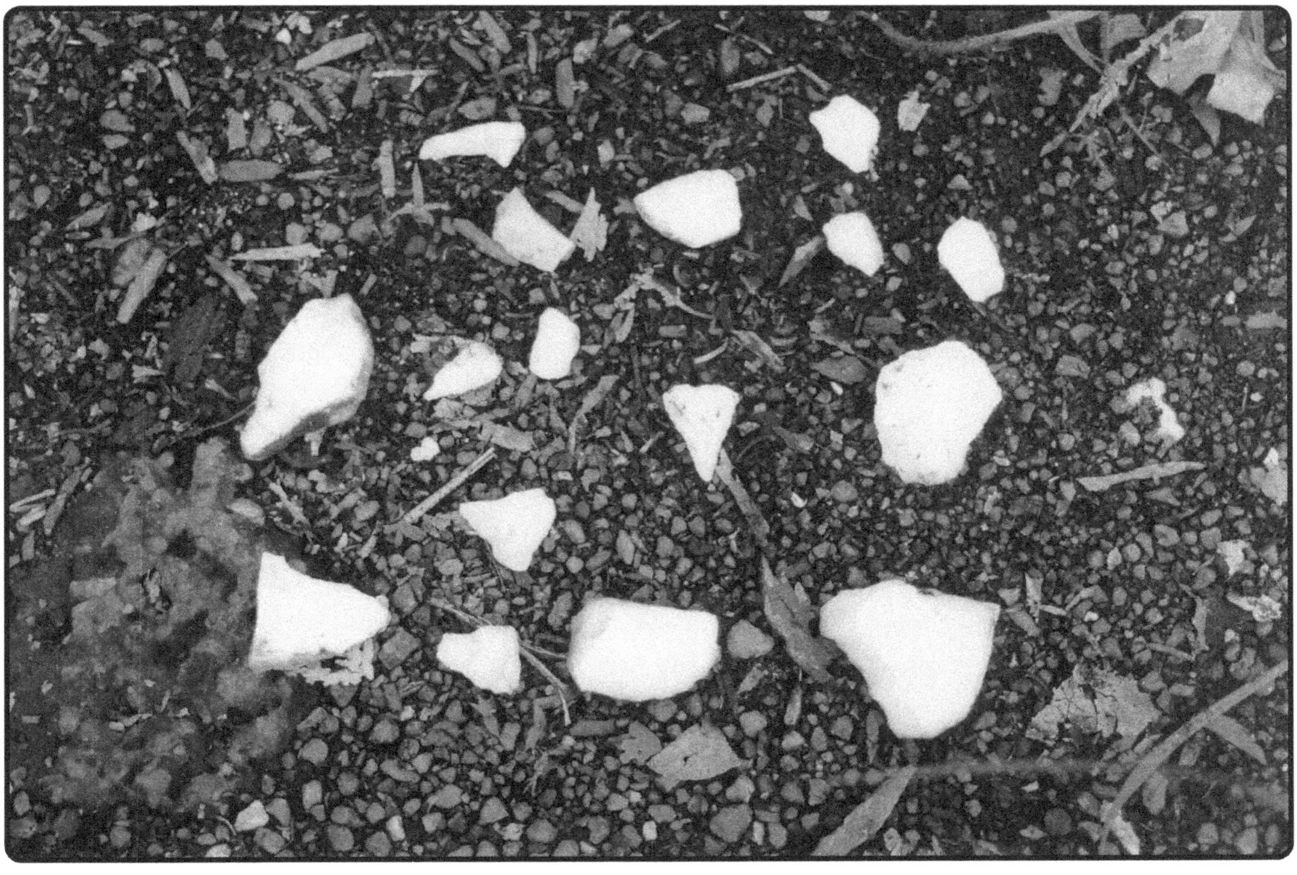

Plate 5.6 Close-up of white quartz stone artefacts in site MA15

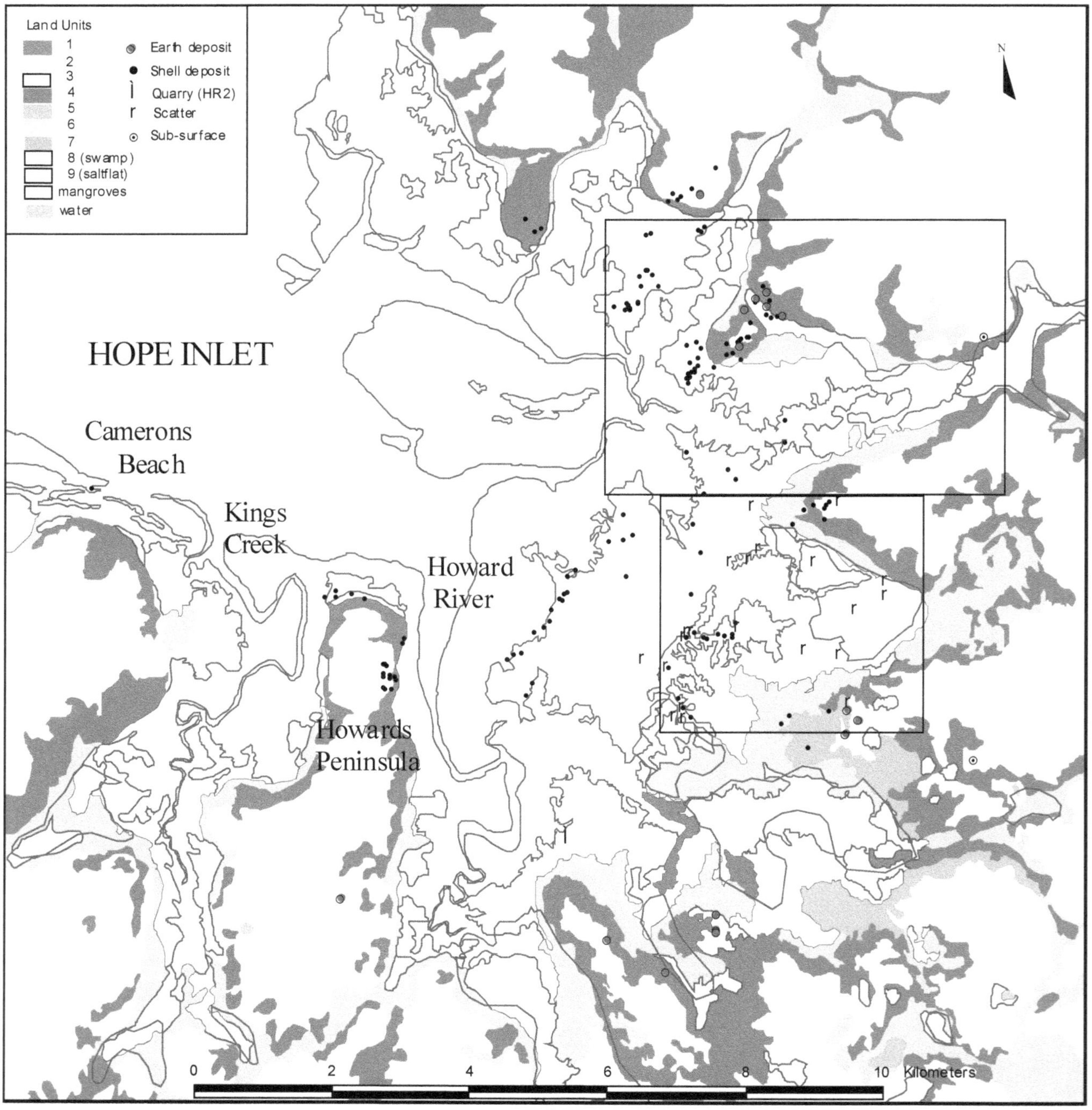

Figure 5.3 Location of sites at Hope Inlet

Plate 5.7 Largest *Anadara* shell mound at Hope Inlet, part of Mound Complex HI81

Plate 5.8 View of second largest *Anadara* mound in Complex HI81, looking northwest along ridge from top of largest mound, with saltflats in background

Plate 5.9 Large *Anadara* mound in open eucalypt woodland at edge of saltflats, seen in background

Plate 5.10 Shell ring in mixed forest inland of large mounds

in lesser numbers edge ground axes manufactured from non-local dolerite and hornfels (Plate 5.15), portable mortars (shallowly hollowed grinding stones commonly of sandstone or quartzite) and large flakes made from exotic quartzite. One large flake with orange and white diagonal stripes and a distinct positive bulb of percussion, measured 160mm long by 40mm wide (Plate 5.16).

Large flakes of the type found at Hope Inlet have been labelled as large blades, "leilira" blades, trigonal points and *ngambi* points (Allen 1997; Brockwell 1989; Jones 1990; Jones and White 1988:52; McCarthy and Setzler 1960:248). The mortars at Hope Inlet are all square and shallowly hollowed, and differ from the mortars with deeper circumscribed depressions also seen in rock outcrops, observed at some other coastal plains sites (see Brockwell 1989:101-5). They also differ from the seed grinding implements of arid Australia, which are more elongated and have long grooves worn into the stone surface (Smith 1988a). This type of mortar at Hope Inlet reflects the generalised and pounding functions of wetlands resource strategies in high rainfall areas as opposed to the desert areas where grinding of seeds is more common (Peterson 1968:568). The wide range of artefact types and raw materials in the Hope Inlet scatters are comparable to those found in wetland sites across northern Australia, often in association with middens, on the South Alligator Rivers woodland margins adjacent to freshwater lagoons and swamps, on Cobourg peninsula and closer to home on the coastal plains of Chambers Bay (Baker 1981:66-7; Brockwell 1989:93-8; McCarthy and Setzler 1960:218, 286; Meehan *et al*. 1985; Mitchell 1994; Woodroffe *et al*. 1988:96).

Scatters of shell and stone artefacts were observed to be mostly concentrated in two areas, and on two different types of substrate. Those found more seaward, including amongst a network of mangrove-lined tidal channels and perennial ponds, that dissect the saltflats midway between the hinterland and seaward mangrove fringe, are situated on a loose sandy substrate. In this area many shell mounds with associated stone pestles also occur. Other stone artefact scatters are situated on a grey cracking clay-pan surface at the landward edge of the saltflats 100-300m from the hinterland margins. These were located on a four kilometre square section of saltflats associated with a large swampy area between two promontories, which is fed by a number of tributaries running into a main tidal channel (Figure 5.4.).

Site disturbance

Different site disturbance processes were apparent in these two areas. In the former area I observed that many of the stone artefacts (mainly small flakes) recorded on a loose sandy substrate during the first survey year were obscured by aeolian deposition the following year. In the latter area artefacts recorded during the dry, consisting of both small flakes and the large ground implements, were exposed on a hard substrate of cracking clays. Some of the smaller

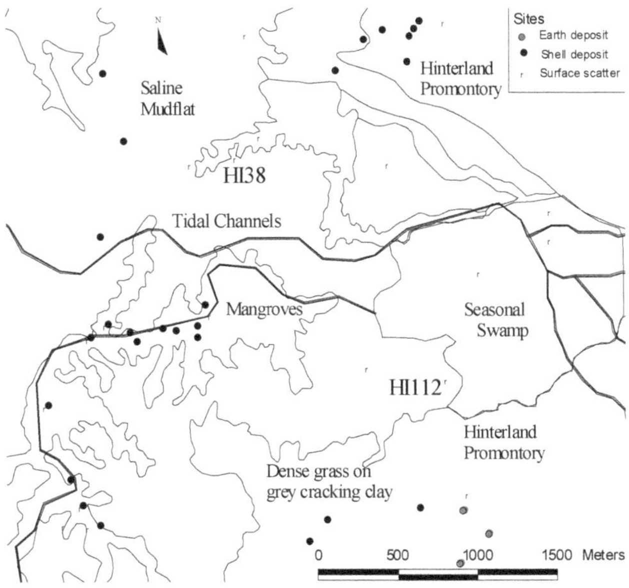

Figure 5.4 Location and environmental context of sites HI38 and HI112 and other shell and stone artefact scatters at Hope Inlet

stone artefacts recorded on this clay surface during the first survey were not visible the following year, probably due to falling down the cracks. This may be due to the pronounced shrink-swell properties of the clay substrate (Wood *et al* 1985:53), which means it is subject to "argilliturbation", a process of soil mixing resulting from alternate shrinking and swelling of the clays during the dry and wet seasons (Wood and Johnson 1978:352). During this process mixing of archaeological materials also occurs through movement of small artefacts downwards and large artefacts upwards (see Baker 1978; Butzer 1982:109; Erlandson and Rockwell 1987; Gregory 1998:111). These disturbance processes affect the ability to accurately record artefact types and densities on these sites and may bias archaeological interpretations of the nature of the sites (Gregory 1998:98-120), and make it difficult to determine the age of the sites through geomorphic data. Examples of sites on different substrates, Site HI38 and HI112, are described below, and their location and environmental context, and that of other shell and stone artefact scatters at Hope Inlet, are given in Figure 5.4.

Site HI38

One example of a shell and stone artefact scatter found on a sandy substrate on the saltflats is Site HI38, located alongside the end of a mangrove-lined shallow tributary of the main tidal channel that dissects the saltflats (Plate 5.17). The creek and surrounding saltflats was dry at the time of survey in the late dry season. There appears to be dieback of the mangroves around these creeks, and this is probably a natural process that occurs around saltpans due to changing freshwater/seawater regimes, strong evaporation and hypersalinity (Galloway 1982:34; Brocklehurst and Edmeades 1996). The site is situated midway between a major tidal channel a kilometre or so to the southwest and the higher ground of a hinterland promontory to the northeast, on which are located a cluster of shell mounds,

Plate 5.11 Stone pestle on surface of shell mound on saltflats

Plate 5.12 Large baler shell found on saltflats midden

Plate 5.13 Shell scatter on Hope Inlet saltflats

Plate 5.14 Large mound on Hope Inlet saltflats, showing typical vegetation

Plate 5.15 Edge ground axe in Terebralia shell scatter on grey cracking clay surface

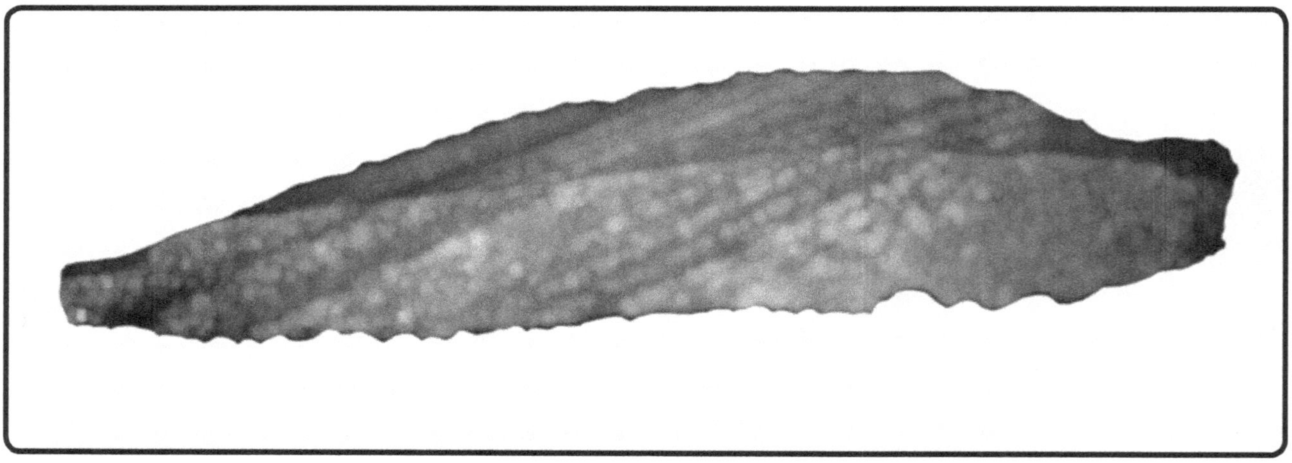

Plate 5.16 Large flake made from exotic orange striped quartzite found in shell and stone artefact scatter

middens and rings. Half to a kilometre south of site HI38 is the network of mangrove-lined tidal channels on the saltflats and shallow perennial ponds described above (Figure 5.4).

Dimensions of this surface scatter include a small 20m x 10m area of *Anadara* shell within a wider scatter over 50m in diameter with other shell taxa including *Terebralia, Telescopium, Volema, Nerita, Cerithidea, Geloina,* and *Ellobium*, and stone artefacts. Identified stone artefact types include mainly small flakes, a few pestles, and a portable mortar and edge-ground axe (Plate 5.18). Raw materials include imported quartz, quartzite, sandstone and dolerite. Stone artefact density varied from five to 30 artefacts per square, counted in two one-metre squares in separate areas of the site. The area around the site appeared quite desolate at the time of survey, with no evidence of resources such as swamp plants nearby, that may have required the use of flaked or ground stone for processing plant resources, as Brockwell (1989, 1996a) describes for the South Alligator wetland sites.

Site HI112

Two kilometres southeast of Site HI38 is an example of a shell and stone artefact scatter on the grey cracking clay surface, designated Site HI112, located at the saltflats edge, about 100 metres from the margins of one of the southern-most hinterland promontories. Fewer and smaller midden sites were recorded on this relatively low-lying promontory, which features large stands of pandanus, than on the three promontories to the north. Site HI112 is positioned on a section of the supra-tidal saltflats that stretches as a narrow tongue between a large seasonally inundated brackish swampy grassland/sedgeland area on its north/north-eastern edge, and a tidally inundated area alongside the hinterland edge to the south-west. The saltflats tongue is of slightly higher topography than these surrounding areas and the scatter is spread over an area some 30m in diameter. Ground surface visibility was poor, due to the lack of contrast between the colour of the stone artefacts and grey clay background, and this has probably hidden this site from collectors (Plate 5.19).

Over 100 stone pestles in a variety of raw materials and oval and angular shapes were counted at this site (Plate 5.20). Other artefact types include small flakes and edge-ground axes. Raw materials include imported quartz, quartzite, sandstone, dolerite and local porcellanite and quartzite. Eight stone artefacts were counted in a representative one-metre square grid, and a small number of *Terebralia* and *Telescopium* shell are present on this site. Like Site HI38, the area surrounding this site appeared quite dry and denuded of vegetation at the time of survey, but presumably the adjacent swamp supported plant resources that the past inhabitants processed by pounding and/or grinding with the large numbers of pestles at this site.

Earth mounds

On surface inspection the Hope Inlet deposits also exhibit a continuum in the relative proportions of shell and earth, from mounds of predominantly shell with very little earth matrix, to earth mounds with only a few weathered shell fragments seen on the surface. The relative proportions of shell and earth are difficult to quantify without excavation. In this study sites are classed as either earth or shell mounds on the dominant component estimated from surface examination. The Hope Inlet earth mounds are generally circular, and occasionally elongated. They range in size from approximately 12m to 25m in diameter and 0.4m to 2m in height.

Similar type earth mounds have been described in northern Australia at Chambers Bay, 90km east of Hope Inlet (Baker 1981), Arafura swamp in Arnhem Land (Peterson 1973) at Milingimbi (Roberts 1991) and near the Blyth River (Meehan 1988; 1991) and South Alligator River (Brockwell 1989:158; Meehan *et al* 1985). Further afield at Aurukun on Cape York Peninsula Cribb *et al.* (1988:68) refer to clusters of small earth "oven mounds" containing termite material located in the vicinity of shell mounds. Stone artefacts are rare on the surface of these earth mounds as Baker (1981:81) also observed for the Chambers Bay earth mounds. Earth mounds recorded by Brockwell (1996b, 1996c) 25km inland from the Darwin coast on the margins of the Adelaide River floodplains, differ in form and composition from most of the Hope Inlet earth mounds. The earth mounds at Hope Inlet that appear most similar to the Adelaide River mounds, which are generally of lower and wider topography, ranging from 20m to 60m in diameter and from 0.20m to 1.5m in height, and containing numerous stone artefacts (Brockwell 1996c:161), are those on the southernmost promontories.

Quarries

Only one quarry (Site HR2) was recorded at Hope Inlet, an outcrop of quartzite rock protruding from a section of saltflats, some kilometres south of the main concentrations of sites, and 600m east of Howard River. An *Anadara* midden and stone artefacts procured from the outcrop, and some manufactured from another unidentifiable material, were scattered on and around the rock. Stone from this quarry appeared similar to raw material recorded on the shell mounds to the north.

Variability between Darwin Harbour and Hope Inlet sites

from information on site numbers, type and distribution obtained through archaeological investigations undertaken by others and for this study (Table 5.1, Figures 5.1-3), it is apparent that there are both similarities and differences in site distribution patterns between the two different coastal environments represented in the study area. In both areas the sites are generally located in clusters at the junction of a number of different land units or ecological zones, incorporating access to the sea, mangroves, freshwater swamps, creeks and woodland habitats. Around the relatively steep shoreline of the Darwin Harbour embayment and associated riverine environment, midden sites occur where the Dissected Foothills and Littoral Complex meet, at the foot of hillslopes, and on the crests of discrete hills with clear views of the surrounding landscape and across the mangroves and harbour. At Hope Inlet the sites are located on a low lying prograding landscape of the flat Coastal Plains and on hinterland promontories formed on the lower slopes of the Northern Plains upland plateau.

Differences can be seen between the two coastal areas in site numbers and frequency of site types (Table 5.1). Although shell middens predominate in both areas, this is more pronounced at Hope Inlet, where large numbers of middens and mounds, including two large mound complexes, and doughnut-shaped shell rings and sub-surface middens, not found around Darwin Harbour, occur. Earth mounds too were only located in the Hope Inlet area, which has a topography of very low relief (Pietsch 1985:1), without the low hills and ridges and outcrops of rock that provide raw materials for the more numerous stone artefact scatters around Darwin Harbour.

Plate 5.17 Site HI38 shell and stone scatter on sandy substrate on saltflats, adjacent to mangrove-lined creek

Plate 5.18 Close-up of portable mortar, pestle, edge ground axe, flaked stone and shells, site HI38

Plate 5.19 Site HI112, scatter on grey cracking clay surface at edge of saltflats

Plate 5.20 Close-up of pestles, edge ground axes, and flaked stone, site HI112

Mound volumes

A frequency distribution table of approximate mound volumes, for a subset of the total number of mounds on which all three measurements - width, length and height - were taken, shows that shell deposits at Hope Inlet are generally larger and more numerous than those around Darwin Harbour (Table 5.2, Figure 5.5). The difference between the two areas in number and size of mounds may simply be due to removal of larger deposits around the harbour through the process of European settlement and development as discussed in Chapter One. Other detailed analyses, such as frequency distribution of mound volumes according to location types ie. saltflats, hinterland margin and woodland, as undertaken by Bailey (1975a:V11:37) for the Weipa mounds, are not attempted here. Investigations suggest that many of the mounds on the saltflats may extend further beneath the level of the saltflats surface, so total shell volume based on surface measurements of mounds on this landscape type is likely to be an underestimate. The estimated volume of Mound HI80 for example, when measurements of its full sub-surface extent are included (see Chapter Eight), is more than seven times that of the volume which can be measured on the surface. Similarly, it is difficult to quantify the total shell volume of the large coalescent composite mounds situated on shell-bearing laterite ridges with other than a rough estimate. Mowat (1995:98-101) found similar problems with estimating volumes of middens on the West Alligator River sites.

Surface scatters

More surface scatters containing only stone artefacts were located around Darwin Harbour, and these are commonly located inland of the middens, on raised eroded areas such as low hills and laterite ridges, close to raw material sources. Shell scatters and combined shell and stone artefact scatters are more common in the Hope Inlet area, and these range in areal dimensions between 20m^2 and 800m^2. The Hope Inlet scatters are generally smaller than but still within the range, of dimensions of 4m^2 to 12000m^2, and of artefact densities, for scatters recorded around Darwin Harbour and further afield in the South Alligator and Mary River regions (Brockwell 1993, 1996a, 1996b; Burns 1996a, 1996b, 1999; Guse 1992:77-78, 1996, 1998; Hiscock 1993b; Hiscock and Mitchell 1991; Navin and Officer 1995; Smith 1981). They are much smaller than the large wetland sites of 10,000m^2 to 25,000m^2 at Kakadu (Jones 1985b:291-2; Meehan *et al.* 1985) but more recent surveys suggest there is just as great a diversity in site dimensions and artefact densities within any one area as there are differences between regions (Guse 1995:17,34-5; Hiscock *et al.* 1992). Possibly, the dimensions of the Hope Inlet scatters are circumscribed

Table 5.2 Frequency distribution of midden volumes at Hope Inlet and Darwin Harbour

Volume m3	Hope Inlet	Darwin Harbour	Total	%
<100	140	25	165	83%
100-499	25	2	27	14%
500-999	3	1	4	2%
>1000	2	0	2	1%
Total	170	28	198	

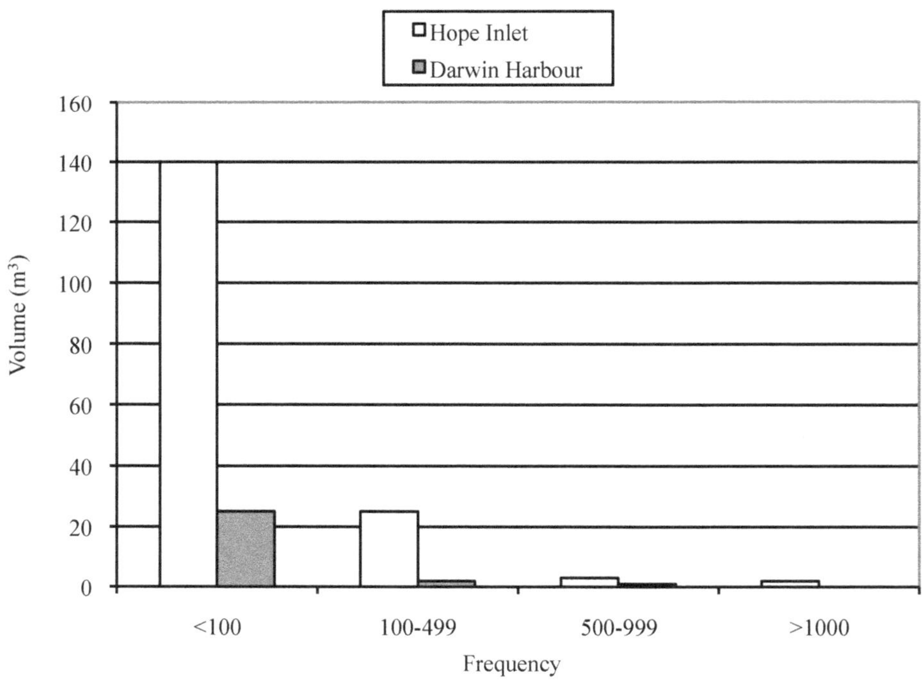

Figure 5.5 Frequency distribution of midden volumes at Hope Inlet and Darwin Harbour

by the nature of their landscape context, placed as they are on small delimited areas of barely discernible rises in topography amongst a network of swamps and tributaries on the saltflats (Figure 5.4).

Differences in stone artefact types and raw materials

Surface scatters exhibit distinct differences in the types of stone artefacts and raw materials represented between the two areas (Tables 5.3 and 5.4). Around Darwin Harbour stone artefact scatters contain mainly flaked stone such as small flakes and bifacial points manufactured predominantly from white quartz, and some quartzite, raw materials which are readily accessible in local outcrops. Flaked stone dominated by quartz was also a common material on the South Alligator River wetland sites, thought to reflect its local availability (Brockwell 1996:93). However, at Hope Inlet, where quartz and quartzite also dominate the flaked stone in shell and stone artefact scatters, only the grey/blue quartzite is available locally from the quarried outcrop described above.

Presumably the source of the white quartz component of these sites is also from the quartz outcrops around Darwin, 25km to the south-west. In addition to the local porcellanite and grey quartzite and Darwin quartz, artefacts at Hope Inlet are made from a variety of other raw materials, including exotic rose quartz, and exotic white or orange-striped quartzite, sandstone, hornfels, dolerite and tuff. The presence of the large flaked leilira blades, and ground stone such as pestles, mortars, and edge ground axes manufactured from this diverse array of materials is noticeable. It is interesting that a couple of the blades are broken in a manner similar to other north Australian blades used for exchange, which are thought by Paton (1994:175-6) to have been deliberately broken.

Other than the locally occurring porcellanite (Pietsch 1985:14), a fine-grained sedimentary rock, and the grey/blue quartzite, sources of these stone raw materials were not found in the Hope Inlet area. Sandstone is also found around Darwin Harbour, in the Berry Springs area (Pietsch and Stuart-Smith 1987:21). The nearest sources of the raw materials hornfels, dolerite and tuff and the exotic quartzite from which the large blades are made, are over 60km kilometres inland and must have been carried for long distances to these sites (Kim Akerman pers. comm.). The relatively rare large leilira blades in this area are similar in form to *ngambi* points reported as manufactured at the Ngilipitji quarry in eastern Arnhem Land, associated with ritual and exchange (Jones 1990:25; Jones and White 1988:52).

Other quarries where blades of this type have been reported are Deaf Adder Creek and Argaluk Hill near Oenpelli (Kamminga and Allen 1973:103-6; McCarthy and Setzler 1960:283, cited in Allen 1997:369). Outside of Arnhem Land quartzite quarries are known from the Daly River and Victoria River districts (Basedow 1907:32; Stanner 1933-4:162); Katherine (Edwards 1967:11-14) and Pine Creek (Cundy n.d. cited in Allen 1997:369). The source of the large Hope Inlet flakes, which are made from an orange-striped quartzite and white quartzite, may not be from Arnhem Land as that stone is reported to be a pinkish-grey colour (Jones 1990:27) or (at Oenpelli) white, pink, mauve, brown and red (McCarthy and Setzler 1960:285).

It is quite possible that the scarcity of larger more noticeable stone artefacts around Darwin Harbour is due to their removal by collectors in what is a populated, urban area compared with the much less accessible Hope Inlet area. Collectors tend to pick up artefacts in readily identifiable form, such as bifacial points, axes or pestles. Oral accounts by locals, the presence of stone assemblages that have since been handed in to MAGNT and the recording of an occasional hammerstone or pestle (eg. Guse 1995:29) and stone axe (eg. Kinhill 1992:4) in the less populated areas, suggests that much collecting by individuals has occurred since first European settlement of the area. Support for this hypothesis is provided in Table 5.3, which shows increased variety, of site types including ground stone artefacts, and of raw materials, in scatters located further out of city areas, at Berry Springs (BS) (see Figure 5.1).

An alternative hypothesis is that the abundance of large heavy stone artefacts suitable for grinding and pounding at Hope Inlet reflects different activities being carried out in this area. Ethnographic studies suggest that one of the uses of large ground stone artefacts such as pestles and mortars was to process plant foods such as the cycad palm nuts, and spike rush (*Eleocharis dulcis*) and waterlilies (*Nymphae spp.*) from the coastal swamps (Russel-Smith *et al.* 1997:171). Numerous coastal swamps that would have supported these plant foods, and stands of cycads, occur both around Darwin Harbour and at Hope Inlet. If similar resources were available for exploitation in both areas it seems likely that similar implements would have been required.

However, these two areas are distinctively different in terms of geology and topography, which underlies the vegetation/environment association (Bowman and Wilson 1986:75; Wilson and Bowman 1986:12). And so, on the limited available ecological data, it would appear that these swamp and lagoon vegetation communities and stands of cycads are more extensive and found in closer spatial association in the gently sloping transitional zone between the Northern Plains lower slopes and Coastal Plains at Hope Inlet. At Darwin Harbour the transition between the mangroves and higher ground of the hinterland is much more sudden (see Christian and Stewart 1953; Baker 1981:15-17; Pietsch 1983, 1985, 1986 and Chapter Two).

According to ethnographic and experimental use-wear studies small, sharp flaked stone artefacts found in both areas could have been used for activities such as scraping or cutting meat and plant material, as well as implement manufacture, spear heads and body scarification (Curr 1886:251; Foelsche 1882:6; Kamminga 1982:24, 32-6, 53-5; McCarthy and Setzler 1960:283). Brockwell (1996:95-6)

Table 5.3 Presence/absence of artefact types and raw materials in stone artefact scatters around Darwin Harbour

Site	Artefact Type		Raw Materials						
	flaked stone	ground stone	quartz	# quartzite	sand-stone	dolerite	tuff	chert	other
WIN1s	F		*						
WIN5s	F, C		*	*					
EA3s	F		*						
EA5s	F		*						
RC1	F, C,		*						
RC2	F, C, FP, B		*						
RC3	F		*						
RC4	F, RF		*					*	
RC5	F, RF		*					*	
RC6	F, FP, RF, BP		*					*	
RC7	F, C		*					*	
RC8	F, C		*					*	
RC9	F, BP, FP, B		*					*	
MS1	F, RF, C		*	*				*	
MS2	F, C		*					*	
MS3	F, BP, C, RF, FP		*					*	
NS1	F, UP		*	*				*	
NS2	F, A		*	*				*	
MC1	F, RF		*	*					
MA2s	F		*					*	
MA13	F, C		*				*		
MA14	F, BP		*	*					
MA15	F, BP, C, RF		*						
MA20q	F, C		*						
MA30s	F		*						
MA12Hs	F, M, RF		*		*				*
RB1q	F, C, RF		*						
BS1	F, C, R, FP, BP		*						
BS2	F, C, RF, FP		*						
BS3	F, C, RF, FP, BP		*						
BS4	F, C, RF, FP		*	*			*		
BS5	F, C, RF, FP, BP		*	*			*		*
BS6	F, C, RF, FP, BP		*	*			*		*
BS7	F, C, RF, FP, BP		*	*			*		*
BS8	F, C, RF, FP, BP		*	*			*		*
BS9	F, C, RF, FP, BP	A, P	*	*		*	*		*
BS10	F, C, RF, FP, BP	A	*	*		*	*		*
BS11	F, C, RF, FP, BP	A	*	*		*	*		*
BS14	F, C, RF, FP, BP, UP		*						
BS21	F, C, BP	M	*	*	*	*	*		
BS22q	F, C, BP, M		*	*					
BS23	F, C, BP		*			*	*		
BS24	F, RF, C, BP		*	*		*	*		
BS25	F, RF, C, FP, BP, UP, M	P, M	*	*	*			*	

WIN=Winnellie, EA=East Arm, MA=Middle Arm (Burns 1994, 1996a, 1999; Heritage Surveys 1997) RC=Rapid Creek, MS=Marrara swamp (Navin and Officer 1995) NS=Noogoo swamp (Stone 1989) MC1=Micket Creek, RB=Rosebery (Hiscock 1993b, 1995b) BS=Berry Springs (Guse 1998; Burns 1999) s= shell present, q= quarry, # quartzite includes local and exotic varieties, F= Flake, C= Core, A= Axe RF= Retouched flake, FP= Flaked piece, BP= Bifacial point, UP= Unifacial point, P= Pestle/pounder M= Manuport

Table 5.4 Presence/absence of artefact types and raw materials in stone artefact scatters at Hope Inlet

Site	Artefact Type		Raw materials					
	flaked stone	ground stone	quartz	# quartzite	sandstone	dolerite	tuff	porcellanite
HR2q	F			*				
HI9	F							*
HI11s	F		*				*	
HI38s	F	GS, P, A	*	*	*	*		
HI39s	F		*	*				
HI40s	F, C, M	P	*	*	*			*
HI41s	F	GS, P	*	*	*			
HI42s	F, M	P	*	*	*	*		
HI44s	F	GS, P, A	*	*	*	*		
HI45s	F, M	P	*	*	*			
HI46s	F	M		*	*			
HI47s	F	M	*		*			*
HI48s	F	P		*	*			
HI51as	F, M		*	*				
HI52as	F, M	P	*	*	*			
HI53s	F		*	*				
HI54s	F, M		*	*				
HI55s	F, M	M	*	*	*	*		
HI112s	F	P, M, A	*	*	*	*		*
HI128s	F, C	P, A	*	*	*	*		
HI129s	F, RF	GS, P, A	*	*	*	*		
HI137s	F, C, B	GS, P, A	*	*	*	*		
HI138s	F, M	P	*		*			
HI139s	F, C, UP	P, A	*	*	*	*	*	*

HR= Howard River HI= Hope Inlet s= shell present, q= quarry, # quartzite includes local and exotic varieties F=Flake, C=Core, RF=Retouched flake, FP=Flaked piece, BP=Bifacial point, UP=Unifacial point, A=Axe, P=Pestle/pounder, GS=Grinding stone, M=Manuport

found in her analysis of similar assemblages on the South Alligator River wetlands that a significant proportion of small flakes exhibited use-polish consistent with their use as processing implements for silica-rich wetland plants. Use-wear analysis may be one avenue of future research that could address some of the questions on the function of these stone artefact scatters. Ethnographic uses of large blades include as hafted spearheads, knives and fighting picks (Davidson 1935:172-8), while edge-ground axes are generally associated with wood-working (McBryde 1984).

Stone artefacts as evidence for trade/exchange

The archaeological evidence, in the form of stone artefact scatters containing a relatively large quantity of exotic stone, is obviously consistent with trade and/or exchange taking place at Hope Inlet, a place bereft of raw material sources for stone artefact manufacture. As noted earlier, quartz may have been brought in to the Hope Inlet area from outcrops of rock around Darwin 25km to the south-west. The nearest sources of other imported raw materials such as hornfels, dolerite, tuff and the exotic quartzite from which the large blades are made, are over 60km kilometres inland and must have been carried for long distances to these sites (Kim Akerman pers. comm.). The interpretation that the presence of exotic stone indicates trade and/or exchange activities at Hope Inlet is bolstered by the voluminous ethnographic literature in which stone artefacts manufactured from exotic materials are associated with ceremonial exchange networks (eg. Berndt 1951; Roth 1904; Spencer and Gillen 1904; Thomson 1949; Warner 1969; and see McBryde 1984). Sourcing of the raw materials found in these sites as undertaken by researchers in southern parts of Australia (eg. McBryde 1984; and see Godwin 1997:301) is one possible avenue for future research that would provide evidence of the distribution networks of stone artefact exchange in the Darwin region.

Trade in large blades documented for other parts of northern Australia (Stanner 1933; Kaberry 1939) also occurred in the Darwin and western Arnhem Land region (Cundy 1980:179-80, cited in Allen 1997:361). Allen (1997:365-8; Allen and Barton n.d.:127) point out that archaeological sites containing large blades in the Kimberley-Victoria River-western Arnhem Land region are restricted in general to those dated more recently than 1000 BP. Allen (1997) argues that large blades in northern Australia had a secular as well as ritual use, secular in the sense that the large blades were part of spear-hafting technology, or used in everyday activities such as women's knives, and that trade/exchange of these items was between related groups. He contends that the ritual or ceremonial status of large blades that existed in central Australia, spread into north Australia with an expansion of regional ceremonial exchange systems, particularly in the last 500 years (Allen 1997:375). The implications of the exotic stone in these sites for the nature of past occupation and ceremonial exchange is explored further in Chapter Nine.

Implications of site distribution patterns

A distinctive pattern of site distribution was revealed through the results of surveys in this region, being that a significantly large proportion of all archaeological site types are situated on or within 300 metres of land unit boundaries (Burns 1996a). This is consistent with a strategy of positioning campsites within easy reach of a variety of ecotones and resources, thought to be characteristic of coastal forager settlement systems (Yesner 1980:730). Minor variations within land units observed on the ground may not be detailed in land unit mapping, which necessarily generalises and is based on interpretations from remote sensing, aerial photographs and ground truthing through field sampling. So even when sites do not appear to be situated on land unit boundaries on a map, site descriptions may indicate there is a localised change in environment in effectively the same manner as land unit boundaries. For example, Guse (1995) provides descriptions of artefact scatters positioned on top of low laterite ridges adjacent to low lying depressions, a boundary that does not appear on the map. This observation bolsters the predictive assessment that the distribution of all prehistoric archaeological site types in the Darwin region is heavily concentrated on land unit boundaries. Such information is useful in cultural resource management as it can be employed in determining survey strategies for archaeological surveys of proposed development areas (Burns 1996a).

Clustering is another feature of the site distribution pattern in both areas, a phenomenon that has implications for forager population density and extent of sedentism. Around Darwin Harbour midden clusters typically occur at the hinterland-mangrove junction on the many promontories which project into the harbour (Figure 5.2). At Hope Inlet, sites are commonly found in clusters of two or more, on the saltflats, chenier ridges and hinterland margins. Clusters of discrete shell and earth deposits on the hinterland promontories projecting into the saltflats, form a pattern where all are situated just within view of each other 50-100m apart (see above paragraph on Shell Middens, Figure 5.3 and Chapter 8, Figure 8.1). A reasonable explanation for the midden clusters, as Bailey (1975a:VII:42-4; 1983:567) argues for mounds at Weipa, is as a result of people placing themselves in the optimal position to exploit both marine and terrestrial resources in a non-linear landscape (Rowland 1994a:155). Bailey (1983:567) relates the pattern of each cluster containing a large shell mound in open ground and smaller middens in the woodland or on the saltpan, to seasonal changes. Thus the position of individual sites within the cluster is thought to relate to resource availability according to season and microenvironment (Bailey 1977:140).

On the margin of the promontories there is a limited areal extent of a relatively open and narrow transitional zone, separating the more thickly vegetated woodlands of the upland Koolpinyah surface and the adjoining seasonally inundated coastal plains. It is likely that the distribution of cultural material in this area is driven partly by the shape of the landscape topography, resulting in exaggeration of the clustering tendency in the form of coalescent composite mounds (see below). Bailey (1975a:VII:43) proposes this explanation for the Weipa mounds, which have similar conditions of an abundance of resources and restricted length of shoreline for occupation. Rowland (1994a:155) has pointed out that in estuarine areas (such as Hope Inlet), resource distribution is non-linear as opposed to the linear resource distribution found along open coastlines.

Explanations for apparent clustering of mound sites in the Late Holocene feed into the "intensification debate" in Australian archaeology, as evidence for more intensive occupation (Lourandos and Ross 1994). Earth mounds are evidently a late Holocene phenomenon in Australia, appearing in the archaeological record in Victoria after 2500 BP (Williams 1988), and in Arnhem Land from 1500 years ago (Meehan 1988). Baker (1981:63) dated two earth mounds at Chambers Bay one to 920±90 (ANU–2888), corrected to 470±90, and one to a modern date (ANU–1887). However, much of the dating of mounds is on single determinations, and there is the recognised problem of assuming that all mounds or even all of one mound were occupied at any one time, which has implications for interpretations of past settlement patterns. Ammerman (1981:77) refers to this as the problem of "overestimated maps" where sites recorded on a map may in fact have been occupied in different time periods and may not have been part of the same landscape.

In southern Australia dating of similarly clustered earth mounds (Williams 1988:94, 139) indicates they are separated in time by up to 1000 years, suggesting sequential rather than concurrent occupation (Bird and Frankel 1991a:8). Elsewhere in northern Australia, on the West and South Alligator Rivers (Mowat 1995:99-123; Woodroffe et al 1988) and in the Weipa area (Bailey et al 1994; Stone 1992:104), evidence from dating of shell middens in clusters is equivocal with regard to their contemporaneity, with some overlap in time periods of accumulation apparent. The question of whether these sites are contemporary is dealt with in the following chapters of this study, using data obtained through excavation and analysis, including radiocarbon dating of samples from the sites (see Chapter 9, Table 9.1). Before examining data from excavated sites, however, in the next chapter I discuss a range of taphonomic factors known to contribute to the formation of archaeological accretions, particularly shell deposits, which it is essential to consider when attempting to discriminate between natural and cultural components of a site.

6 Considerations in midden analysis

Introduction

Researchers have long expounded the need to identify and consider taphonomic processes when constructing theories of past human behaviour based on data from archaeological sites (Behrensmeyer and Hill 1980; Binford 1981; Hughes 1983; Nash and Petraglia 1987; Schiffer 1987; Solomon *et al.* 1990; Wood and Johnson 1978). Schiffer (1987) provides a synthesis of the cultural and environmental (natural) processes that transform the archaeological record into its contemporary form. Cultural processes include patterns of use and reuse, discard and curation behaviour and site reoccupation and disturbance. Environmental processes include deterioration or destruction of archaeological remains through chemical, physical and biological agents and the influence of climatic and geomorphological conditions.

Cultural and environmental processes involved in the formation of coastal middens include the types of occupational and discard behaviour that influence the spatial patterns of archaeological remains, and the ecological conditions that determine availability of molluscs (Bailey 1983; Meehan 1982a; Waselkov 1987:142-6). In Australia, development of criteria for distinguishing between different types of deposits and agents of accumulation, stimulated by debate on the human or natural origins of shell mounds (Bailey 1991, 1993, 1994; Bailey *et al.* 1994; Burns 1994; Cribb 1991; Mitchell 1993; Stone 1989b; 1992) has drawn attention to the myriad of factors contributing to formation and post-depositional disturbance. Post-depositional processes that affect shell middens may also be either cultural or environmental. Cultural disturbance occurs through reoccupation of sites, and mining of shell deposits for industrial, gardening and other uses (Ceci 1984; Hiscock and Hughes in press; Meehan 1982a:114,166; Waselkov 1987:146-150). Environmental processes that transform coastal archaeological sites range from the more dramatic cyclones in the tropics, erosion and sedimentation, to the more subtle yet cumulative degrading effects of temperature, rainfall and vegetation, and bioturbation by animals (Bird 1992; Hughes and Lampert 1977; Hughes and Sullivan 1974; Muckle 1985; Nash and Petraglia 1987; Rowland 1989; Stein 1992; and see below).

Post-depositional taphonomic processes may transform archaeological assemblages, but as some have pointed out, information may be gained as well as lost from patterns seen as a result of these processes, which assists in reconstructing the cultural contribution to formation of archaeological deposits (Gould 1980:113; Hiscock 1985, 1990; Yellen 1977). This chapter discusses factors contributing to formation of archaeological deposits and the manner in which they affect patterns of distribution of cultural material. It is argued that information gained from patterns resulting from taphonomic processes will assist to reconstruct past events in the interpretation of archaeological data from excavations, discussed in following chapters. Identification of taphonomic patterns also helps to address the problem identified by Frankel (1988), of explanations for the nature of cultural change being derived from the construction and definition of temporal analytical units (aggregated excavation units), which assume an equivalent time span for each unit and a constant rate of accumulation of deposits. In order to tease out cultural and natural formation and post-deposition processes a number of features of shell and earth deposits in the Darwin region will be examined for evidence of any gross changes, patterns or trends over time, both within and between sites. These include:

1. Changes in stratigraphy and the nature and distribution of macroscopic organic matter such as shell, bone and charcoal, and stone artefacts, through the deposit.

2. The rate of shell fragmentation in the deposits.

3. Changes in relative frequency of taxa and in three Hope Inlet shell mounds changes in shell size.

Stratigraphy

Stratigraphy in archaeology is derived from geological concepts of strata, or layers, which represent geologically distinct events. The human accumulation of refuse in a discrete period of occupation is considered such an event, as are any intersecting periods of natural deposition (Stockton 1980:128). In Australian midden analysis, stratigraphy has traditionally been used to refer to the physical attributes of unconsolidated sediment and the matrix (Stein 1992b:77; Attenbrow 1992:5; Bailey *et al.* 1994:76; Mitchell 1993:186; O'Connor and Sullivan 1994b:23).

The matrix is a generic term used by archaeologists to refer to the accumulated fine organic matter in cultural deposits which surrounds archaeological material such as shells, bone, charcoal and stone artefacts (Stein 1992c:137). Minor changes in the colour of the matrix (other than the presence of ashy organic lenses associated with hearths) as noted in these and other tropical mounds (eg. Mitchell 1993:186) are now generally agreed to be more likely due to natural processes such as heavy rainfall runoff, groundwater saturation and associated chemical processes, rather than changes in cultural activities and content as sometimes thought in the past (Ford 1992:315; Hughes and Lampert 1977:137; Stein 1992a; Veitch 1996:82).

The nature and distribution of matter in cultural deposits

there are two sources of the organic material, natural and cultural, in deposits associated with prehistoric occupation, which require differentiation (Stein 1992d:194). Cultural organic matter, such as shells, bones, charcoal, seeds and decomposed residues, are described by Stein (1992d:195) as *sediment*, which is transported by an agent (usually human) to a site for subsistence and manufacture requirements. Stein (1992d:195) defines *soil* as organic matter which is the product of biological weathering of parent material (sediment) on a stable surface. Only the upper portion of a deposit acts as a parent material for soil production, so the thickness of the deposit determines the extent to which it is affected by weathering and the accumulation of soil.

Thus thick deposits such as shell mounds which form rapidly are removed from the processes of soil formation and post-depositional effects of the surface zone of weathering, such that the source of the organic matter under this zone is primarily *sediment* (cultural deposit). When only moderate quantities of artefactual material are deposited, the entire thickness of the deposit is affected by weathering and accumulation of soil organic matter (Stein 1992d:197-8).

Vertical decay processes

It is now accepted that the distribution of shell in cultural deposits generally follows a characteristic pattern resembling a skewed "normal curve" distribution, which is related to taphonomic processes of decay, as discussed by Sullivan (1984:5-6) and others (Brennan 1977:125; Stein 1992a:11-15). In this pattern, as seen in deposits analysed by these researchers and in this study (Chapters Seven and Eight), shell density first increases in the upper to middle layers of the deposit, relative to a slightly lower density in the top 5-20cm surface layers, then decreases again gradually with depth to the base. The lower density in the surface layer can be explained by weathering processes and higher fragmentation of surface material (see below).

In the upper levels of the deposit below the weathering zone, shell is relatively well preserved in an alkaline environment. With depth of the deposit, the proportion of shell to sediment decreases, a phenomenon observed in previous analyses of shell deposits, especially in areas of the world with heavy rainfalls (eg. Stein 1992; Sullivan 1984:5). This is explained as a process in which carbonic acids present in all rain and groundwater dissolves the calcium carbonate in shell. In this way shell is leached out from the lower levels of shell deposits by rain and groundwater percolation, or by seawater saturation (Hughes & Lampert 1977:136; Stein 1992b; Waselkov 1987:149). Factors that influence rates of shell decay include alkalinity levels, temperature and moisture content of a deposit (Claassen 1998:59-61). There is evidence that these factors also affect the preservation of other organic material in cultural deposits (see below and Hughes 1980:4-6), and may explain the characteristic "taphonomic curve" pattern seen for the distribution of bone and charcoal in the excavated Darwin region sites as shown in Chapters Seven and Eight.

For bone it is thought that decay occurs through a similar process to leaching of shell, where exchangeable calcium from bone is transported in solution (Linse 1992; Waselkov 1987:149; Sawbridge and Bell 1972:848). Bone comprises around two-thirds mineral matter and one third organic tissues, mainly protein with some fats (Butzer 1972:255). Processes of decay of bone may involve the removal of the mineral component, the organic tissue, or both (O'Connor 1990:111). O'Connor (1990:111) found that in bone, which appeared physically preserved, from deposits in the Koolan Shelter in the Kimberley, the mineral component was preserved but most of the organic material had been replaced. Research has shown that hydroxyapatite, the inorganic component of bone, is soluble in alkaline as well as acidic environments. As pH levels increase above 7.8 the Ca/Phosphorous ratio is affected so that more phosphorous leaches from the bone (Linse 1992). Thus the presence of bone fragments and high phosphorous levels in a shell midden may indicate that destruction of bone has occurred (Coutts 1970:46).

Charcoal from archaeological deposits can also be classed as organic matter, as it is often not highly carbonised and inert as sometimes assumed. Thus, like humus, charcoal would be subject to biochemical decay by oxidation, and may not be present where expected in cultural deposits, as Hughes (1977:216) observed decades ago in a geomorphological study of several shelter sites in New South Wales (O'Connor 1990:112). As Hughes (1977:213) noted, even though they are alkaline and favourable for the preservation of the mineral component of bone, in shell deposits that are moist and aerated, organic material (including shell, bone and charcoal) will be oxidised by biochemical attack. Decomposition processes of organic material are likely to occur at different rates spatially across a deposit, dependent on extent of water passing through the sediments (Clarke 1999).

Of particular interest when I first analysed shell mounds on Middle Arm Peninsula was that the distribution of stone artefacts exhibits a similar characteristic pattern to the organic material (see Chapter 7; Figure 7.5). Some researchers (eg. Hiscock 1990; Lilley 1993:39) have commented on this type pattern in distribution of stone artefacts in rockshelter deposits. Hiscock (1990) has pointed out the similarity between taphonomic processes acting on bone and those that alter stone. A study by Hiscock (1985, 1990) on taphonomic processes affecting stone artefacts shows progressive decay of chert artefacts related to the depth of the deposit. Once artefacts are "buried by a sufficient depth of sediment" (Hiscock 1990:43) a range of post-depositional chemical processes result in weathering of artefacts, including leaching, discolouration, encrustation and increased breakage and disintegration (Hiscock 1985:92, 1990:49).

Schiffer (1987) discusses the decay processes which act on buried stone, including chemical, physical and biological agents. Water, present in open midden sites, and in rockshelter deposits, as attested by the ubiquitous "dripline" mentioned in most excavation reports, acts as a medium for chemical reactions and dissolves atmospheric gases to form weak acids. Organic decay, of waste products of human activity and decomposition of plant material to form humic acids, adds to a lowering of the pH of a deposit. Acids may react with the surface of stones with chemical reactions such as ionic dissociation, hydrolysis and oxidation, resulting in deterioration of the stone through the removal of ions. Cycles of wetting and drying increase the rate of deterioration (Schiffer 1987:151-5).

In the case of deposits around the Darwin coastal area, quartz is the predominant material. Despite this stone material being extremely resistant to weathering, its inherent features may mean it is susceptible to disintegration. Quartz has a crystalline structure with numerous flaws and micro-fractures, making it a brittle material that breaks up under pressure through knapping or other processes into very small fragments (Cotterell and Kamminga 1990; Dickson 1977; Kamminga 1982:24). It is possible that much of the debris resulting from working quartz is not readily recognisable under normal archaeological excavation procedures.

Vertical movement

Other site formation processes that may contribute to the distribution patterns seen in middens are diagenesis and vertical movement of components of the deposit. Recent studies in rockshelter sites overseas suggest that volume reduction may occur in archaeological deposits with a high ash and charcoal content, associated with decomposition processes of the ash due to dissolution of ash-derived minerals (diagenesis). This results in compaction and increased concentrations of artefacts in layers reduced in volume and downward movement of younger lithics and bones into older sediments (Karkanas *et al.* 1999:1179; Schiegl *et al.* 1996).

Experiments by Stockton (1973) investigating vertical dispersal of stone artefacts in sandy deposits subjected to human trampling, which have been replicated by others (eg. Gifford-Gonzalez *et al.* 1985; Villa and Courtin 1983) gave results of a pattern approximating a "normal curve" in vertical distribution of artefacts, similar to that seen in midden deposits (Figure 6.1). Very short periods of trampling (less than half a day) resulted in movement of artefacts downwards to a depth of up to 16cm, distributed in a this typical curve pattern. During archaeological excavations of midden deposits, intrusive material such as glass is quite often recorded down to depths of 15cm, and occasionally as much as 45cm below the surface, in levels otherwise dated by radiocarbon analysis to thousands of years BP (eg. Jones and Johnson 1985:50-2; Schrire 1982:196; Sullivan 1984:10).

As an explanation for penetration of glass slivers into apparently undisturbed lower levels, Stockton (1973) suggests that tremors from footfall above may shake small particles which have moved below the depth of soil actually being disturbed, so they sink deeper among spherical objects such as sand grains. Based on observations from an excavation at King's Table shelter in the Blue Mountains, where charcoal pieces dated to similar ages were distributed vertically though 60cm of deposit, Stockton (1977:49, 1980:128) puts forward the idea of differential vertical movement of material such as charcoal and stone artefacts, in a deposit that behaves like a fluid medium. In this process, sorting of artefacts, charcoal and sand grains is related to density, size and shape, so the higher an object's area relative to mass, the more it tends to rise. Small flaked artefacts tend to move downwards (sink), while charcoal found in association, which is often used to date the artefacts, is lighter and tends to rise. Gifford-Gonzalez *et al.* (1985:816) suggest a similar mechanism, whereby processes such as bioturbation or wetting/drying cycles may act on a buried layer of objects to raise some and lower others, producing a decrease in artefact frequencies on either side of a mode.

In sandy rockshelter deposits it has been demonstrated through conjoining that post-depositional disturbance may result in vertical movement of artefacts without visible signs of disturbance in the matrix surrounding the artefacts (Villa 1982:278). Researchers have shown through conjoining of artefacts and experimental studies that considerable vertical, horizontal and differential sub-surface movement of artefacts occurs in sandy deposits after deposition, catalysed by wetting and drying and biogenic activity. This work demonstrates that there is not necessarily any direct relationship between dated charcoal pieces and stone artefacts at the same depth (eg. Cahen and Moeyersons 1977; Hofman 1986; Richardson 1996b). It does not support precise interpretations of chronological sequences of changes in stone artefact types, such as the "Small Tool Tradition", developed from radiocarbon dates obtained on charcoal found in association with the stone artefacts.

Nor in midden analysis, as Brennan (1977) stated over 20 years ago, can the distribution of cultural material in a shell deposit be assumed to be in a contemporary context. Brennan tested the proposition that cultural contents found in shell deposits, such as stone artefacts, bone and charcoal, may be "*intrusive into them, deriving from later periods of occupation when the midden surface had weathered down to an occupiable condition*" (emphasis in Brennan 1977:122). From excavations of oyster shell middens along the Lower

Hudson River, New York, Brennan (1977:127) found that there was no correlation between diagnostic stone artefacts, of known provenance, and their depth within a dated midden (cf. Sanger 1981; Spiess 1988). The artefacts were all in a "humus-shell zone" - a layer typically found capping the surface of shell deposits to a depth of 5-20cm, formed through the process of soil formation on the upper portion of a deposit as discussed above (Brennan 1977:125; Hughes and Lampert 1977:136; Sawbridge and Bell 1972; Sullivan 1982:118). Trampling by humans or animals may also contribute to the formation of a compacted surface capping of fragmented shell and soil (see below).

Site formation processes and patterns of material distribution

An understanding of the effects of natural site formation and post-depositional processes, which can often be seen as regular patterns, allows these factors to be taken into account when seeking to explain archaeological patterns in terms of human behaviour (Schiffer 1987:22-3). In shell deposits, shell is the dominant sediment. If rate of shell deposition parallels rate of discard of other material, such as bone and charcoal as well as stone artefacts, then this may reasonably be explained as increased use of site while the nature of occupation remains the same (Hiscock 1984:134).

When shell discard rate decreases during increases in discard of bone, charcoal and stone artefacts, the nature of occupation has changed. Since shell is by far the dominant material, the pattern that arises when deposition of this dominant component ceases or slows is one in which the other proportionally minor components, - bone, charcoal and stone artefacts - increase markedly in their manifestation. And while natural processes may cause decreases in density of cultural material correlating with increased fragmentation, it can be argued that increases in density of cultural material that correlates with increased fragmentation is related to intensity of human activity.

Shell fragmentation and bioturbation

An understanding of the causes of shell fragmentation is another important consideration in midden analysis. Fragmentation may reduce the volume and porosity of deposits, and increase both post-depositional movement of shell and shell decomposition (Muckle 1985:32). Differential rates of fragmentation according to shell taxa may bias the recovery of shell types during excavation leading to inaccurate reconstructions of the relative abundance of the original range of taxa in a deposit. The thinner and more fragile a shell, the more likely it is to be fragmented. Fragmented shell generally means smaller sized shell, which have a large surface area to mass ratio and are therefore affected to a greater extent by post-depositional chemical dissolution (Ford 1989:168; Stein 1992c:138; Waselkov 1987:149).

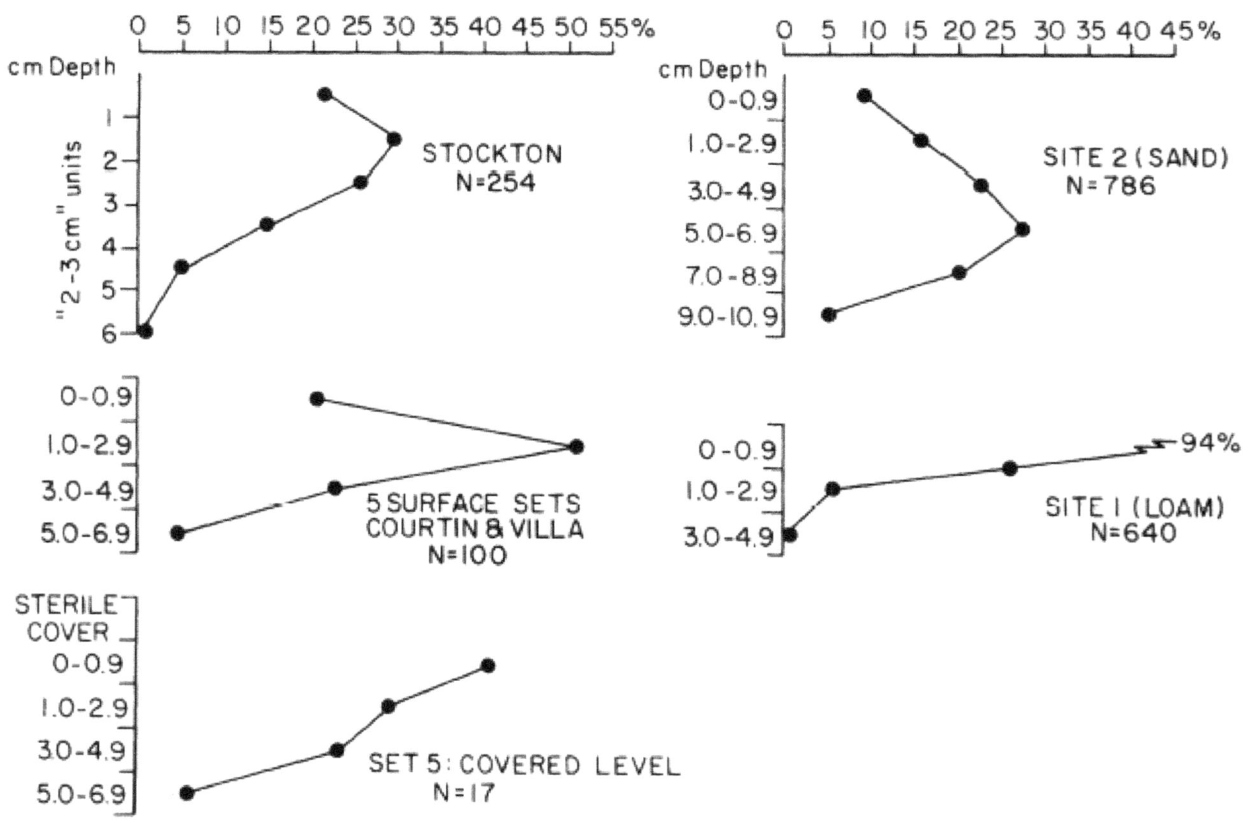

Figure 6.1 Vertical distribution of objects trodden into sand, showing typical pattern approximating a normal curve [Extract from Gifford-Gonzalez *et al.* (1985:814) with data from Stockton (1974) and Villa and Courtin (1983)]

It is well known that shell fragmentation in middens is caused by both cultural and natural processes that occur during midden formation or post-depositionally. Cultural behaviour that may contribute to shell fragmentation during midden formation includes heating through cooking (Robins and Stock 1990:89; Spennemann and Colley 1989:56), use of discarded shell as hearth material (Jones 1980a:155; Meehan 1982a:114), and occupational disturbance (trampling, dropping rocks brought to the site, and movement of shell by people during daily activities) (Matthews 1965, cited in Hughes and Lampert 1977:135; Meehan 1982a:112-7; Muckle 1985:33, 69).

Experimental work by Robins and Stock (1990) and Spennemann and Colley (1989) found that heating (cooking) differentially affects the extent to which different shell taxa are fragmented and removed by post-depositional chemical and physical processes of decay. This is dependent on differences in shell fragility between taxa, which is determined by shell composition and structure (Robins and Stock 1990: 87; Waselkov 1987:146). For example, the bivalve shell *Marcia* would conceivably be more susceptible to post-depositional destruction after cooking, as it is a relatively fragile thin layered shell in comparison with the robust bivalve *Anadara*, which is strengthened by its thick external ribbing (Mowat 1994). Similarly, thin-walled gastropod shells such as *Cassidula* and *Nerita*, would be more susceptible to destruction than robust thick-walled gastropods such as *Telescopium* and *Terebralia*.

Post-depositional natural taphonomic processes that increase shell fragmentation include physical processes such as natural bushfires, animal trampling and bioturbation, and chemical processes related to inundation of the deposits by groundwater or seawater (see above, Stein 1992a:10; Sullivan 1984:5). It has been demonstrated that short periods of inundation by sea water increases the rate of shell loss from middens (Claassen 1991b:268). Other researchers have noted increased fragmentation of fragile shell such as *Marcia* on the surface of north Australian mounds in comparison with the subsurface material. They provide explanations of exposure to extremes of climate (Baker 1981:79) or a combination of mound antiquity, topography and recent trampling by animals such as buffalo (Mowat 1994:207; O'Connor and Sullivan 1994b:22).

Bioturbation by birds (especially megapodes), goannas and dogs (Burns 1994:72, 75; Dwyer *et al.* 1985; Meehan *et al.* 1985:147; Stone 1989:61), turtles, rodents and humans (Ceci 1984:64) pigs and crabs (Roberts 1991:58-62; Specht 1985), termites (Wood and Johnson 1982:545) and tree roots (Waselkov 1987:149) may cause some fragmentation, but more significantly in terms of dating sites, may result in horizontal and vertical movement of shells, bone, charcoal and artefacts. Erlandson and Rockwell (1987) identify radiocarbon reversals resulting from differential movement of different sized objects through natural disturbance processes in archaeological deposits on the California coast. During surveys of the Darwin region I commonly observed holes and disturbance in mounds consistent with being dug up by goannas.

In his experimental studies on the effects of trampling on shell deposits, Muckle (1985:75) found that degree of shell fragmentation generally reflects intensity of human activity. Muckle's study showed that initial deposition by tossing or pouring shells resulted in either no fragmentation or breakage into only large pieces, which as suggested by Claassen (1998:58), indicates that highly fragmented shell in deposits are likely to be the caused by activities that take place after initial discard. Muckle argues, based on experimental trampling on shell deposits, that very intense trampling (several thousand passes of a person of weight 65kg, wearing hard, rubber-soles shoes, walking over the deposit, with one foot striking the shells) would be required to reduce shells in a shallow deposit to a point where roughly half the shell weight would not be retained by 4mm sieves (Muckle 1985:68, 77). In this analysis, a more sensitive measure of shell fragmentation is used. All shell pieces retained by both 6.4mm sieves and by 3.2mm sieves, which are less than half the whole shell, are classified as fragmented. However, a check of calculations of fragmentation rate based only on pieces in the 3.2mm sieves gave very similar results.

Hiscock (1985:89-90) demonstrates an inverse relationship between breakage of stone artefacts and rate of sediment accumulation, which also applies to accumulation of shell middens. When the rate of shell deposition is slow, material including shell, bone and stone artefacts remains on the surface for longer periods of time and is subject to a higher degree of fragmentation than when deposition is rapid and the cultural material is exposed for a short time. As noted in the following chapter, in the excavated mounds MA1 and MA7, which radiocarbon dates indicate accumulated over hundreds, perhaps up to 700 years, the proportion of fragmented *Anadara* shell is between 30-40% throughout each deposit (Figure 7.6). As will be demonstrated below, attention to rates of fragmentation in shell deposits allows us to go some way to resolving the problem identified by Frankel (1988) of assuming equivalent time spans for arbitrarily defined excavation units or their aggregates.

Changes in shell size and relative frequency of taxa

Interpretations of environmental change, or of regional cultural change in strategies of resource procurement, in population demography or social organisation have been derived from midden analysis. Interpretations are often based on perceived changes in rates of shell discard or the relative frequency of molluscan taxa through time, related to depth of the deposit (eg. O'Connor and Sullivan 1994b:24; Schrire 1982:233-4; cf. Allen and Barton n.d.:88, 104-6). However, taphonomic processes have long been noted to cause problems for the unwary in midden analysis (Muckle 1985:47,76,93-95; Sanger

1981:40). Relative frequency of molluscan taxa according to depth in a deposit may be the result of differential post-depositional decay of shell taxa, as demonstrated for the Middle Arm mounds in the next chapter.

In Chapter Ten this study also examines changes in shell size of the dominant mollusc within and between three Hope Inlet shell deposits, following previous research elsewhere on cultural shell deposits, which uses shell size as an indicator of extent of human exploitation of molluscan resources and/or environmental change over time (eg. Spennemann 1987; Swadling 1976). A variety of environmental and biological factors are known to affect shell size as well as extent of human exploitation, including predation by animals other than humans, specific location of the shell beds according to tidal zonation, seasonality and salinity (Claassen 1998:47-9,133-7; Spennemann 1987 and see Chapter Ten). These are take into account by considering shell size in conjunction with a range of other data, such as shell fragmentation and relative proportions of other mollusc taxa and other fauna and on different temporal and spatial scales, at both intra-site and inter-site levels (Frankel 1988:43).

In the following chapters detailed analysis of excavated sites on Middle Arm Peninsula, Darwin Harbour and at Hope Inlet, Shoal Bay are presented respectively, with one chapter devoted to sites in each area. These analyses examine trends over time in rates of shell discard or the relative frequency of mollusc taxa, and of other organic material and stone artefacts as exhibited vertically through deposits, in the light of taphonomic information demonstrating differential preservation and shell fragmentation (Muckle 1985:93-5).

7 Inter and intra-midden variability on Middle Arm Peninsula

Introduction

This chapter compares the results of archaeological excavations and analysis of three *Anadara* shell middens located on Middle Arm Peninsula, Darwin Harbour. Each of the middens, designated MA1, MA7 and MA10, was first excavated for different purposes in 1994 and 1995 respectively. Excavation of midden MA1 was undertaken as part of my honours thesis and the excavation report is given in Burns (1994). Middens MA7 and MA10 were excavated partly to test the proposition that they were older than MA1, based on the more weathered appearance of surface shell, and to augment the somewhat limited knowledge of the chronology of sites in the region. In this study I examine gross chronological patterns exhibited in the data, on stratigraphy, vertical distribution of organic matter and stone artefacts, together with rate of shell fragmentation, obtained through these excavations and explore various explanations for inter and intra-midden variability seen in middens such as these in close proximity.

Environment and location of Middle Arm middens

Middle Arm peninsula, the largest of many promontories jutting into Darwin Harbour, lies approximately 12km south of Darwin (Figure 5.2). Today, like the rest of the harbour shoreline, a fringe of mangroves hundreds of metres wide borders the peninsula. The mangroves merge into estuarine mudflats that extend into the harbour and inland along the tidal sections of the Elizabeth and Blackmore Rivers. The topography of Middle Arm peninsula is described as an undulating low ridge 15-40 metres high, with ephemeral wet season creeks cutting through it (Caldwell 1983:44-9). Vegetation consists of mainly open eucalypt woodland typical of the northern monsoonal coast, with patches of monsoon vine forest (MVF).

In 1993 eleven shell middens and mounds were recorded during an archaeological survey of the peninsula as part of an honours thesis (Burns 1994). All of the middens recorded in this area are composed predominantly of *Anadara granosa* shells. Most are situated on the hinterland margin at the foot of the raised hillslopes, or on the crests of the low hills and rocky knolls overlooking a narrow sandy strip and the mangrove zone, which extends hundreds of metres to the mudflats and sea. Vegetation around the mounds consists of open grassland and nearby eucalypt woodland with a few pandanus shrubs.

Site MA10 is a non-mounded shallow shell deposit (less than 30cm in depth) situated on the northern side of the coastline on Middle Arm peninsula. Sites MA1 and MA7 are mounded deposits located less than a kilometre apart, and are two of nine middens dotted along a 3km stretch of the southern shoreline of the peninsula (Figure 5.2). Although MA1 and MA7 are both mounds situated on a terrestrial surface of silty clay and rocks at the hinterland margin, they exhibit different topographic characteristics. Mound MA1 sits on the lower hinterland boundary of a continuous gentle slope from the low inland hills to the sandy shoreline. Mound MA7 sits on a 1.5m high rocky ledge at the hinterland margin, which drops down sharply to the sandy strip separating the hinterland and mangroves. Densely wooded forest and discontinuous tracts of MVF occur along a line of low hills immediately inland of the mounds. Another 17 shell mounds were also identified a few kilometres south of these sites, at Haycock Reach (Hiscock and Hughes in press).

Description of middens

On preliminary surface inspection, mounds MA1 (10mx9mx0.45m) and MA7 (7mx9mx0.48m) appear similar in size, shape and composition, comprising roughly circular, discrete shell deposits, and while MA10 is not mounded in form, it too appears similar in composition. All are dominated by the bivalve shell *Anadara granosa*, with lesser numbers of other taxa, including *Marcia hiantina*, *Saccostrea echinata* (oyster), and the gastropods *Chicoreus capucinus*, *Nerita sp.*, *Telescopium telescopium*, *Terebralia semistriata* and *Volema cochlidium*. Observed on the surface of the mounds were a few stone artefacts, consisting of flakes manufactured from white quartz, and on each of the middens MA7 and MA10 a large specimen of *Syrinx aruanus* (false trumpet shell).

An apparent difference observed between the middens was the more weathered condition of the shell on MA7 and MA10 than on MA1, which was thought to possibly indicate that these deposits were older than MA1. The deposits exhibited some bioturbation, with goanna holes and some mixing of the deposit evident. Whilst MA7 is a single mound, MA10 is found 50m from a mounded shell deposit (MA6) and MA1 is found in association with a cluster of archaeological features within a 20m radius. These include another shell mound (MA2), a sparse stone artefact scatter of mainly flaked pieces manufactured from white quartz, and on nearby outcropping sandstone, a shallow shell midden (MA9) adjacent to an Aboriginal engraving (Plates 7.1 and 7.2).

The excavations

Descriptions of the excavations of MA7 and MA10 and analysis of the remaining spits of mound MA1 not examined previously in my honours project are given below. In order to facilitate comparison of these three sites, Tables containing the results of analysis of each site are presented together on the same page. Thus the quantitative data on the components of the excavated samples for Sites MA7, MA1 and MA10, are given in order in Tables 7.1, 7.2 and 7.3 respectively. The total identified shell taxa for each of the samples from MA7, MA1 and MA10, are listed in descending order of MNI frequency, with Weight and Wt/MNI, in Tables 7.4, 7.5 and 7.6 respectively.

Site MA7

Site MA7 is a single discrete, roughly circular mounded shell deposit located on the southern coastline of Middle Arm peninsula, registered as MAGNT site No. 5072-0087. The mound sits on a 1.5 metre high rocky ledge at the seaward edge of eucalypt woodland, overlooking a ten metre wide sandy strip to the mangroves (Plate 7.3). Vegetation on the mound then consisted of grass, cycad palms and a few other shrubs. A small seasonal creek flows from hills inland of the mound, spreading to a heavily grassed swamp adjacent to the eastern side of the rocky ledge (Figure 7.1). Amongst the mangroves that front the mound today may be found the gastropods *Nerita*, *Telescopium* and *Terebralia*. Tidal mudflats that may have formed the ideal habitat for the mudflat bivalves *Anadara* and *Marcia* occur seaward of the dense mangroves. But according to local Aboriginal people who collected these molluscs as children, few still occur here today.

Stratigraphy and chronology

A 0.5m square pit was excavated near the centre of the mound to a depth of 0.5m. Two main stratigraphic layers were identified: a shell layer being the mound itself, and a silty clay soil layer, with almost no shell, on which the mound rests. Stratigraphic profiles of mound MA7 and midden MA10 are given below in Figures 7.2 and 7.3 respectively. A thin layer of weathered and fragmented *Anadara* shell caps the shell layer surface of MA7. This layer is comprised of predominantly densely packed *Anadara* shell in a matrix of dark greyish brown clumpy sediment. The lower layer was composed of compacted yellow/brown clay and rocks with little shell.

During the excavation, two shell/charcoal pair samples consisting of *Anadara* shell and adjacent charcoal pieces were collected from 5-15cm (spit 2) below the surface, and from the basal levels between 30-40cm depth (spit 4). Radiocarbon dating of the surface samples returned an

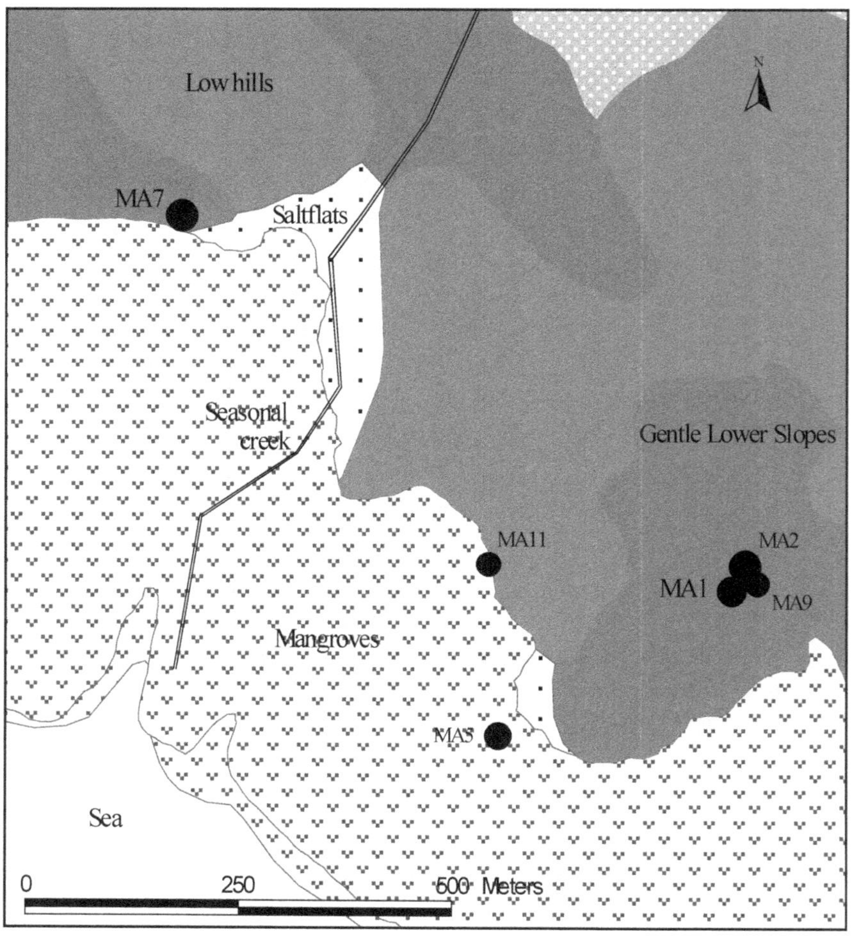

Figure 7.1 Location plan of Sites MA7 and MA1

Plate 7.1 Excavated mound MA1, showing sandy strip and mangroves in background

Plate 7.2 Engraving of bird's nest and tracks on sandstone at edge of sandy strip, looking inland toward mound MA2 at edge of woodland

Plate 7.3 Excavated mound MA7, overlooking sandy strip and mangrove zone

Plate 7.4 Excavated midden MA10 on rocky knoll overlooking sandy strip and mangroves

estimate for the shell of 1870±70, calibrated to 1395 calBP and 850±80 for the charcoal, calibrated to 480 calBP. For the basal levels, an estimate was obtained for the shell of 1220±60 calibrated to 742 calBP, and 1070±80 on charcoal calibrated to 637 calBP (Table 7.7). The shell dates appear to be inverted, which may be due to vertical movement of shell through post-depositional disturbance of the deposit by goanna activity, or it may be that the surface shell date Beta-95257, is anomalous. The dates on the shell and charcoal samples from the base of the mound correspond closely. Nonetheless, these radiocarbon estimates suggest a slow deposition rate, with an accumulation period of around 200 years for this small mound, or around 700 years if the inverted surface shell date is accepted.

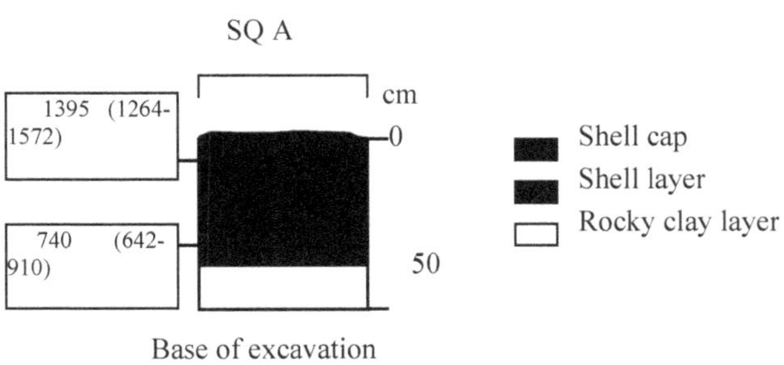

Figure 7.2 Stratigraphic profile of MA7

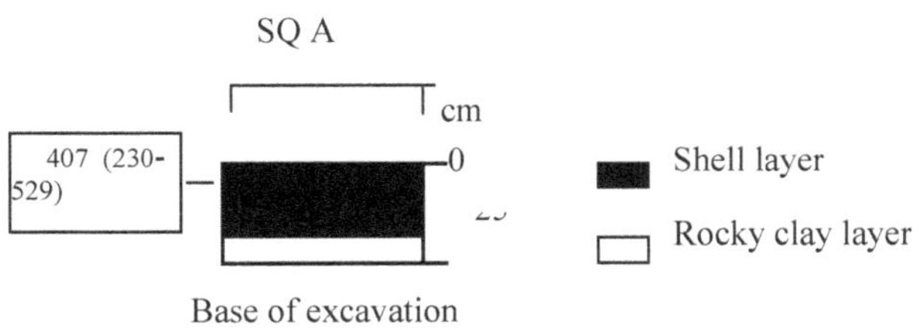

Figure 7.3 Stratigraphic profile of MA10

Shell

Sixteen shell taxa were identified in the sample from mound MA7 (Table 7.4). In mound MA7 the mudflat bivalve *Anadara* is the dominant shell taxon for both MNI and weight, representing 38% by MNI, and 65% by weight, of the total shell. The next most common taxa are the bivalves *Marcia*, representing 19% by MNI and 4% by weight, and the oyster *Saccostrea*, at 11% by MNI and 4% by weight. The mangrove gastropods *Nerita sp*., *Cassidula angulata*, *Terebralia semistriata* and *Chicoreus capucinus* together comprise another 21% by MNI and 11% by weight. All other mollusc taxa are in frequencies of less than 1%, excepting juvenile *Anadara*, which comprise 8% by MNI and 0.2% by weight. Like the few chenier shells (*Cerithium*) and barnacles recovered from the base of the deposit the juvenile *Anadara* shells are probably incidental additions.

Other organic material

Fragments of crab (13.2g) and bone (total 16.2g, density 24g/m^3) were recovered (Table 7.1), some burnt. Three fish otoliths were identified by Dr. Helen Larson of MAGNT as fork-tailed catfish (*Arius sp*.). Other fish bone recovered includes the jaw and vertebrae of small fish. Mammal bone, two teeth of the large Antilopine kangaroo (*Macropus antilopine*) and two small macropod teeth, two possum teeth and snake bone fragments, including vertebrae, were also recovered (Dirk Megirian pers. comm.). A relatively large quantity of charcoal (94.2g) was recovered, total density 0.9gm/kg of deposit.

Stone assemblage

Analysis of excavated material found large numbers of laterite rocks, some white quartz rocks, ochre and stone artefacts (Table 7.1). A total of 107 stone artefacts weighing 27.5g was recovered from the 0.125m^3 excavated sample, mostly (89%) from the 3.2mm sieve. For comparison with samples from the other excavated sites, this is standardised to an estimated average density of 856 artefacts for one cubic metre. Artefact types include flakes (n=100), flaked pieces (n=4), and two retouched flakes. The stone artefacts were all manufactured from white quartz, with the exception of one retouched flake manufactured from a fine-grained sedimentary rock and a small (<0.1g) granite (acid-igneous) manuport.

Analysis of Site MA1

The excavation report for this site is given in Burns (1994). Below is the analysis of all material excavated, from a 0.5m square pit 2m from the western edge of the mound, including spits not analysed for my honours thesis. Two radiocarbon estimates obtained on *Anadara* shell from the south wall of the excavation pit at a basal depth of 25cm, and from 5cm below the surface, gave results of 2430±90, which calibrated to 2056(1829-2311) calBP, and 1780±60, which calibrated to 1305(1197-1469) calBP, respectively (Table 7.8). These dates are interpreted as indicating that mound MA1 most likely accumulated between 2050 and 1300 years BP.

Shell

Fourteen mollusc taxa were identified in MA1 (Table 7.5). *Anadara* shell is clearly dominant, representing 60% by MNI, and 83% by weight, of the total shell. The next most common taxa are the mangrove gastropods *Nerita*, which makes up 10% by MNI and 1% by weight, and *Cassidula*, which makes up 7% by MNI and 0.2% by weight. The proportion of juvenile *Anadara* is very similar to MA7, comprising 7% by MNI and 0.1% by weight. The presence of a certain percentage of juvenile *Anadara* is possibly related to the season of collection (Richard Willan pers. comm.). Most of the shell in both samples was recovered in the 6.4mm sieve.

Table 7.1 Quantitative data, Site MA7

XU	Mean depth	XU wt	Shell	Charcoal	Bone	Otoliths		Crab	Stone artefacts			Quartz rocks	Ochre	Laterite rocks >6mm
	(cm)	(kg)	wt(g)	wt(g)	wt(g)	n	wt(g)	wt(g)	n	wt	material	wt(g)	wt(g)	wt(g)
1	6	14.7	4210	6.6	2.3	2	0.3		14	2	q	112	2.4	2232
2	15	23.2	6687	29.1	4.9			3.5	27	6.7	q, g, fs	46		2000
3	25	23.5	6948	27.3	3.5			8.7	32	9.5	q	81		1735
4	40	29.5	6603	24.2	4.8			1	27	8.5	q	56	0.7	3722
5	48	9.3	1933	7	0.7	1	0.2		7	0.8	q	66		4738
Total		100.2	26381	94.2	16.2	3	0.5	13.2	107	27.5		361	3.1	14427

Table 7.2 Quantitative data, Site MA1

XU	Mean depth	XU wt	Shell	Charcoal	Bone	Otoliths		Crab	Stone artefacts			Quartz rocks	Ochre	Laterite rocks >6mm
	(cm)	(kg)	wt(g)	wt(g)	wt(g)	n	wt(g)	wt(g)	n	wt	material	wt(g)	wt(g)	wt(g)
1	6	19.274	6426.8	0.1		1	0.3		16	43.7	q, t	220	1.2	3509
2	9	10	3428.9	0.2					10	6.4	q	87		1392
3	15	16.7	4753.6	0.7	0.1				8	6.9	q	420		1487
4	21	16.846	4944.9	1.5	0.1	2	0.9		10	0.5	q	40		1248
5	27	13.978	3502.1	2.7		1	0.2		5	1.7	q	170		2060
6	32	12.02	2138.2	1.6					2	0.5	q	97		1747
7	41	23.727	3119.2	2.5	0.1				4	0.7	q	59		5152
8	44	21.959	724	0.8					1	0.2	q	50		13556
Total		134.504	29037.2	10.1	0.3	4	1.4		56	60.6		1143		30151

Table 7.3 Quantitative data, Site MA10

XU	Mean depth	XU wt	Shell	Charcoal	Bone	Otoliths		Crab	Stone artefacts			Quartz rocks	Ochre	Laterite rocks >6mm
	(cm)	(kg)	wt(g)	wt(g)	wt(g)	n	wt(g)	wt(g)	n	wt	material	wt(g)	wt(g)	wt(g)
1	2	19	921	0.4					7	2.9	q	274		584
2	7	16.5	2523	0.6	0.5				21	3.3	q	380		692
3	14	17	2435	0.8	0.3				27	58	q	261		684
4	20	26.4	1039	0.5	0.6				14	4.1	q	739		1007
5	23	12.4	57		0.1							121		350
Total		91.3	6976	2.3	1.5				69	68.3		1775		3317

q= quartz
t= tuff
g= granite
fs= fine grained sedimentary

Table 7.4 Total mollusc taxa MNI and Weight, Sq.A Site MA7

Shell Taxa	MNI	MNI%	Wt(g)	Wt%	Wt/MNI
Anadara granosa	907	38.2%	17028.9	64.6%	18.8
Marcia sp.	446	18.8%	1053.9	4.0%	2.4
Saccostrea echinata	259	10.9%	1093.2	4.1%	4.2
Juvenile Anadara	190	8.0%	57.2	0.2%	0.3
Nerita sp.	197	8.3%	472.9	1.8%	2.4
Cassidula angulata	158	6.7%	60	0.2%	0.4
Terebralia semistriata	110	4.6%	1667	6.3%	15.2
Chicoreus capucinus	42	1.8%	637.9	2.4%	15.2
Melina ephippium	20	0.8%	119.6	0.5%	6.0
Ellobium aurisjudae	16	0.7%	38.1	0.1%	2.4
Telescopium telescopium	12	0.5%	645.1	2.4%	53.8
Fragum hemicardium	5	0.2%	5.7	<0.1%	1.1
Volema cochlidium	4	0.2%	97.3	0.4%	24.3
Cerithidea sp.	2	0.1%	1.8	<0.1%	0.9
Cerithium sp.	2	0.1%	<0.1	<0.1%	<0.1
Placamen sp.	1	<0.1%	<0.1	<0.1%	<0.1
Xanthomelon sp.	3	0.1%	11.0	0.0%	3.7
Unidentified shell			3391.3	12.9%	
Total	2374		26380.9		

Table 7.5 Total mollusc taxa MNI and Weight, SQ.A Site MA1

Shell Taxa	MNI	MNI%	Wt(g)	Wt%	Wt/MNI
Anadara granosa	1246	60.1%	24128.7	83.1%	19.4
Nerita sp.	211	10.2%	295.4	1.0%	1.4
Cassidula angulata	139	6.7%	49.1	0.2%	0.4
Juvenile Anadara	136	6.6%	18.3	0.1%	0.1
Saccostrea echinata	86	4.1%	302.3	1.0%	3.5
Terebralia semistriata	79	3.8%	1023.3	3.5%	13.0
Chicoreus sp.	70	3.4%	740.8	2.6%	10.6
Ellobium aurisjudae	57	2.7%	46.8	0.2%	0.8
Telescopium telescopium	20	1.0%	753.7	2.6%	37.7
Marcia sp.	10	0.5%	15.9	0.1%	1.6
Volema cochlidium	8	0.4%	69.9	0.2%	8.7
Cerithidea sp.	8	0.4%	1.2	<0.1%	0.2
Geloina sp.	1	<0.1%	7	<0.1%	7.0
Turbo cinereus	1	<0.1%	1.7	<0.1%	1.7
Xanthomelon sp.	1	<0.1%	2.6	<0.1%	2.6
Unidentified shell pieces			1580.6	5.4%	
Total shell	2073		29037.3		

Table 7.6 Total mollusc taxa MNI and Weight, SQ.A, Site MA10

Shell taxa	MNI	MNI%	Wt(g)	Wt%	Wt/MNI
Anadara granosa	376	75.0%	6169.5	88.4%	16.4
Juvenile Anadara	60	12.0%	7.6	0.1%	0.1
Terebralia semistriata	45	9.0%	398.7	5.7%	8.9
Telescopium telescopium	7	1.4%	90.4	1.3%	12.9
Chicoreus capucinus	3	0.6%	19.6	0.3%	6.5
Nerita sp.	3	0.6%	8.8	0.1%	2.9
Xanthomelon sp.	3	0.6%	2.8	<0.1%	0.9
Ellobium aurisjudae	2	0.4%	1.0	<0.1%	0.5
Volema cochlidium	1	0.2%	9.5	0.1%	9.5
Marcia hiantina	1	0.2%	2.6	<0.1%	2.6
Unidentified shell			266.0	3.8%	
Total	501		6976.5		

Table 7.7 Radiocarbon estimates, Site MA7

Material	Depth below surface (cm)	Lab Code	dC13	Date (BP)	Calibrated Date (calBP)
Anadara shell	5-15	Beta-95257	1#	1870±70	1395 (1264-1572)
Charcoal	5-15	Beta-95256	-24	850±80	480 (296-609)
Anadara shell	30-40	Beta-87872	0#	1220±60	742 (642-910)
Charcoal	30-40	Beta-87873	-25#	1070±80	637 (504-771)

Assumed dC13 values

Table 7.8 Radiocarbon estimates, Site MA1

Material	Depth below surface (cm)	Lab Code	dC13	Date (BP)	Calibrated Date (calBP)
Anadara shell	5	Beta-76830	-2.6	1780±60	1305 (1197-1469)
Anadara shell	25	Beta-76831	1#	2430±90	2056 (1829-2311)

Assumed dC13 values

Other organic material

Non-shell components recovered from the MA1 sample (Table 7.2) include a very small quantity of highly fragmented bone (0.3g) and charcoal (10g), and four fish otoliths, identified as forktailed catfish (*Arius* sp.) (Helen Larson pers. comm.).

Stone Assemblage

A total of 56 stone artefacts weighing 60.6g were recovered from the 0.1125m^3 of excavated sample (Table 7.2), which gives an estimated average density of 498 artefacts per cubic metre. Locally available white quartz is the raw material from which all except two of these artefacts were manufactured. Artefact types include flakes (n=46), flaked pieces (n=7), a retouched flake, a bifacial point and a small (0.2g) acid igneous manuport. The only formal implement, a bifacial point, was recovered during excavation from just below the surface, suggesting that it is most likely to be intrusive in the deposit. This artefact, measuring 70mm x 20mm and weighing 13.7g, is manufactured from a fine-grained non-local material, probably Gerowrie tuff. The nearest sources of this raw material are many kilometres inland of the mounds, where numerous scatters comprising mainly stone artefacts of this type have been recorded (Burns 1996b; Guse and Gregory 1994; Roddom 1997).

Site MA10

Site MA10 is a non-mounded shallow shell deposit situated on the northern side of the coastline on Middle Arm peninsula (Figure 5.2). This midden is located on top of a discrete rocky knoll around four metres high overlooking a bare sandy strip and the mangroves (Plate 7.4), 50m north-east of a mounded shell deposit (MA6) on lower outcropping rock (Figure 7.4). At the time of survey, vegetation on the site consisted of a few eucalypt and other trees and shrubs and the surface was covered with large rocks, leaves and ash. Monsoon Vine Forest covers the neighbouring rocky knolls, and inland of the sites the ground rises along a gentle laterite slope covered in eucalypt woodland.

These sites are situated almost midway between a small tidal inlet to the west off East Arm, and the mouth of the Elizabeth River to the east, both less than a kilometre away, but today separated from the sites by a dense 500m wide mangrove fringe. Mudflats that may have provided a suitable habitat for the bivalves found in the midden could be accessed via these tidal waterways. It is also possible that the bare sandy interface between the middens and the mangrove zone was once an intertidal mudflat providing a suitable habitat for *Anadara* bivalves.

Stratigraphy and chronology

In the excavation of a 0.5m square toward the centre of midden MA10, to a depth of 0.25m, two main stratigraphic layers were identified. The upper layer consists of the shell midden itself, which rests on a silty clay soil layer, with almost no shell (Figure 7.3). The dominant constituent of the upper layer was densely packed *Anadara* in a matrix of dark greyish brown clumpy sediment. The lower layer was composed of compacted yellow/brown clay and rocks with little shell. The shell content decreased gradually through to bedrock. During excavation, numerous laterite rocks and tree roots were evident in the deposit; the number of rocks increasing with depth. A sample of *Anadara* shell collected for radiocarbon dating from the north wall 4cm below the surface returned an estimate of 750±90, calibrated to 407 calBP (Beta-94184) (Table 7.9). This estimate places this shallow midden as the most recently accumulated *Anadara* deposit of sites dated in the Darwin Harbour area.

There is a breakage rate between 25-40%, as would be expected from a shallow deposit such as this. Human trampling during accumulation is a reasonable explanation for this breakage, given that the deposit is situated on the crest of a high rocky knoll, not readily accessible to large animals such as cattle and buffalo.

Shell

A total of nine shell taxa were identified in the sample from MA10 (Table 7.6). *Anadara* is the dominant mollusc taxon for both MNI and weight, representing 75% by MNI, and

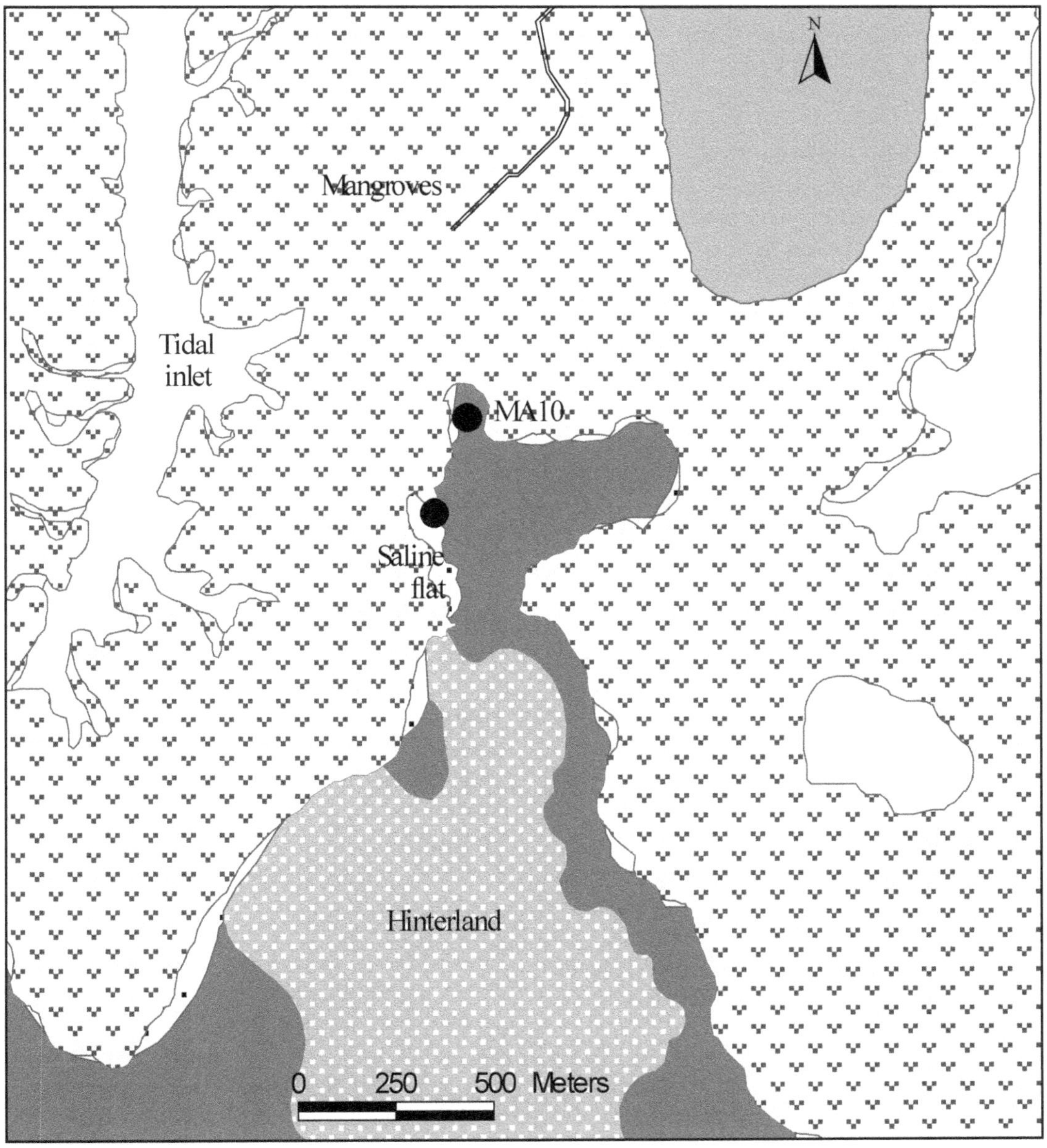

Figure 7.4 Plan of site MA10

Table 7.9 Radiocarbon estimates, Site MA10

Material	Depth below surface (cm)	Lab Code	dC13	Date (BP)	Calibrated Date (calBP)
Anadara shell	4	Beta-94184	1#	750±90	407 (230-529)

Assumed dC13 value

88% by weight, of the total shell weight. The gastropod *Terebralia* is the next most common taxa, representing 9% by MNI and 6% by weight.

Other organic material

A very small quantity of bone fragments was recovered (1.5g) (Table 7.3). Mammal bone, and one Antilopine kangaroo (*Macropus antilopine*) molar tooth were identified. No fish otoliths were identified in the sample, and it was not possible to determine if any of the highly fragmented bone was fish. The dearth of bone is possibly due to lack of depth of this shallow deposit, compared with the thicker shell mounds which provide a more protective alkaline environment for bone. Poor preservation could also have contributed to the small quantity of charcoal (2.3g) recovered.

Stone assemblage

Large numbers of laterite and white quartz rocks were present and a total of 69 stone artefacts weighing 68.3g were recovered from the 0.0625m^3 excavated sample (Table 7.3). Compared with the samples from MA7 and MA1, this gives a relatively high estimated average density, of 1104 artefacts per cubic metre. Most of the stone artefacts (87%) were recovered from the 3.2mm sieve residue, and all artefacts are either flakes (n=67) or flaked pieces (n=2) manufactured from white quartz.

Discussion: Comparison of midden contents and chronology

Analysis of the excavated material from 0.5 x 0.5m test pits in three Middle Arm sites confirmed that the shell deposits are broadly similar in composition to each other. Stratigraphy in each mound comprised two main strata: a shell layer, and a silty clay soil layer. No other discrete strata was identified, other than a thin surface cap on each of the shell mounds, identified on the basis of the degree of weathering and fragmentation of shells and compaction of sediment. In each midden the density of shell was observed to gradually decrease with depth, merging into yellow/brown clay sediment and rocks, and degrading bedrock at the base of the excavations. This results in a characteristic pattern resembling a skewed normal curve distribution (Figure 7.5), related to taphonomic processes of decay as discussed in Chapter Six.

Sixteen and fourteen shell taxa were identified in the samples from mounds MA7 and the MA1 respectively, and nine shell taxa in midden MA10. Some difference was revealed in the relative frequency of shell taxa between the mounds MA7 and MA1, and this is discussed below. The apparent similarity in midden composition must be qualified by the fact that the total volumes of the excavated samples (0.1125m^3 for MA1 and 0.125m^3 for MA7 and 0.0625m^3 for MA10) are small samples (less than 1%) of each total volume, of around 14m^3, 12m^3, and 7m^3 respectively. It is recognised that interpretations based on this data are limited by the proviso as to how representative these samples are, given that shell mounds may be heterogenous, and contain a range of spatially discrete remains from a variety of occupation episodes (O'Neil 1993; Waselkov 1987:143). However, observation of other sections of the mounds exposed through disturbance by goannas suggested that the bulk samples excavated generally represent gross overall mound content.

In each midden the dominant constituent was observed to be the bivalve shell *Anadara granosa*, densely packed in a matrix of dark greyish brown clumpy sediment. Less abundant shell discernible during excavation included those seen on the surface, including *Marcia, Chicoreus, Nerita, Telescopium*, *Terebralia, Ellobium* and *Volema.*

Eight of the marine and estuarine taxa listed above and land snail are common to each deposit and make up over 98% of the total shell numbers in each sample (Tables 7.4, 7.5, 7.6). Only one taxon, *Melina* (mangrove tree oyster) not common to both samples, occurs in any significant quantity, comprising 1% of shell in MA7 but absent from MA1 and MA10. The meat from all these mollusc shells is edible and most of the shells are much larger than 6.4mm (large enough to provide a reasonable amount of flesh for human consumption). A proportion of shells is broken in both samples consistent with ethnographic methods of extraction of the mollusc flesh (Meehan 1982a:86:105-9). Non-shell components recovered (mostly from the 3.2mm sieve) from the middens include faunal remains such as crab, mammal, snake and fish bone (some burnt) and fish otoliths of fork-tailed catfish, macropod teeth of both small and large species and possum teeth. Charcoal, vegetation such as roots, ochre and stone artefacts, and laterite and white quartz rocks were also recovered. The distribution of bone, charcoal and stone artefacts through each deposit exhibits a similar pattern to that seen for shell (Figure 7.5), and this is discussed below.

According to my interpretation of radiocarbon determinations on samples from Sites MA7 and MA10, these deposits are most likely to have accumulated around 1400 to 700 years BP, and around 400 years BP respectively. These two sites are not older than the excavated Site MA1, which radiocarbon estimates suggest accumulated around 2050 and 1300 years BP. Thus the more weathered appearance of the surface shell material

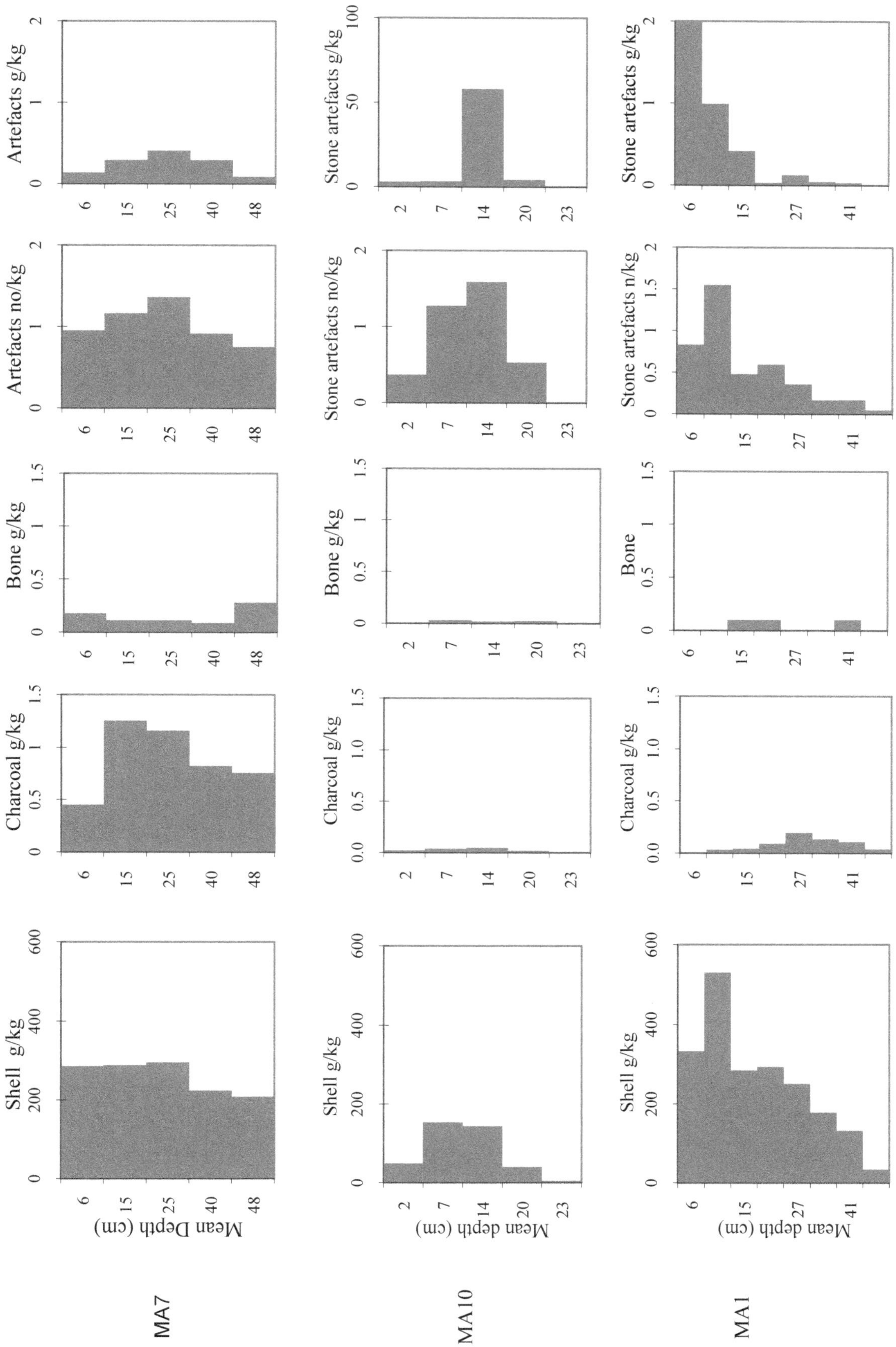

Figure 7.5 Distribution of shell, charcoal, bone and stone artefacts in mounds MA7, MA10 and MA1

on Sites MA7 and MA10 compared with that of Site MA1 is not a reliable indicator of their relative ages. The degree of weathering of shell may be influenced by other factors such as topography of the site, and related extent of inundation and chemical processes, and extent of post-depositional disturbance (cf. Mowat 1994:207).

These sites, although there may be some overlap in time between MA1 and MA7, on the evidence do not appear to be contemporary in terms of human lifespans, and probably represent sequential occupation of sites by small groups over time. It may be pertinent at this point to mention Meehan's (1982a) calculation from ethnographic observation that a group of 40 Anbarra gatherers would have collected approximately $8m^3$ of molluscs in one year. Assuming a similar size group and collection strategy here, by these figures the estimated volume of $12m^3$ for MA7 would only have taken around 18 months to accumulate. Meehan (1982a:166) notes, however, that the shell remains were distributed over three base camps and numerous processing sites, and no such hypothetical midden of $8m^3$ volume accumulated.

Thus it is possible that a mound the size of MA7 would take a group that size hundreds of years to accumulate if it represented only a fraction of the total quantity of discarded shells. Radiocarbon estimates (Table 7.8) indicate that mound MA1, with an estimated volume of $14m^3$, similar to MA7, accumulated over some 700 years. If both mounds accumulated over 650-700 years, this would give a deposition rate, assuming a constant rate, of $0.02m^3$ per year. Even allowing that the rate of deposition is unlikely to be constant, these figures are consistent with contemporary accounts of shell discard by small mollusc collecting groups in Arnhem Land.

Thus, investigations suggest that these mounds are the remains of a variety of activities, undertaken by relatively small groups over many centuries. These include subsistence activities such as mollusc gathering, fishing, hunting of terrestrial fauna, cooking, and manufacture of stone artefacts as well as possible ceremonial activities. The dominance of *Anadara* in the mounds indicates that gathering of mollusca focussed on the tidal sandy mudflats of the shoreline, with light foraging of the muddy substrates of the mangrove forest. Such a scenario is consistent with ethnographic accounts of mollusc gathering in Arnhem Land, where *Marcia* and *Anadara* are staple foods, collected in large quantities, with variety added by the addition of a few other gastropod taxa from the mangroves (Meehan 1982a:69). Ethnographic records suggest the large single specimens of *Syrinx aruanus* (false trumpet shell) on some sites may have been used for as water receptacles as well as a food resource, as observed in contemporary Aboriginal groups (eg. Cribb 1986:141; Davis 1985:302; Meehan 1982a:61,69).

Despite the small quantity of bone remains in the midden samples, their very presence is evidence of utilisation of resources other than mollusca, such as forktailed catfish and crabs, from the estuarine river system and/or shallow intertidal waters, as well as terrestrial resources such as kangaroos, possums, and snakes from the coastal lowlands and woodlands. The range of cultural material from a variety of activities, and high shell fragmentation rates of these mounds is consistent with the type of debris that would form from patterns of disposal recorded by Meehan (1982a:114-6) in "home base" camps. In this pattern the mounds accumulate gradually over time as piles of shell and other material left over from repeated consumption episodes in a base camp are scraped up to form middens around a camp.

Stone artefacts recovered from these middens consist of mostly flaked stone manufactured from white quartz, which can be procured from outcropping rock that occurs sporadically in the local Darwin region. This provides support for conclusions from consultancy surveys in the area that locally available rocks were predominantly exploited as sources of raw material for stone artefact manufacture (eg. Hiscock and Mitchell 1992). However, the bifacial point manufactured from non-local Gerowrie tuff found in MA1, indicates some movement, of artefacts from tens of kilometres inland, where outcropping tuff occurs, to the coast (Burns 1996b; Guse and Gregory 1994).

Although excavation supported surface impressions that these two mounds were broadly similar, some qualitative differences were revealed. Densities of bone, charcoal and stone artefacts are much higher in the more recent mound MA7, than in MA1 (Tables 7.4 and 7.5). Analysis also revealed variability in the shell composition between the samples from mounds MA7 and MA1. This is of interest considering that the mounds lie less than a kilometre apart on the same stretch of coastline in similar environmental contexts. Similar variability in shell composition between middens in close proximity has been described in the coastal Kakadu region by Hiscock and Mowat (1993), who argue that an understanding of the diversity found in middens contributes to theories on past use of coastal wetland landscapes. In the following discussion, the economic interpretation of these mounds is augmented by an exploration of the potential contribution of cultural, environmental and taphonomic factors to the observed variability between the mounds.

A tale of two mounds: inter-midden variability on Middle Arm Peninsula

Introduction

Variation in composition between these two mounds consists of:

1. A change in the relative proportions of the dominant shell taxa, ie. a much less marked dominance of *Anadara granosa* and corresponding greater proportion of *Marcia hiantina* and *Saccostrea echinata* in MA7 than in MA1 (Tables 7.4 and 7.5).

2. A chronological difference in the period spanned by the mounds: between 2000 and 1300 years BP for MA1 and 1400 and 600 years BP for MA7.

The second variation probably explains the first, ie. that inter-midden variability in the molluscan assemblages is probably linked to the different periods of accumulation of mound MA7 and the older mound MA1. This explanation may be further elaborated as due to one or more of three factors, cultural, environmental or taphonomic:

Cultural factors

One explanation for the dearth of *Marcia* in MA1 could simply be that the source shell beds were outside the foraging area where the mound was formed, but ethnographic reports on mollusc gathering behaviour do not support this idea. For example, Meehan (1982a:66) documents regular trips undertaken to obtain mollusca from a source more than 3km away. Since MA1 and MA7 are within a kilometre of each other on the coastline, this explanation seems unlikely.

However, ethnographic data from Arnhem Land (Meehan 1982a) and the Torres Strait Islands (Bird and Bliege-Bird 1997) demonstrates that cultural collection, cooking and discard patterns do result in differential proportions of shell taxa accumulating in a shell deposit. This data also indicates seasonal changes in exploitation of molluscan resources, determined not only by availability but by condition of, and cultural preference for (whether "fat" and tasty) particular mollusca in Arnhem Land (Davis 1985:300; Meehan 1982a:55, 1982b:108, 1983). *Anadara* and *Marcia* were staple foods in the wet season (from the late dry through to the early wet). In historic times, preference for *Marcia* meant that this mollusc was collected, whenever the shell beds were exposed by the strong tides of most lunar months, at the expense of other taxa, including *Anadara,* even though the Anadara beds were located closer to the base camps (Meehan 1982a: 70, 78, 142).

Discard patterns, resulting from processing small quantities of *Marcia* at "dinnertime camps" prior to returning to "home base" (Meehan 1982a:112-6), would explain a dearth of Marcia in MA1, which could reasonably be presumed part of a "home base", given the cluster of archaeological features at this site (see above). Bird and Bliege-Bird's (1997:52) observations on contemporary shellfish gathering strategies among the Meriam of the Torres Strait islands provide support for this hypothesis. They observed that mollusc taxa which are collected from a close source and relatively difficult to field process were transported back to "home base", while mollusc taxa which are relatively easily processed in the field and collected from a source far from the "home base" were usually field processed.

Marcia is relatively easier to process than *Anadara* (Meehan 1982a:87,97), so is more likely to be processed in the field if collected from a more distant source. If the *Marcia* shell beds were further from MA1 than from MA7 during their accumulation, then discard practices could explain their differential molluscan proportions. Possible tests of this hypothesis would include investigations of the chronology and relative molluscan proportions in the two other shell deposits (MA2 and MA9) located within 40 metres of MA1.

Environmental factors

Short-term localised climatic events such as cyclones, and/or greatly increased rainfall, which were observed on the Arnhem Land coast to destroy the *Marcia* shell beds for short periods, also destroyed the *Anadara* beds (Meehan 1982a:163-5; 1991). So such events would not explain the variability in frequency of these two molluscs. However, variation in the locations and dominance of *Anadara* and *Marcia* shell beds over short periods of time, in response to local environmental change in relative amounts of sand, river and sea have been documented on the Arnhem Land coast (Meehan 1982a:62). Short-term localised environmental change such as this would provide an explanation for differential collection of the two taxa, but represent a much shorter time-scale event, of years, rather than a span of centuries as indicated by the radiocarbon estimates.

The longer time-scale represented by the accumulation period of these two mounds more closely corresponds with the time span of "intermediate frequency" environmental change (Chappell 1982:77). Examples of this "intermediate frequency" environmental change include the episodic construction of chenier ridges, and associated variation in abundance of inter-tidal molluscan resources (Chappell 1982:77; Chappell and Grindrod 1984:221-7; O'Connor and Sullivan 1994:25; see discussion in Chapter Two).

However, if the relative frequencies of different molluscs are viewed from a different perspective following ethnographic data a different picture emerges. Meehan (1982a:69) quantified the Anbarra collection strategies over a year as heavily weighted toward bivalves, which consistently made up 98% of molluscs harvested. Thus another way of looking at the molluscan assemblage in the middens is to group together the bivalve molluscs that inhabit mudflats and the gastropods that live in the mangrove forests. Grouping together the mudflat bivalves *Anadara* and *Marcia*, and the mangrove gastropods common to each midden, *Telescopium*, *Terebralia*, *Nerita*, *Cassidula*, *Ellobium* and *Chicoreus*, gives very similar ratios in each sample (see Chapter 9, Table 9.9). The relative proportions of mudflat bivalves to mangrove gastropods for the Middle Arm mounds MA1 and MA7 is 60% to 28% and 57% to 23% respectively. The relatively larger proportion of mangrove gastropods in MA1 and MA7 when compared with the contemporary Arnhem Land collection may be related to temporal and/or spatial ecological variation in availability of mollusc taxa in these two regions. Nonetheless, the focus in these mounds too would appear to be mainly on bivalves, consistent with the customary cultural collection strategy as exhibited by the Anbarra.

The relatively similar ratio of mudflat bivalves to mangrove gastropods exhibited by the Middle Arm mounds MA1 and MA7 over the hundreds of years spanned does not suggest major long term environmental change. Within this context, an increase in *Marcia* at the expense of *Anadara* is consistent with the theory of people with a cultural preference for *Marcia* taking advantage of a short-term variation in the locations and dominance of *Anadara* and *Marcia* shell beds over time, in response to local environmental change as discussed above. It is also consistent with the notion that the mounds accumulated through a series of short-term depositional events within the span of centuries.

Taphonomic factors

It was thought possible that the under-representation of the relatively fragile *Marcia* and *Saccostrea* shells in mound MA1 compared with mound MA7 could be due to differential shell preservation between the two mounds and subsequent differential recovery of shell (Muckle 1985). Research by Mowat (1994) in the West Alligator region found that shell fragmentation (specifically of *Marcia sp*) is high on the surface of shell mounds, and has a direct relationship with the topographical characteristics of the mound. Her research demonstrates that both mound topographic characteristics such as height, and antiquity are important factors in shell preservation.

Although mounds MA1 and MA7 are both situated on a terrestrial surface of silty clay and rocks at the hinterland margin, they exhibit different topographic characteristics. Mound MA1 sits on the lower hinterland boundary of a continuous gentle slope that allows easy access to animal traffic from the low inland hills to the shoreline. Mound MA7 sits on a 1.5m high rocky ledge at the hinterland margin, which drops down sharply to the sandy strip separating the hinterland and mangroves. Due to the greater antiquity and lower height of mound MA1 compared with MA7, the nature of post-depositional processes would possibly have varied between the two mounds.

This includes possible increased animal traffic (trampling) of the older, lower mound MA1, and a greater likelihood short periods of inundation by sea or ground water, which has been demonstrated to increase the rate of shell loss from middens (Claassen 1991b:268). The combination of such factors would provide a reasonable explanation of differential preservation for the decreased survival rate of *Marcia* and *Saccostrea*, which are relatively fragile (brittle in the case of *Saccostrea*), thin-layered shells lacking the thick external radial ribbing that acts to strengthen the more robust bivalve shell *Anadara* (Boyd 1998:254; Mowat 1994:204, 1995:29-30, 35-6). However, if this were the case then we would expect the thin-walled gastropods like *Cassidula* and *Ellobium* that are also relatively fragile (with generally only the flange surviving in archaeological deposits (see Chapter 9), to be similarly affected. But *Cassidula* and *Ellobium* shells, like the more robust thick-walled *Telescopium* and *Terebralia* shells, occur in roughly similar or slightly higher proportions in the older mound MA1 than in MA7 (Tables 7.4 and 7.5).

In addition, a measure of the proportion of all broken shell in each of these mounds, which is known to affect rates of decay, gave similar results of around 30-40% throughout both deposits (Figure 7.6). Research has shown that shell fragmentation may be related to intensity of human activity and/or rate of deposition of shell (see Chapter Six). Similar rates of shell fragmentation in these deposits is consistent with similar levels of intensity of activity and/or slow deposition rates of the deposits as indicated by the radiocarbon dates, but is not consistent with differential survival of shell taxa through taphonomic processes of decay. There is also a lower quantity of other organic remains such as bone and charcoal in the older mound MA1 than in MA7, which would be consistent with decay processes as a causative factor in differential shell preservation. However, this may also be explained as a reflection of the extent of subsistence activities at these sites (cf. McNiven 1989:46; Walters *et al.* 1987:92).

Discussion

Analysis of the excavated material from mounds MA1 and MA7, which appeared similar on surface inspection, revealed variability in chronology and in relative frequency of the molluscan taxa. Three factors: cultural, environmental and taphonomic, potentially contribute to the inter-midden variability between the two mounds, which is probably linked to their different periods of accumulation. The ethnographic data suggest that *Marcia* is a favoured resource that will be collected in preference to other mollusca if it is available. Cultural discard patterns, linked to short-term local environmental change, with movement of shell beds and difference in distance to the resource source, provide the most likely explanations which is consistent with the archaeological data.

With regard to long-term regional environmental change, the longer time-span of many centuries represented by the accumulation of these two mounds corresponds with the "intermediate frequency" time span of environmental change such as episodic periods of construction of chenier ridges (Chappell 1982:77), and corresponding variation in inter-tidal molluscan resources. Thus the composition of these mounds would represent the long-term average assemblage of macro-mollusc communities in the locality during their formation period (Claassen 1998:134).

Lower proportions of other organic remains such as bone and charcoal in the older mound MA1 are consistent with differential post-depositional destruction through chemical processes facilitated by inundation. But taphonomic processes involving differential preservation of shell taxa over time as explanations for the variability between the mounds seems unlikely, given that the more fragile gastropods are not similarly affected and the deposits exhibit similar levels of shell fragmentation. The initial impression

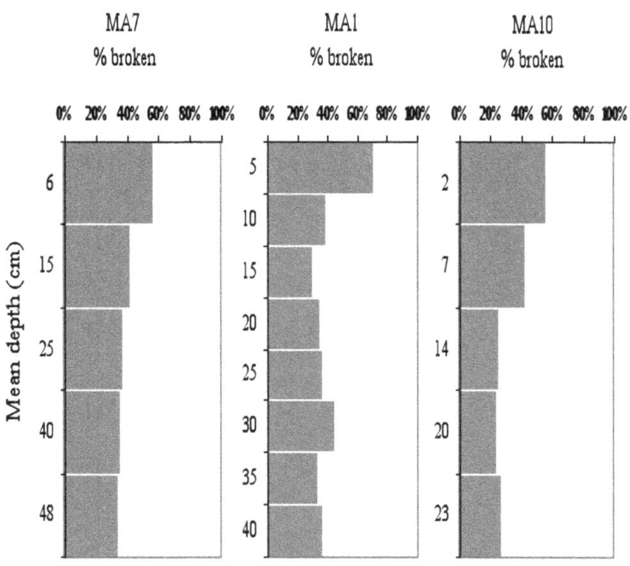

Figure 7.6 Proportion of *Anadara* shell breakage vs depth for Sites MA7, MA1 and MA10

that MA7 was the older mound, may have been because it contained a greater proportion of the more fragile shell *Marcia* on the surface, and thus appeared more weathered generally. Whilst inter-midden variability between these two mounds appears on the available evidence to derive from environmental change over time, with associated changes in collection and/or discard behaviour, intra-midden variation within each of the mounds is a different story.

Intra-midden variation in Middle Arm mounds

Interpretations of chronological cultural or environmental change are often based on perceived changes in rates of shell discard or in relative proportions of shell taxa through time, as determined by radiocarbon dating, in individual middens. This also applies to excavations of midden deposits in rockshelters, where changes in relative proportions of mollusc species with depth are interpreted as reflecting changes over time in mollusc exploitation. A well-known example is Schrire's (1982) interpretation of the midden deposit at Malangangerr in Arnhem Land. Schrire (1982:89) noticed that the large mangrove bivalve *Geloina* and gastropod *Telescopium* were dominant (by weight) in the lower levels of the midden but that the small gastropod *Cerithidea* dominated in the upper levels. She first explains this as the result of environmental change, in the form of devegetation of coastal flats resulting from buffalo trampling in the last 200 years, which reduced availability of the preferred *Geloina* and *Telescopium* gastropods.

The interpretation of increased abundance of *Cerithidea* relative to *Geloina* over time is challenged by Allen and Barton (n.d.:88), who point out that using MNI instead of weight as a measure gives a different result, in which *Cerithidea* remains most abundant throughout the deposit. In a later synthesis of Holocene coastal occupation in Arnhem Land, Hiscock (1999) disputes Allen and Barton's (n.d.88) claim, arguing that the change from *Cerithidea* to *Geloina* dominance is apparent even when the weight measures are converted to minimum numbers (Hiscock 1999:94-6). Hiscock puts forward an explanation of mollusc procurement correlating with environmental change in the late Holocene as indicated by geomorphic data from the Alligator Rivers region (Woodroffe *et al.* 1985; 1988). In this model the shell remains are said to reflect mollusc procurement during the transition from moist *Rhizophora*-dominated mangrove forests favoured by *Geloina* and *Telescopium*, to drier *Avicennia* and *Bruguiera* forests favoured by *Cerithidea*.

Thus the intra-midden variation, related to the depth of the midden, is said to reflect the changes in local habitat and availability of molluscs over time. However, when Schrire's data is converted to MNI numbers and plotted against depth without using log calculations (Table 7.10), which emphasise the smaller numbers represented by *Geloina* (Hiscock 1999:95), a slightly different pattern of the relative proportions of *Cerithidea* and *Geloina* can be seen, in which *Geloina* is clearly never numerically abundant (Figure 7.7).

Problems with interpretations based on changes in relative proportions of different mollusc taxa in middens over time, as represented by vertical distribution of shell, are exposed by the following analysis of intra-midden variation in the two Middle Arm mounds MA1 and MA7. In this analysis the arbitrariness and difference between spit boundaries of two excavated mounds, MA1 and MA7 was adjusted for comparison of the two deposits. The spits were apportioned to regular 5cm depth intervals, which allows comparative calculations of shell MNI from a given volume in each mound. Using this method reduces well-known problems of variation in MNI counts of faunal remains, which are affected by how the sampling (excavation) units are aggregated (Grayson 1984). This was done even though the problem is less critical for molluscs (than for vertebrate remains) because these calculations are usually based on only one non-repetitive durable element such as a shell hinge or spire, and more accurately represent the number of actual individuals in the sampling unit (Mason *et al.*1998:309; cf. Glassow 2000; Horton 1984).

As shown in Figure 7.8, the graphs exhibit a replicated pattern, characterised by an increase in proportion of shell content at round 5cm to 25cm depth, relative to the surface spits, and an overall trend of decreasing shell content with depth of the deposit. This pattern resembling a skewed distribution, a feature consistently seen in analyses of open shell midden sites (eg. Hughes 1977:158), is explained by Sullivan (1984:5-6) and others (Brennan 1977:125; Stein 1992a:11-15), as related to taphonomic processes of decay (see below). The pattern of decreasing shell density with depth occurs for each individual shell taxa, and is more marked for bivalves and more fragile thin-walled shell such as *Nerita* and *Cassidula* than thick-walled shell, such as the robust gastropods *Telescopium* and *Terebralia* (Figure 7.9).

While the fragile gastropods exhibit a similar pattern to the bivalves, the robust gastropods do not necessarily follow this pattern. As can be seen in Figure 7.8, whilst the actual numbers in each spit of a robust gastropod such as

Telescopium stays very low through the mounds vertically, the proportion relative to the more abundant bivalve *Anadara* increases markedly, simply because the *Anadara* decreases greatly in abundance with depth, through decay processes. *Cerithidea* is a relatively thin shelled gastropod which would be more susceptible to decay than the thick robust shells like the large bivalve *Geloina* or gastropod *Telescopium* (see Healy and Wells 1998:724). Comparing densities at different depths, of the relatively fragile *Cerithidea*, which decreases markedly with depth, with thick robust shells like the bivalve *Geloina* or gastropod *Telescopium*, which stay relatively constant in low numbers with depth, gives a biased picture of the relative frequencies over time. This is exacerbated when combined with a large difference in the numbers being compared, as is the case here (Claassen 1998:117). I argue that the patterns seen in this example are the result of differential post-depositional decay of shell taxa. The problem that this taphonomic process poses for midden analysis has long been noted (eg. Sanger 1981:40) and must be taken into account when interpreting changes in relative molluscan proportions over time.

Thus, taking these taphonomic patterns into account, the midden at Malangangerr rockshelter could be interpreted as the remains of different molluscs deposited together, which may then be subject to post-depositional chemical decay processes. In this explanation large quantities of *Cerithidea* from the *Avicennia* and *Bruguiera* section of the mangrove forests, along with small numbers of *Geloina* and *Telescopium* from the *Rhizophora*-dominated areas would have been collected during the same foraging expedition and their shells deposited in the rockshelter. The pattern shown in Figure 7.7 is consistent with an interpretation of two depositional periods, each of relatively short duration, separated by a hiatus in deposition. In this interpretation, the upper 20cm, or four spits, which contain freshwater shell and is dated by Schrire (1982:85) to 370±80 BP, was deposited sometime after the deposition of the lower (up to) 60cm deep deposit. The implication remains that the molluscs in the midden represent discard of what was available (in part) within the locality of the rockshelter, but possibly over centuries, or even years, rather than millennia. As Schrire (1982:236) says, using Meehan's (1982a:166) calculations of $8m^3$ per year for the Anbarra (see above), all of the midden volume at Malangangerr rockshelter could have accumulated in nine years. Moreover, the evidence suggests that vertical distribution of different shell taxa in this midden is partly a result of taphonomic processes and can not be taken as a direct reflection of environmental change (cf. Hiscock 1999).

Table 7.10 Conversion of Shrire's (1982:88) percentage data to weight and MNI figures for Malangangerr sample2

Spit	Cerithidea		Geloina	
	Wt(g)	MNI	Wt(g)	MNI
1	70	50	29	2
2	128	91	164	14
3	25	18	18	2
4	27	19	42	4
5	334	239	167	14
6	149	106	56	5
7	77	55	45	4
8	18	13	50	4
9	13	9	107	9
10	2	1	32	3
11	0.6	1	86	7
12	0.3	1	105	9
13	0	0	61	5
14	0	0	3	1
15	0	0	0.1	1

Of interest during analyses of the Middle Arm sites was that the characteristic skewed normal curve pattern resulting from decay processes seen for shell distribution through the deposits is also exhibited by other organic material such as bone and charcoal, and this too was generally replicated in each deposit. Even more interesting was that the distribution of stone artefacts also roughly exhibited similar patterns (Figure 7.5). Not that this is something new. This characteristic pattern can be seen in innumerable archaeological reports on excavations of both open midden and rockshelter deposits (eg. Barker 1991; Lilley 1993; Morwood and Oste-Brown 1995), and factors that may contribute to this are discussed in the previous chapter.

Conclusion

Whilst inter-midden variability in frequencies of mollusc taxa between the two Middle Arm mounds is consistent with local environmental change, intra-midden mollusc variation within each mound follows a pattern best explained as due to taphonomic processes. This is not to say that molluscan variation within any midden is always due to taphonomy. Recognition that a characteristic pattern exists for shell taxa in a midden, which varies with their morphology, allows differentiation between patterns due to cultural or environmental change and those resulting from taphonomic processes. Significant deviations from the characteristic pattern can be identified as

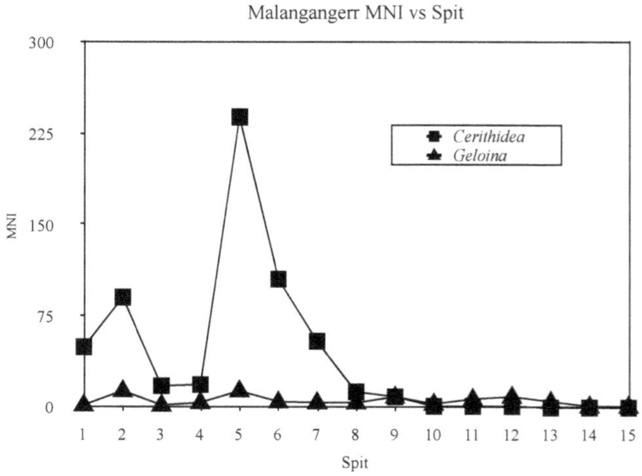

Figure 7.7 Graph showing relative patterns of decreasing density with depth for Cerithidea and Geloina at Malangangerr. Data from Schrire (1982:88-9)

representing cultural and/or environmental change and not taphonomic processes. Examples can be seen in the change from dominance of the fragile shell *Marcia* (*Tapes*) in the lower levels to the relatively more robust shell *Anadara* in the upper levels of the Idayu mound on the Kimberley coast (Veitch 1996:85), and the vertical distribution of *Anadara trapezia* in the middle layers of the Disaster Bay midden in New South Wales (Colley 1997:12-13).

For the Middle Arm mounds, the data are consistent with an interpretation of a relatively stable environment over many centuries, with some local changes in relative amounts of sand, creeks and sea along the southern coast of the arm producing a favourable habitat for *Marcia* shell beds over short periods. During this time small foraging groups, in a series of discrete occupational episodes, selected a limited variety of available macro-molluscs from a number of different habitats, and deposited the shells on the mounds. This analysis has shown how consideration of taphonomic processes allows a more informed assessment of the cultural contribution to the observed distribution of remains in midden deposits, and this is further discussed and applied in the analyses of excavated deposits at Hope Inlet.

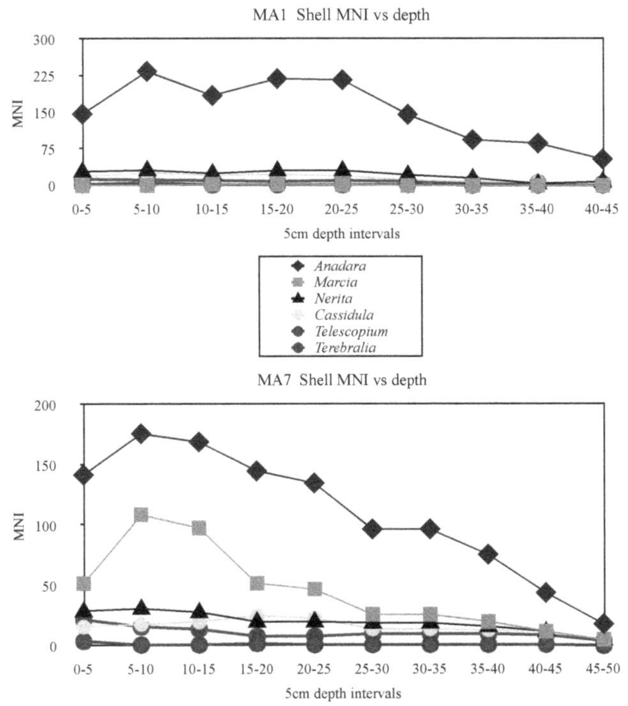

Figure 7.8 Graphs showing relative patterns of decreasing density with depth for different shell taxa in mounds MA1 and MA7

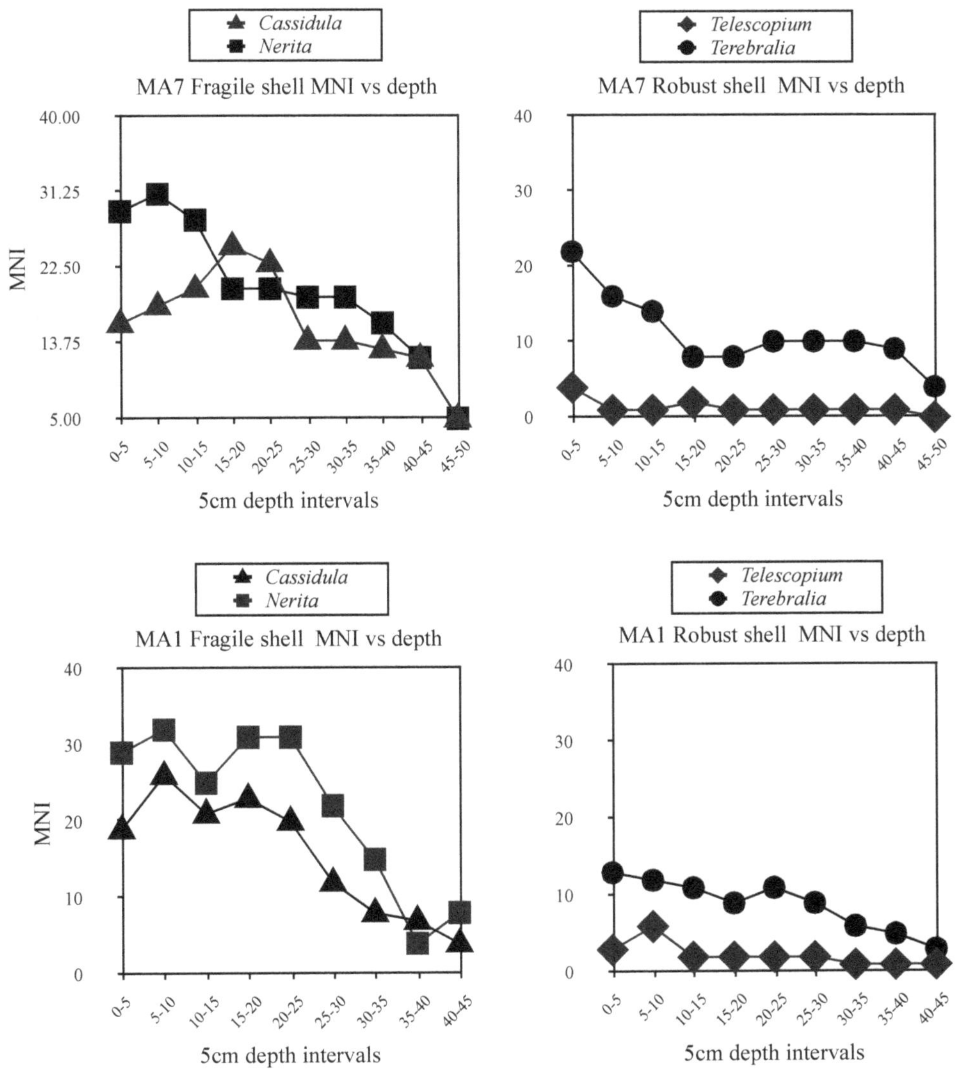

Figure 7.9 Graphs showing patterns exhibited for relatively fragile gastropod shell, and for robust gastropod shell, when MNI plotted against 5cm depth intervals, in sites MA7 and MA1

8 Variability within and between archaeological sites at Hope Inlet

Introduction

This Chapter discusses the results of excavation of three shell mound sites, Sites HI83, HI81, and HI80, one subsurface shell midden, Site HI66, and an earth mound containing shell, Site HI97 at Hope Inlet. The excavations were undertaken in order to compare data from a number of stratified sites with potentially contrasting and complementary deposits within the same local area across different microenvironments. As pointed out by archaeologists such as McBryde (1974:16-17) and Bailey (1993:15), this allows a more comprehensive assessment of the role of individual sites and local inter-site comparison when reconstructing a cultural sequence for that region. Radiocarbon estimates obtained on shell and charcoal samples from each of these sites to determine the chronology of occupation of the area, found that some, or part of, the Hope Inlet sites may be contemporary with each other, and with Darwin Harbour sites discussed in Chapter Seven. Features of these deposits examined for evidence of any gross changes over time within and between sites, include stratigraphy and the distribution of macroscopic organic matter (shell, bone and charcoal), and stone artefacts, through the deposit, rates of shell fragmentation, shell size and relative frequency of mollusc taxa. In this way the relative contribution of cultural and natural formation and post-deposition processes, reviewed in Chapter Six, that contributed to the present day form and nature of these archaeological remains, are disentangled.

Thus the sites were selected for excavation for comparative analysis, of contents, stratigraphy, and chronology of different site types, and of sites located in adjacent ecological zones (Figure 8.1). Each of the hinterland promontories and an associated saltflats area are partitioned from the others by tidal channels, mangroves and swamps. Sites selected for excavation are those representative of different site types in one of these partitioned areas, across the different environmental zones; Site HI80 on the saltflats, Site HI81 on the hinterland margin overlooking the saltflats, and the more landward Sites HI83, HI66 and HI97 (Figure 8.1). Excavation methods are given in Chapter Four. The environmental setting, history, structure and composition of each individual excavated deposit at Hope Inlet are considered in detailed descriptions below.

Description and analysis of excavated sites

Site HI83

Environmental setting

Site HI83 is the most inland of the excavated shell mounds at Hope Inlet, situated on the footslopes of the Northern Plains upland plateau, which forms a narrow promontory projecting onto the saltflats. It lies a few kilometres inland from the present day shoreline and approximately 300m inland of the large mound complex, Site HI81 discussed below, which borders the promontory margin. The saltflats, less than 300m to the south, are visible from this mound. This site is one of a dozen shell middens clustered along the centre of the promontory, 50-100m apart, most just visible from one another through the trees (Figure 8.1). Some of these middens are mounded, some are doughnut-shaped shell rings that are devoid of shell in the centre, and 100m east of HI83, is one large elongated earth mound. Site HI83 is a discrete, low, roughly circular *Anadara* mound measuring 16m north-south by 14m east-west and approximately 70cm above the surrounding ground surface (Figure 8.2; Plate 8.1). Other than a rounded fist-sized quartzite pestle and lateritic rocks, no stone artefacts were observed on the mound surface.

Vegetation covering the 500 metres or so wide promontory is mixed forest, probably what Wilson *et al.* (1990:32) refer to as mixed species closed-forest or monsoon vine forest thicket that occurs on sand dunes or lateritic soils adjacent to the coast. Species such as *Terminalia ferdinandiana* (Billy-goat plum) *Erythophleum chlorostachys* (Ironwood) *Syzygium suborbiculare* (Red bush apple) *Acacia auriculiformis* (Black wattle) *Dioscera transversa* (Long yam) *Planchonia careya* (Cocky apple) *Ficus opposita* (Sandpaper tree) and *Flueggea virosa* (White currant) were recorded by Larrakia ethnobotanist Lorraine Williams in this forest in 1998 as plants still utilised by the Larrakia locals. (The relevance of plants to past subsistence strategies discussed in Chapter 9, and see Table 9.8). The common weed horehound (*Hyptis*) also occurs in dense patches across the promontory and on the mounds. Water would have been readily available during the wet from seasonal creeks and and shallow drainage lines that dissect the hinterland in this area and from shallow wells in the dry season.

Stratigraphy, shell fragmentation and chronology

Excavation of a one metre square pit toward the southeast edge, two metres from the centre of the deposit, to a depth of approximately one metre, revealed two main strata, being the shell mound itself (Stratum I) and the clay layer on which it rests (Stratum II). Two layers were distinguished

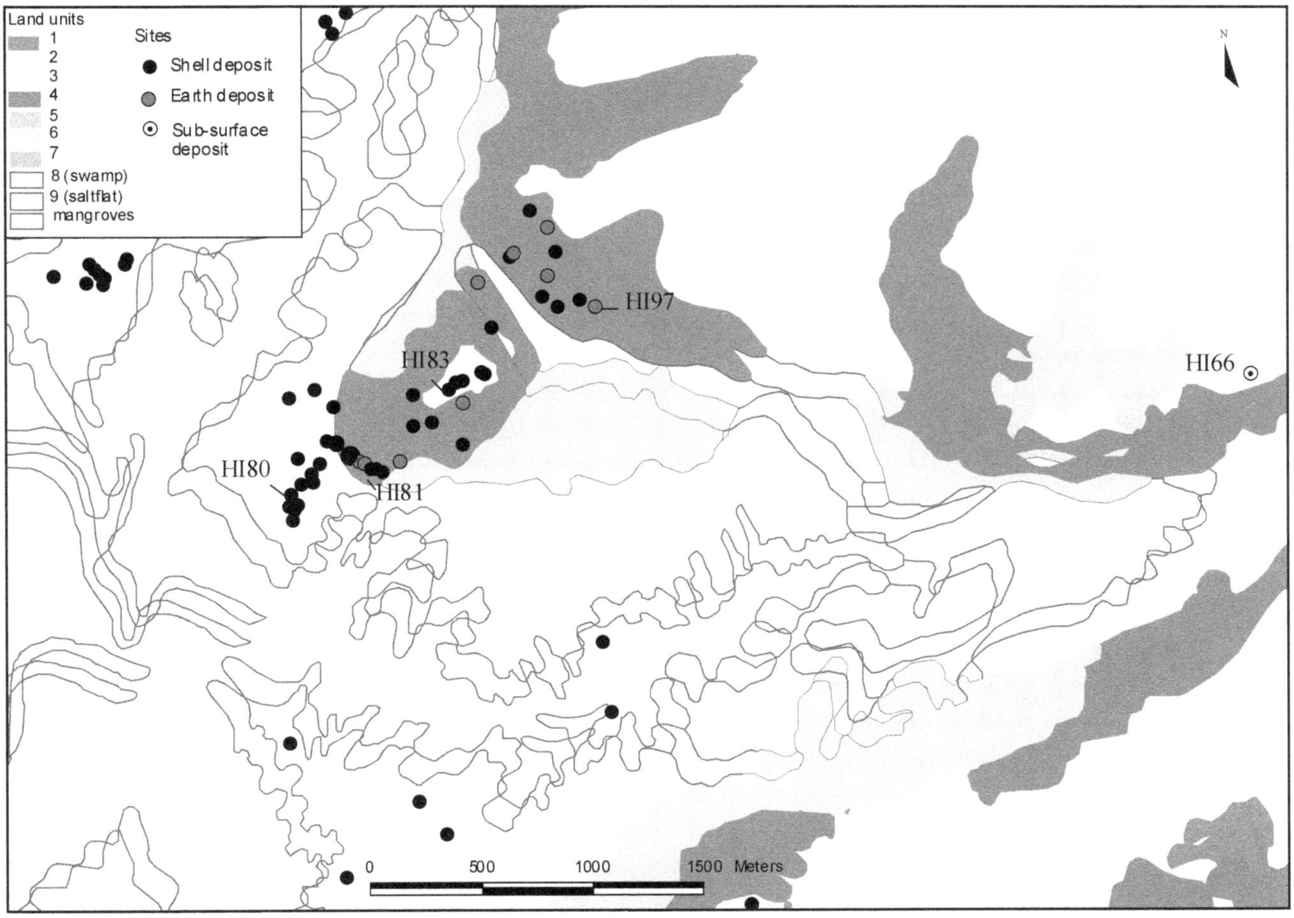

Figure 8.1 Location of excavated sites at Hope Inlet

within Stratum I: an upper shell layer of mainly densely packed whole *Anadara* shells in a sparse matrix of dark brown greyish ashy silt and a compacted 5-10cm thick cap of highly fragmented shell and soil. Numerous roots were observed in the top 10cm of deposit (the weathering zone), which contains a higher proportion of soil than the rest of the shell deposit. Large quantities of charcoal and ash, and thin, fine ashy lenses were evident throughout the shell layer during excavation. The pH levels were high between 9.5 and 10 throughout the deposit. The shell layer extends to a depth of around 65cm, resting on Stratum II, a lower layer of compacted light brown clay with numerous rocks and few shells. As the excavation proceeded the lower layer became more compacted with decreasing shell content and increasing rocks and lateritic gravel, until a rock hard clay base was reached.

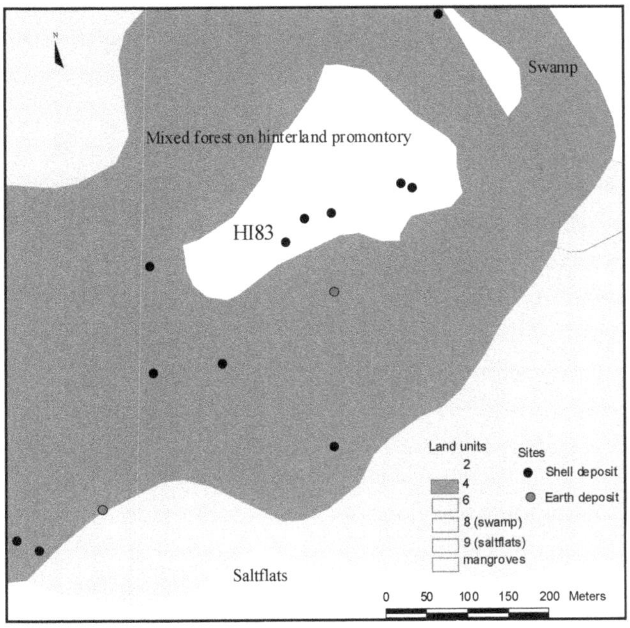

Figure 8.2 Location plan of shell mound site HI83

During excavation, at around 35cm depth it became apparent that due to the dearth of matrix in the mound a frame was required to prevent loose shell falling inwards. Thus at spit 15 a plastic cover was placed in the base of the pit. At the start of spit 16 the walls were cleaned back and a frame to support the walls was placed in the pit, before excavation resumed (Plate 8.2). On removing the frame after the excavation was complete, the walls collapsed inwards. Therefore the stratigraphic profile is based on observation during the excavation (Figure 8.3). Samples of *Anadara* shell collected for dating during the excavation from Section A, spit 9, 16cm below the surface, returned a radiocarbon estimate of 2020±90, which calibrated to 1561 calBP. A large piece of charcoal cemented with *Anadara* shell *insitu* at the basal depth of 67cm in spit 29 Section B of the shell layer, collected as a shell-charcoal pair, returned an estimate for the shell of 1910±70, calibrated to 1448 calBP, and 1850±70 for the charcoal, calibrated to 1369 calBP (Table 8.1).

Plate 8.1 *Anadara* mound HI83 in mixed forest

Plate 8.2 Excavation pit with frame insitu, mound HI83

Radiocarbon age estimates of the shell samples from the surface and basal levels are statistically indistinguishable at the 95% level (Alan Hogg pers. comm.). The condition of shells from the surface was distinctively less weathered than shells from the base of the mound, which lead me to argue with confidence that the surface shell sample dated had not been moved up from the lower levels. With the close correspondence of dates on the stratigraphically secure shell/charcoal pair from the base of the mound, this suggests that the older age obtained for the surface shell may be due to the inherent error of radiocarbon dating, related to exchange of carbon with the environment (*cf.* Head *et al.* 1983; Higham and Hogg 1995). The surface date has a more negative stable isotopic composition (dC13) than the basal shell, and may represent increased freshwater influence (Head 1991). Taking these factors into account, I interpret the radiocarbon estimates as indicating that mound HI83 accumulated rapidly around 1450 years BP.

Rate of shell fragmentation in this mound is low relative to that observed for the Middle Arm mounds (Figure 8.4). At the base of the deposit there is a higher proportion of broken shell, which is consistent with taphonomic decay processes as discussed above. The proportion of broken shell then decreases and remains low at mostly less than 10% throughout the deposit, with two peaks of slightly increased breakage between the depths of 40-45cm and 60-65cm, which may indicate short periods of hiatus of shell deposition. In the top 15cm of the deposit, the rate of fragmentation increases progressively to 80% on the surface. It may be that the high degree of shell fragmentation for the surface layers of this mound reflects the level of intensity of human activity on the mound site after deposition (Muckle 1985:75). However, given the low topography of this mound in an area where livestock were previously run (Christian and Stewart 1953:116; Pietsch 1985:1), it is possible that the high surface level fragmentation is due to recent trampling by cattle or buffalo, as Mowat (1994:207) argues for mounds near the West Alligator River.

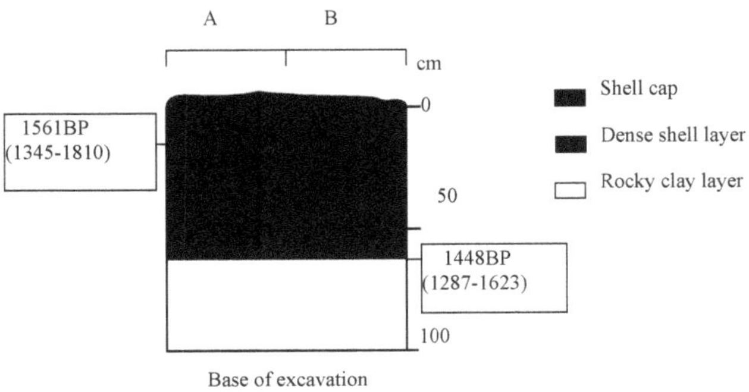

Figure 8.3 Stratigraphic profile, south wall, HI83. Provenance of dated shell shown

Distribution of organic material shell, charcoal and bone in this mound generally follows the characteristic pattern (Table 8.2; Figure 8.5), that is related to taphonomic processes of decay, as discussed above. However, within this mound there are also apparent alternating layers of the main component, shell, and other material. The highest density of bone, charcoal and stone artefacts occurs in the upper level of the deposit between 30cm and 10cm depth, coinciding with a decrease in shell density. There are also small peaks in the density of organic materials other than shell, which coincide with the small peaks in shell fragmentation (Figures 8.4 and 8.5). Given the low fragmentation rate and distribution of organic matter, the stratigraphy is consistent with the dates, which suggest relatively rapid accumulation of this mound, with at least three main discrete depositional events.

Shell

Fifteen macro-mollusc taxa were identified in the excavated material from HI83, and these and the non-molluscan components are listed in Tables 8.3 and 8.4 sequentially. The bivalve *Anadara* clearly dominates the material, representing 92% by MNI and 99% by weight of the total shell content of the sample. Juvenile *Anadara* makes up 6% numerically and 0.2% by weight, and the mangrove gastropod *Cassidula* 1% and 0.1% by weight of the shell component. The presence of juvenile *Anadara* is possibly related to season of collection (Richard Willan pers. comm.). Land snail (*Xanthomelon sp.*) is more numerous in this deposit compared with the more seaward mounds HI81 and HI80. This is consistent with the notion that land snails are intrusive post-depositional agents, which may burrow into shell deposits seeking calcium necessary for their shell formation (see David and Stanisic 1991:20; Claassen 1998:143). Also not considered part of the humanly harvested molluscs are isolated specimens of *Placamen*, *Cerithidea* and *Nassarius,* and small numbers of the diminutive gastropod *Cerithium*, associated with cheniers, recovered from the lower half of the shell layer. Frequency of the bivalve shells *Marcia* and *Saccostrea* and gastropods *Nerita, Terebralia, Chicoreus* and particularly *Cassidula,* is higher in the upper 20cm of the excavated deposit (Table 8.5, Figure 8.6).

Other organic material

Fragmented and for the most part unidentifiable bone (101.3g) was recovered from throughout the shell deposit (Table 8.2). This was distributed vertically in the characteristic pattern related to taphonomic processes of decay, as discussed above

Table 8.1 Radiocarbon estimates, Site HI83

Material	Depth below surface (cm)	Lab Code	dC13	Date (BP)	Calibrated Date (calBP)
Anadara shell	67	Wk6526	-2.3	1910±70	1448 (1287-1623)
Charcoal	67	Wk6527	-25.3	1850±70	1369 (1259-1530)
Anadara shell	16	Wk8252	-3.2±0.2	2020±90	1561 (1345-1810)

(Figure 8.5). Some pieces were identified as fish, mammal, rodent and bird bone. A few macropod and possum teeth and a rodent incisor were recovered from spits 7 to 14, between 15 and 36cm depth, and spits 27 (65cm depth) and 39 at the base of the pit. Two macropod incisors, 27mm and 34mm in length were recovered from spits 13 and 14 respectively, at depths of 33cm and 36cm below the surface (Table 8.5). Both incisors were broken at the tip, similar to a wallaby incisor excavated by Schrire (1982:93-4) at Malangangerr rockshelter in Arnhem Land. A macropod metatarsal bone, probably from an Antilopine kangaroo (*Macropus antilopine*) was noted in Section B, spit 5, 10cm below the surface. These are of interest given ethnographic reports of the use by coastal Northern Territory and north Queensland Aboriginal groups, of kangaroo incisors, teeth and "knuckle-bones" as ornaments attached to the hair (Basedow 1907:41; Foelsche 1882:14; Roth 1984b:25), and incisors mounted on sticks and used as engraving tools (Mulvaney 1975:90, Plate 28).

Fish appeared to comprise a significant proportion of the bone remains by volume, but only a small quantity was identifiable to taxon level. This includes the otoliths of fork-tailed catfish (*Arius* sp.) and a wrasse tooth, recovered from the top 20cm of deposit, and one broken otolith of black jewfish (*Protonibea tiacanthus*), from spit 24 at a depth of 57cm. Unidentified vertebrae of small fish were also present. Numerous crab taxa are known to inhabit these estuaries, but due to the extent of fragmentation and small quantity of crab exoskeleton (5.4g), it was only possible to tentatively identify the claw of the mud crab (*Scylla serrata*).

A total of 76g of macroscopic charcoal was recovered from the excavated sample, also exhibiting a vertical distribution characteristic of the pattern related to taphonomic processes of decay, as discussed above. Small peaks in the density of charcoal correlate with peaks in bone and stone, and in shell fragmentation. In the upper layers, increased density of charcoal, bone and stone corresponds with a dip in shell density, exhibiting a pattern of alternate accumulation, of thick shell layers with lesser amounts of charcoal and other faunal remains, alternating with layers of less shell and equivalent higher density of charcoal and other faunal remains (Figure 8.5).

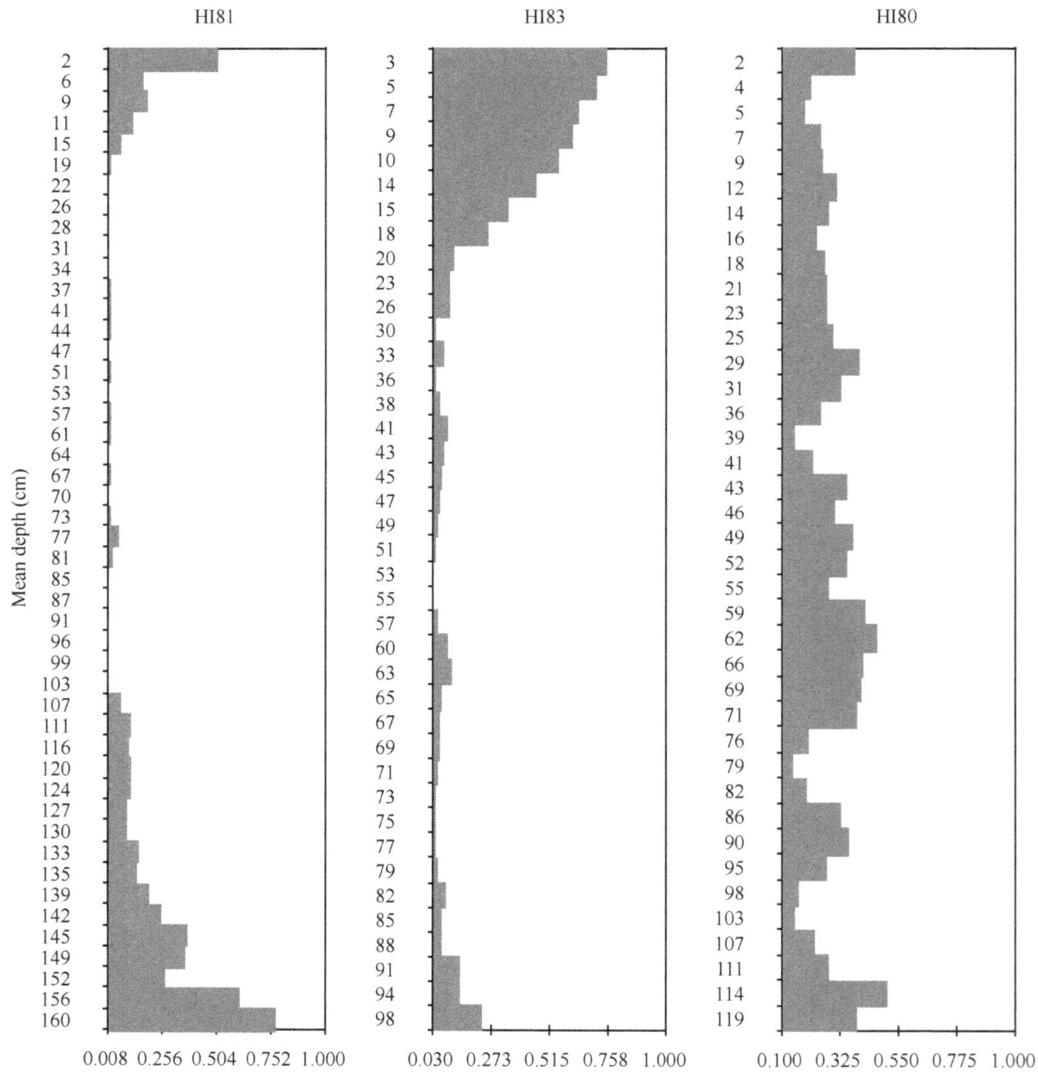

Figure 8.4 Rate of shell fragmentation of mounds HI81, HI83 and HI80

Table 8.2 Quantitative data, Section A, Site HI83

XU	**Mean depth**	**XU wt**	**Shell**	**Char.**	**Bone**	**Otoliths**		**Crab**	**Barn.**	**Stone artefacts**			**Ochre**	**Rocks >6mm**
	(cm)	(kg)	wt(g)	wt(g)	wt(g)	no.	wt(g)	wt(g)	wt(g)	no.	wt(g)	mat.	wt(g)	wt(g)
1	3	18	11543	0.6	0.9					4	4.7	q	2.0	149
2	5	7	4509	0.4	0.7									53
3	7	9	6124	1.1	2.2	1	0.3		0.6					134.5
4	9	12	7693	1.8	2.9				0.4	1	0.1			407.4
5	10	13	8781	3.8	9.2	2	0.4	0.4	0.8	2	1.2	q d	0.5	119.4
6	14	12	8094	4.7	8.6	2	0.4	1.7		2	0.2	q		43
7	15	14	9529	6.3	16.5	1	0.2		0.2	1	0.1	q		21.8
8	18	14	10041	7.2	14.7			0.8					0.6	31
9	20	11	8983	5.1	8.1					2	0.3	q		10.2
10	23	11	8832	4.3	5.0									9.9
11	26	11	8664	2.9	4.7									9.3
12	30	10	8580	2.0	1.5									90.5
13	33	12	9534	2.4	3									267
14	36	9	8177	2.7	2.3					2	0.1	q		64.9
15	38	9	7730	1.5	1.0									73.4
16	41	14	11174	2.0	2.2			1.8						191.4
17	43	12	8919	2.0	3.6			0.4	0.1					165
18	45	12	8483	2.7	2.9									228
19	47	12	8743	1.7	1.8								0.3	474
20	49	10	6774	0.7	1.1									352
21	51	11	7190	1.8	0.7									364.4
22	53	10	6712	0.7	0.9			0.3		1	8.7	qe	0.3	507
23	55	16	7160	1.0	1.0									1913
24	57	14	6337	0.9	0.4	1	0.6			2	0.1	q, t		1646
25	60	17	7688	2.3	0.9								0.9	2466
26	63	17	8118	2.1	1.0									1717
27	65	16	7067	1.1	0.8									1311
28	67	16	6438	1.6	0.7					1	0.2	q		1385
29	69	17	5516	1.7	0.4									2339
30	71	18	2936	0.4	0.1									3501
31	73	19	2578	0.6	0.1									3598
32	75	21	3017	1.0	0.2									4388
33	77	20	2894	1.1	0.4									3878
34	79	18	2321	0.8	0.3									3206
35	82	22	2122	0.4	0.1									4737
36	85	20	1622	0.8	0.1									3261
37	88	21	1306	0.2	0.1					1	0.2	t		4640
38	91	22	846	0.9										4835
39	94	22	306	0.4	0.1									6281
40	98	24	210	0.2	0.1									6768
Total		588	253290	75.9	101.3	7	1.9	5.4	2.1	19	15.9		4.6	65636.1

q = quartz d = dolerite Char. = charcoal mat. = material qe = quartzite t = tuff Barn. = barnacles

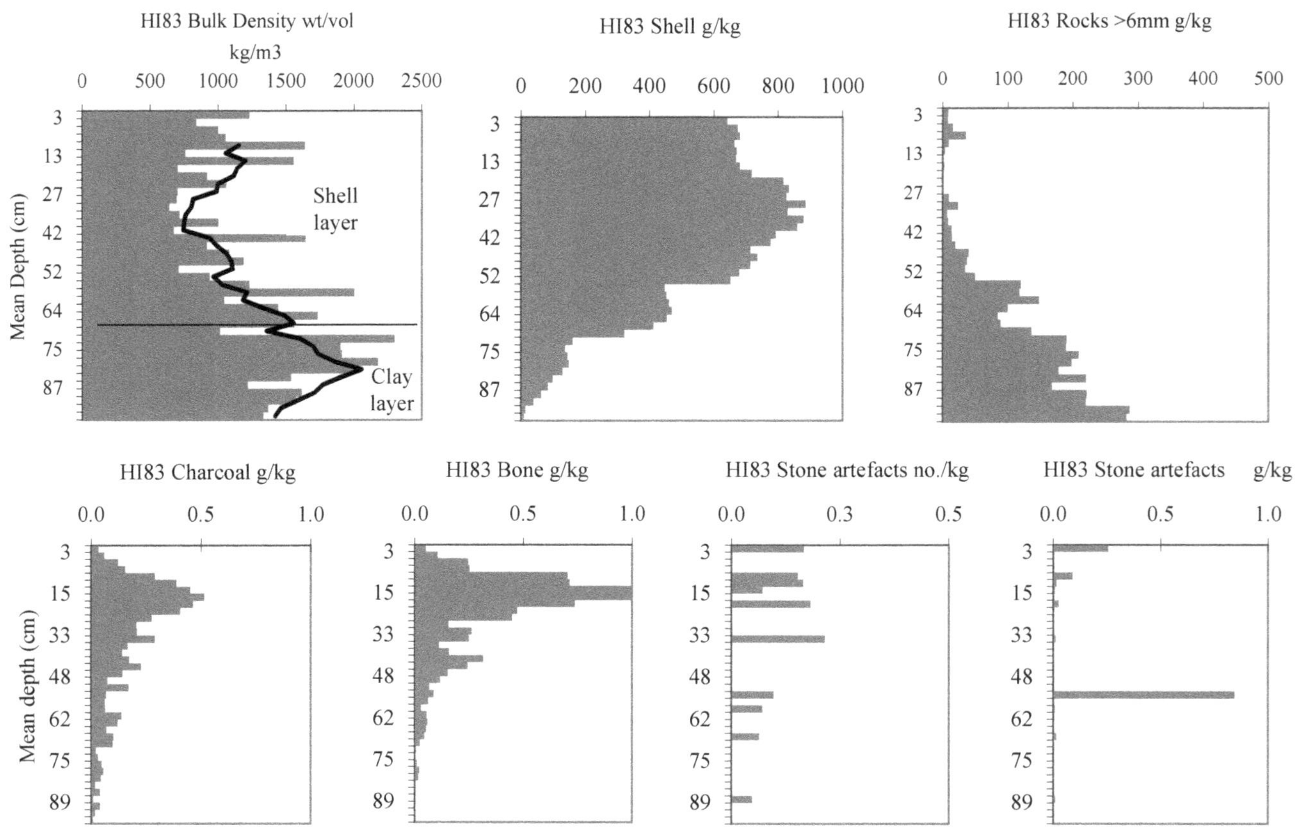

Figure 8.5 Bulk density and distribution of shell, rocks, bone and stone artefacts, Section A, Site HI83

Table 8.3 Total mollusc taxa identified in Section A, Site HI83

Taxon	MNI	Wt(g)	%MNI	%Wt	Wt/MNI
Anadara granosa	28641	250159	91.7%	98.8%	8.7
Juvenile *Anadara*	1953	429.7	6.2%	0.2%	0.2
Cassidula angulata	273	136	0.9%	0.1%	0.5
Nerita sp.	103	187.7	0.3%	0.1%	1.8
Xanthomelon sp.	77	35.3	0.2%	<0.1%	0.5
Terebralia semistriata	60	289.7	0.2%	0.1%	4.8
Chicoreus capucinus	35	164.4	0.1%	0.1%	4.7
Marcia hiantina	28	44.1	0.1%	<0.1%	1.6
*Cerithium coralium**	24	1.1	0.1%	<0.1%	0.0
Ellobium aurisjudae	12	8.3	<0.1%	<0.1%	0.7
Telescopium telescopium	11	113.9	<0.1%	<0.1%	10.4
Volema cochlidium	10	45.4	<0.1%	<0.1%	4.5
Saccostrea echinata	9	17.8	<0.1%	<0.1%	2.0
Geloina coaxans	4	54.2	<0.1%	<0.1%	4.5
Melo amphora	3	69.9	<0.1%	<0.1%	23.3
Cymbiola flavicans	3	18.4	<0.1%	<0.1%	6.1
Placamen	2	1.1	<0.1%	<0.1%	0.6
Nassarius sp.*	1	0.4	<0.1%	<0.1%	0.4
Cerithidea sp.*	1	0.1	<0.1%	<0.1%	0.1
Unidentified shell		1512.4		0.6%	
Total	31250	253288.9			

* Chenier shell

Table 8.4 Identified components of non-shell faunal remains, Section A, HI83

Type	Wt(g)	n	Av. wt
Total Bone: Mammal, Rodent, Bird, Fish (Fork-tailed catfish, Black jewfish, Wrasse)	101.3		
includes: Macropod incisors	1.8	2	
Macropod teeth	0.5	3	
Possum teeth	0.1	2	
Rodent incisor	<0.1	1	
Total Otoliths:	2	8	
includes: Fork-tailed catfish	1.3	6	0.2
Black jewfish	0.6	1	
Crab	5.4		
Barnacles	2.1		

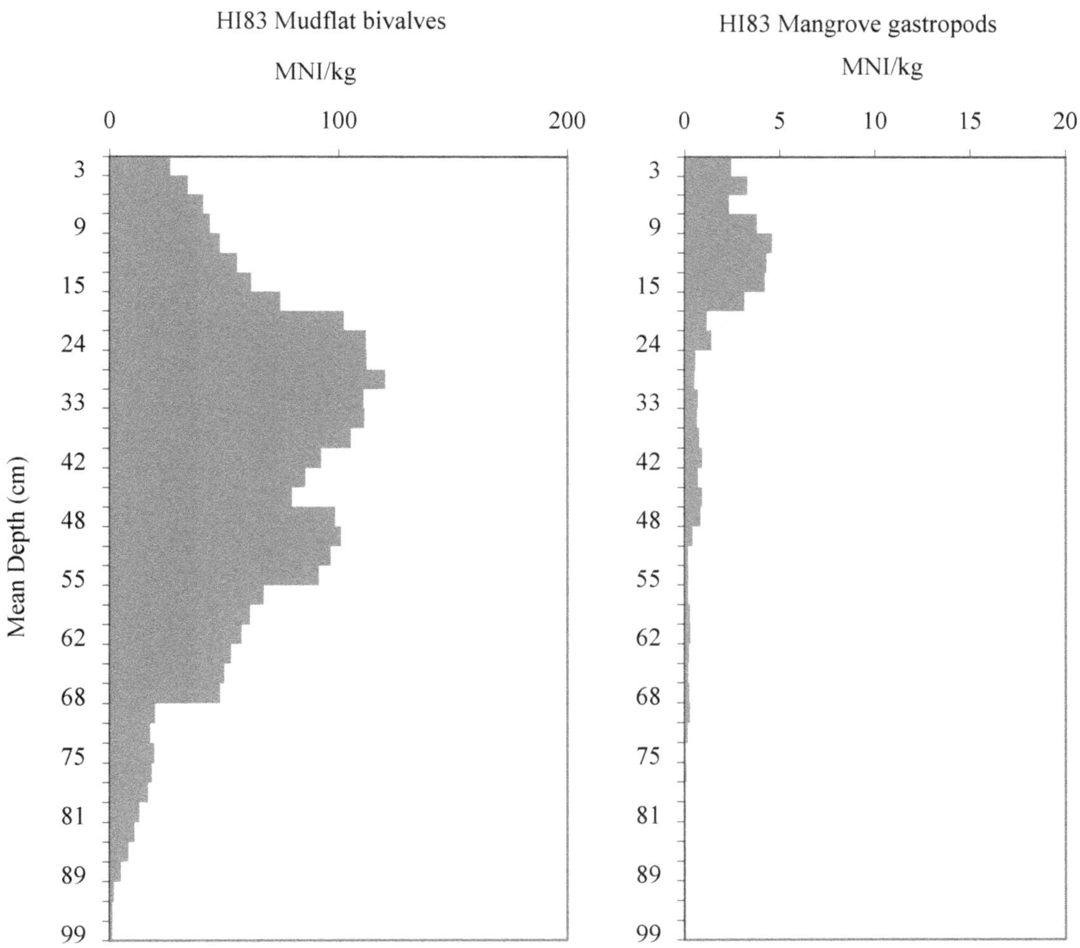

Figure 8.6 Graph showing relative densities (MNI/kg) vs depth of deposit HI83, of mudflat bivalves Anadaraand Marcia and the mangrove gastropods Cassidula, Nerta, Chicoreus, Terebralia,Telescopium nd Ellbim

Table 8.5 Distribution of shell taxa with depth in Section A, Site HI83

XU	Mean depth (cm)	*Anadara*	Juvenile *Anadara*	*Cassidula*	*Nerita*	*Terebralia*	*Chicoreus*	*Marcia*	*Ellobium*	*Saccostrea*	*Telescopium*	*Volema*	*Geloina*	*Cymbiola*	*Melo*	*Cerithium**	*Xanthomelon*
1	3	476	98	23	7	6	6	1		1	2						
2	5	228	48	9	7	2	2	1			2						2
3	6	367	96	12	4	2	2	1			1	1	1				5
4	9	507	169	28	5	7	2	3	1	1							5
5	10	630	213	38	9	10	2	2	1	2		1					4
6	13	672	172	32	9	7	3	4	1	2							7
7	15	864	176	36	10	7	4	4	2	1							7
8	19	1044	163	22	7	9	3	5	3	1					1		5
9	22	1125	69	7	3	1	2	4				1					4
10	24	1185	59	9	3	2	1	2									4
11	27	1177	32	3	2		1	1								4	3
12	29	1166	20	2	2		1					1	1				
13	33	1273	50	3	3		1				1				1		2
14	36	1033	15	1	2	1			1		1	1	1	1			
15	37	947	33	4	1	1					1	1					3
16	42	1301	66	6	3	1	1		1		1	1					4
17	43	983	62	3	3	1	1					1		1		9	1
18	46	948	32	6	4		1						1				5
19	48	1172	41	6	2				1	1	1				1		2
20	49	959	24	1	2		1									5	2
21	52	1023	34	1	1												1
22	55	940	20	1	1							1		1			1
23	57	1075	18	2	1												3
24	59	857	20	1	2				1								2
25	62	962	29	3	2							1				6	1
26	64	917	30	1	2						1						
27	66	781	33	2	1												
28	68	752	24	1	3												
29	71	620	19	3	1		1										
30	73	365	6	2	1												
31	75	338	9	1													
32	77	410	6	2													
33	79	361	33	1													
34	81	303	4														3
35	84	280	10	1													
36	87	213	6														
37	89	173	5														1
38	92	109	4														
39	95	42	2														
40	99	27	3														
41		36															
		28641	1953	273	103	57	35	28	12	9	11	10	4	3	3	24	77

* Chenier shell

Stone assemblage

A total of 19 stone artefacts weighing 15.9g were recovered from 0.5m^3 of excavated deposit (Table 8.2), which gives an estimated average density of 38 artefacts per cubic metre. Artefacts comprise mostly small quartz flakes from the 3.2mm sieve residue (83%), with a maximum dimension between 2.0mm and 10.0mm, which places them in the category of small sized lithic debitage (cf. MacDonald 1991:12). Quartz is the dominant material of the assemblage numerically (79%), but only makes up 37% by weight, as one of the three larger flakes retained in the 6.4mm sieve, made from quartzite, weighs 8.7g. Discounting the quartzite flake, average weight for artefacts in this deposit is 0.1g.

In all, four varieties of raw material were identified. Quartz was recovered from throughout the shell layer, a dolerite flake from the top 10cm, a large quartzite flake from the basal shell levels, and fine tuff material from the lower clay layer. The pattern of distribution of stone artefacts, which are concentrated in the upper 20cm and between 50 and 70cm depth of the deposit with a hiatus in between, supports the interpretation above of discrete occupational episodes during accumulation of this mound (Figure 8.5). Ochre, mostly red, but also one yellow piece, was recovered from a few spits in the shell layer of the deposit. Lateritic rocks >6.4mm totalling 65.7kg occur throughout the deposit, concentrated in the lower layers. These are another feature of tropical shell mounds, and may have been used as hearthstones (see Veitch 1994).

Discussion

Geomorphic and archaeological data and radiocarbon dating indicate that around the period 1450 years BP, the occupants of this area gathered large quantities of bivalve molluscs and other resources and cooked and deposited the remains in the form of mound HI83 on a ground surface of rocky clay. The predominant mudflat bivalve *Anadara* was gathered either from the supra-tidal flats fronting the promontory, which were possibly at the time inter-tidal, or from the mudflats seaward of the mangrove zone, which could have been accessed with canoes via the tidal channels. Ethnographic data shows that it is possible for large quantities of midden material to accumulate quickly, and this appears to be the case for this mound, judging by the statistically indistinguishable dates and low rate of shell fragmentation.

The pattern of shell fragmentation is consistent with the proposition that (at least) three discrete depositional events formed this mound. Using Meehan's (1982a:166) figures for the Anbarra of 8m^3 per year, the estimated volume of 57m^3 for HI83 could have taken around 7 years to accumulate, assuming a similar size group. However, this assumes a different discard practice than that observed ethnographically, of depositing all mollusc remains in one site. A similar discard practice of distributing remains over many sites, would take longer to accumulate a mound this size, or alternatively, a larger group size could accumulate this volume over an even shorter time span.

A low rate of shell fragmentation indicates a short accumulation period, as shell remains on the surface for only short periods of time, before being covered by a protective layer of more shell. Mounds MA1 and MA7 on Middle Arm peninsula, with smaller estimated volumes of 14m^3 and 12m^3 respectively, which accumulated over many hundreds of years, each exhibit a rate of fragmentation at around 40% (Figure 7.6). Moreover, radiocarbon estimates indicate that mound HI80 (see Table 8.11 below), with an estimated volume of 73m^3, and higher rate of fragmentation at 20-40% throughout (Figure 8.4), accumulated over approximately 200 years. Thus the data suggests that mound HI83, with a rate of fragmentation less than 10%, accumulated relatively rapidly in human terms.

The noted pattern, of shell-thick layers accumulated alternately with layers containing less shell but increased other organic remains such as fish, bird, mammal, and charcoal, was also detected by Koike and Okamura (1994) in the Satohama shell mound on Miyato Island, on the northern Pacific coast of Japan. Koike and Okamura interpret this cyclical deposition pattern as reflecting alternating deposition of large quantities of shell during the intensive shell gathering period, and harvesting of other resources during periods of low shell gathering activity, which they suggest is related to seasonality of exploitation activities. If this were so, the pattern may be a type of seasonal banding, like growth rings in a mound (Ian Walters pers. comm.).

Peaks in density of bone, charcoal and stone artefacts at depths also coinciding with peaks in the shell fragmentation rate supports such an interpretation for this mound, and is consistent with accumulation of the mound through (at least) three separate depositional periods. In addition, within the periods of overall decrease in shell density, there is an increase in density of molluscs other than the dominant bivalve *Anadara*, particularly the mangrove gastropods. There is also a higher frequency of other bivalve shells *Marcia* and *Saccostrea* and gastropods *Nerita, Terebralia, Chicoreus* and particularly *Cassidula,* in the upper 20cm of the deposit (Table 8.5, Figure 8.6). It is possible to explain this as due to differential chemical destruction with depth of the deposit as discussed above, especially with regard to the relatively fragile shells *Marcia, Saccostrea* and *Cassidula,* though not for the more robust *Terebralia,* and *Chicoreus* shells. An alternative explanation is that a decrease of 15 to 20% in the relative frequency of *Anadara* to the other taxa in the top 30 to 10cm of deposit reflects a relative increase in quantity of molluscs other than *Anadara* (in particular mangrove gastropods) collected during the terminal stages of accumulation of this mound (Figure 8.6).

I interpret the data as indicating that the other organic and stone material were the remains of continued activity on what was then the surface of the mound, during a temporary

Plate 8.3 *Anadara* mound HI81 on laterite ridge at edge of saltflats

Plate 8.4 Excavation square mound HI81 overlooking saltflats

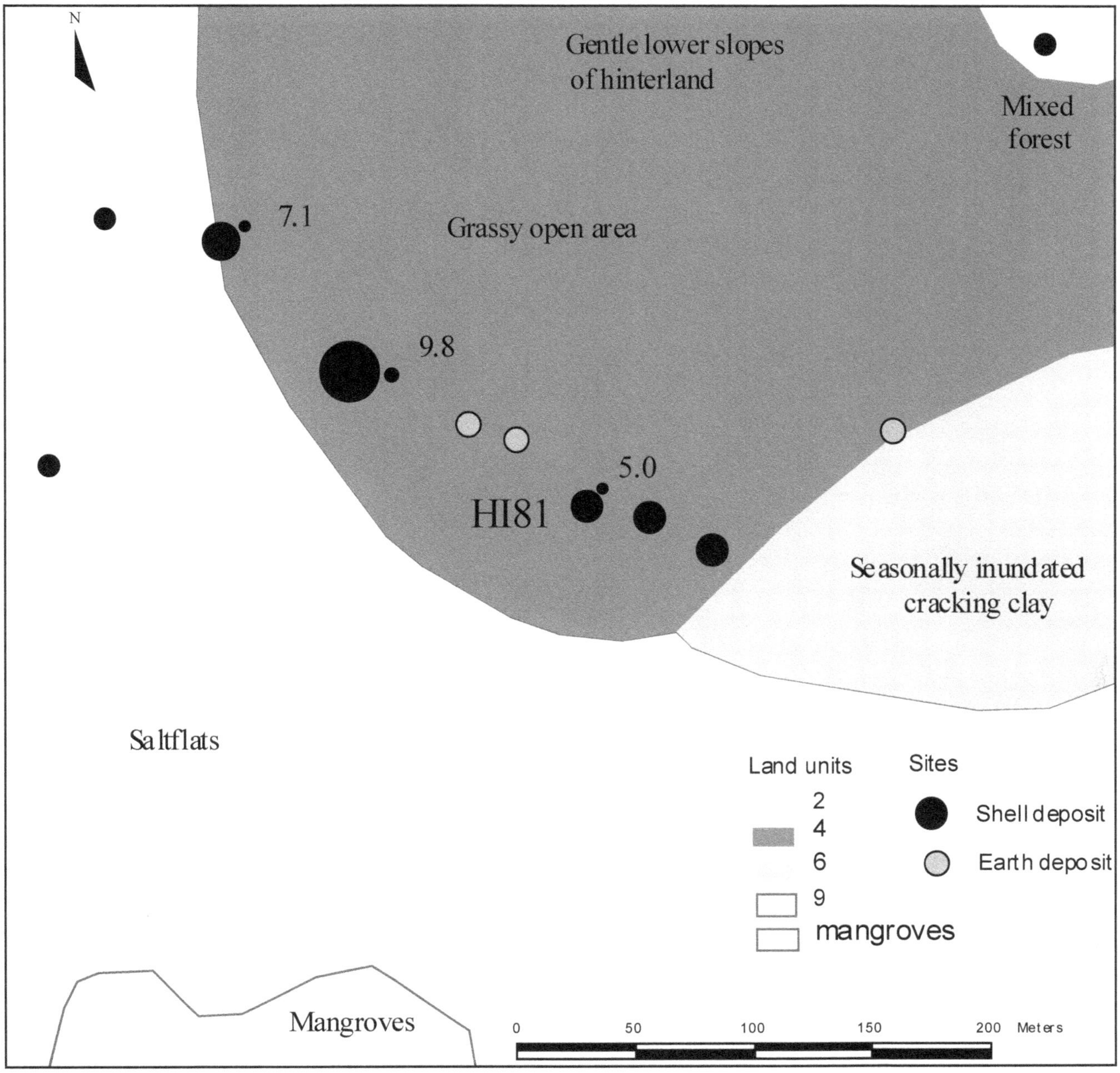

Figure 8.7 Plan of Site HI81

hiatus or slow down in deposition of the dominant shell. These activities included cooking a variety of resources including molluscs other than the staple *Anadara*, such as the mangrove gastropods, and fish, macropods, rodents, possums and birds as well as use of red and yellow ochre and maintenance and manufacture of artefacts. The numbers of artefacts from this sample are too small for any meaningful analysis, other than to say that the presence of debitage indicates knapping of imported stone was carried out on this site.

If the depositional episodes are alternate, perhaps seasonally, and there are only around three, this suggests a much faster rate of accumulation during periods of deposition than seen ethnographically, which implies a larger group size, more intensive harvesting and/or different discard practices. Such an inference must be qualified by the knowledge that the small sample size of excavated material may not necessarily be representative of the remainder of the mound. As discussed in Chapter Four, shell mounds may accumulate through many occupational episodes and contain spatially discrete and heterogenous contents (eg. O'Neil 1993; Waselkov 1987:143), though here, as observed for the Middle Arm deposits, sections of the mound exposed by goanna disturbance suggests that the bulk samples generally represent gross overall mound content.

Site HI81

Environmental setting

Site HI81 is part of a large mound complex or "composite mound site" (Cribb 1996:160), extending for some 300m along a lateritic ridge that borders the seaward edge of a narrow hinterland promontory (Figure 8.1). This promontory projects onto the saltflats midway between

Plate 8.5 Soot-covered stone hearth on laterite ridge just outside excavation pit

Plate 8.6 Section profile of exposed wall of mound HI81 after cave-in, showing stone hearth at base

two small mangrove fringed tidal channels around 500m to the north and south, which bifurcate a kilometre southwest of the site from one of the main channels leading to the shallow seas of Hope Inlet. The mound complex comprises a series of coalesced shell and earth mounds, some of which appear similar to what Cribb (1996:160, Plate 9.2) refers to as "double mounds", while the largest two mounds in this complex at Hope Inlet (Plates 5.7 and 5.8) exhibit distinct foothills and outworks. Although shell appears to be incorporated along the whole length of the ridge, five discrete shell mounds are discernible. Two earth mounds are situated along the landward edge of the laterite ridge and two more some 100m southeast from the eastern end of the ridge. On the landward side of this ridge an open area extends some 100m to the edge of the mixed forest that covers most of the promontory. Numerous smaller shell middens and mounds are found both seaward of the complex on the saltflats and in the forest.

To minimise disturbance, one of the smaller shell mounds that make up this large composite site was excavated (Figure 8.7; Plates 8.3 and 8.4). This mound measures 14m north-south across the width of the ridge by 15m east-west. It lies approximately 100m south along the ridge from the largest mound (HI81a) of seven metres high. Shell and earth material along the ridge between these two mounds has been heavily bioturbated by scrubfowl and goannas. Situated between the excavated mound and the edge of the ridge approximately 50m south are two smaller mounds. Rising a metre or so above the level of the ridge on which it sits, the mound overlooks and abuts the mudflats. At its highest point it is 2.3m above the level of the mudflats, and 1.8m above the ground surface on its landward aspect.

Patches of monsoon vine thicket (MVF) cover some sections of the ridge, including around but not on the excavated mound and were identified by Larrakia ethnobotanist Lorraine Williams in 1998 as species utilised by the traditional owners of this area. Species include *Dioscera transversa* (long yam), *Ficus opposita* (sandpaper tree), *Planchonia careya* (cocky apple), *Sterculia quadrifola* (bush peanut), *Abrus precatorius* (poisonous, decorative red and black seed) and *Flacourtia territorialis* ("good fruit") (Lorraine Williams pers. comm.). The nearest small creek is about two kilometres north-east of the site, but as noted above water would have been available in the area from shallow wells.

Stratigraphy, shell fragmentation and chronology

Excavation of a one metre square pit toward the centre of the mound (Plate 8.4), to a depth of 1.6m revealed two main stratigraphic layers: the shell mound itself (Stratum I) and the lateritic ridge on which it rests (Stratum II), also containing shell (Figure 8.8). Two layers can be distinguished within Stratum I. The upper shell layer consists of densely packed predominantly *Anadara* shell in a sparse, fine matrix of dark brown grey ashy sediment, interspersed with organic ash and silt lenses, and extends to a depth of approximately 110cm. A thin surface layer

Table 8.6 Radiocarbon estimates, Site HI81

Material	Depth below surface (cm)	Lab Code	dC13	Date (BP)	Calibrated Date (calBP)
Anadara shell	5	Wk6524	-1.6	1900±70	1427 (1281-1612)
Anadara shell	140	Wk6523	-2.4	2220±70	1814 (1614-1986)

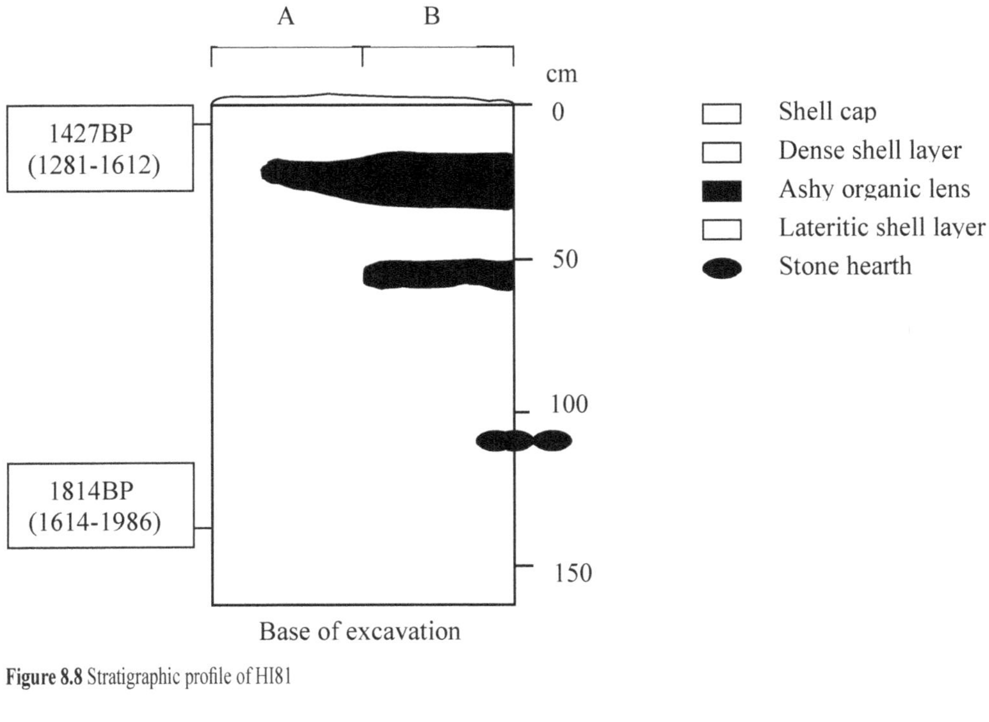

Figure 8.8 Stratigraphic profile of HI81

Table 8.7 Quantitative data, Section A, Site HI81

XU	Mean depth	XU wt	Shell	Charcoal	Bone	Otoliths		Crab	Stone artefacts			Ochre	Rocks >6mm
	(cm)	(kg)	wt(g)	wt(g)	wt(g)	n	wt(g)	wt(g)	n	wt(g)	material	wt(g)	wt(g)
1	2	18	13317		1.9				1	0.2	q		368
2	6	19	16452	0.1	2.8	1	0.3		2	0.1	q	0.2	22
3	9	17	14431	0.3	4.3	5	0.7		2	0.1	q		45.8
4	11	15	14248	0.3	2.3	2	0.1		2	0.1	q		4.5
5	15	15	13609	0.1	1.3	1	0.1		3	0.4	q		92.4
6	19	16	15732	0.2	0.2								
7	22	15	14603	1.2	0.9								13
8	26	14	13508	0.1	0.1				1	0.1	q		100
9	28	14	13805	0.1	0.1								0.8
10	31	14	12944	0.1	0.2	1	0.1						3.8
11	34	13	12519	0.3	0.5	1	0.1		1	0.1	q		5.9
12	37	13	12946	1.0	0.5								21.8
13	41	16	14759	1.7	0.7	1	0.1	0.2					23.8
14	44	14	13472	1.3	0.4	2	0.7	0.1	1	0.3	q	0.1	28.1
15	47	16	15241	0.6	0.5								15
16	51	15	14301	1.2	0.8								28.9
17	53	14	13499	0.9	0.1			0.1					2.5
18	57	15	14716	2.0	1.0			0.1					4.5
19	61	15	13424	1.2	0.8							1.2	20.7
20	64	15	14637	0.5	0.2								13.8
21	67	15	13760	1.0	0.4	2	0.5					0.1	9
22	70	15	14239	1.4	0.2								3.1
23	73	15	13951	1.2	0.6								
24	77	16	13658	1.5	0.8	2	0.3						6.6
25	81	16	13895	0.9	0.8				2	0.1	q	0.1	3.4
26	85	16	14411	0.6	0.2								
27	87	16	14403	0.4	4.3								6
28	91	16	15352	0.8	0.1								
29	96	16	14818	0.9	0.1								37.5
30	99	15	14636	0.6	0.1								3.8
31	103	15	14373	2.2	0.1								46
32	107	16	14543	3.8	1.2			3.2				13.4	260
33	111	20	15615	4.7	7.2	1	0.6	1.7				1.6	613
34	116	24	14327	11.3	5.4	5	1.1	0.6					979
35	120	26	10437	7.8	4.5	2	0.5	1.1					1808
36	124	29	8782	2.3	0.6				1	0.1	q	1.0	2755
37	127	31	7869	2.0	0.6				2	0.1	q		3894
38	130	31	8099	1.2	0.4							159.2	4341
39	133	30	5624	1.8	0.4	1	0.1	2	1	3.4	c	8.0	4374
40	135	34	4380	2.3	0.5				4	0.4	q	3.7	6415
41	139	32	3249	1.9	0.4			0.3	3	0.4	c	61.4	7490
42	142	31	1492	0.7	0.2			0.1				34.7	7465
43	145	31	1003	1.0	0.4				1	0.3	q	0.9	5214
44	149	31	827	0.6	0.5	1	0.2		2	0.2	q		4796
45	152	30	666	0.5	0.4				1	0.3	q		2006
46	156	32	61	0.1	0.1				1	0.1	q		1744
47	160	31	39						2	0.1	q	0.7	1843

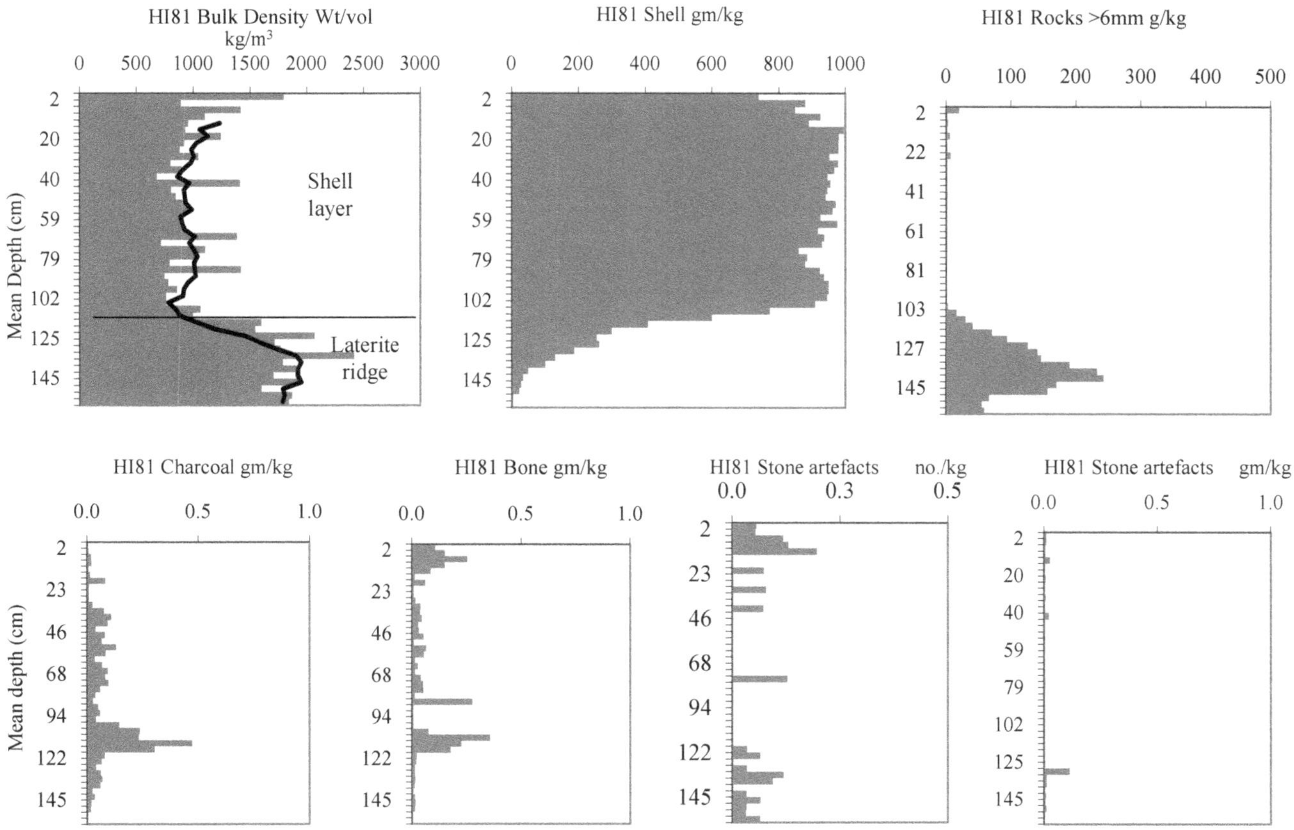

Figure 8.9 Distribution of material HI81

less than 5cm thick of compacted highly fragmented *Anadara* shell and soil caps this shell layer. Stratum II, the lower layer, consists of the more consolidated lateritic ridge comprising dark brown red gravelly soil with shell content decreasing gradually with depth. Lighter medium brown sandy patches appear at a depth of around 130cm, and the sediment in this ridge then grades to a more sandy, lateritic gravelly soil that continues to the base of the pit. The condition of shell in the lower layer is much more weathered than in the upper layer. Throughout the deposit the pH level remained constant at 8.5.

Due to the dearth of matrix in this mound, loosely bound shell from the walls began to fall inwards at around 10cm depth during the excavation, and this was rectified by placing a frame in the pit. On removal of the frame after the excavation was complete, the wall caved in to reveal a stone hearth, consisting of fifteen soot covered lateritic rocks, sitting on the laterite ridge at the base of the pit, just outside the excavation square at a depth of around 110cm (Plate 8.5). The stratigraphic profile presented in Figure 8.8 (Plate 8.6) is of the exposed wall remaining.

Anadara shell for dating was collected during the excavation from the wall 5cm below the mound surface, and from spit 42 in the basal lateritic layer 1.4m from the surface. Results of the radiocarbon age estimates are presented in Table 8.6. Shell from the base of this deposit returned the oldest dates for sites dated at Hope Inlet.

The age estimates suggest this site accumulated between approximately 1800 and 1450 years BP. Some movement of shell was shown in this mound, by the presence of a few individual shells stained red in spit 6, some 80cm above the lower lateritic layer containing the bulk of red stained shell, presumably from the iron content of the soil.

There is a very low rate of shell fragmentation in the upper shell layer of this mound relative to all other excavated mounds, even lower than that for HI83, which radiocarbon dating shows accumulated rapidly. The proportion of broken shell is highest in the lower stratum (laterite ridge), decreasing from 80% at the base to 10% in the top layers of the ridge. Shell fragmentation rate then decreases rapidly at the base of the upper shell layer, remaining very low at less than 5% throughout other than a small peak at around 75-80cm depth, until the top 15cm, when there is a gradual increase up to 50% in the surface spit (Figure 8.4).

The pattern of shell fragmentation of this mound is consistent with an initial slower shell deposition period, and/or relatively high level of intensity of human activity on the lateritic ridge, and/or increased decay processes in the lower levels. This is followed by continuous rapid deposition of the upper shell layer, with a possible hiatus at around 75-80cm depth. As discussed above, the increased shell fragmentation for the surface layers would reflect post-depositional human or animal activity on the mound site and/or weathering processes (Muckle 1985:75).

The distribution of shell and bone is consistent with the notion that there may have been two major separate depositional periods for this mound (Table 8.7; Figure 8.9). That is, the taphonomic pattern typically exhibited in the distribution of organic material in cultural deposits is replicated in the lower and upper strata of this mound for both shell and bone. This provides support for the interpretation of at least two major depositional periods, possibly separated in time by up to four hundred years.

Shell

Twenty-one molluscan taxa were identified in the excavated material from HI81, and these and the non-molluscan components are listed in Tables 8.8 and 8.9 respectively. Four of the mollusc taxa are very small chenier shells (see below) associated with a sandy substrate, recovered from the 3.2mm sieve residue, and comprise <0.1% of the shell content. Another is the land snail *Xanthomelon*, found in most shell deposits but considered in this analysis to be a post-depositional intrusive agent. Of the macro-molluscs, *Anadara* is clearly the dominant material, representing 97% by MNI and 99% by weight of the total shell content of the sample. Juvenile *Anadara* makes up 2% numerically and 0.1% by weight of the total shell component.

Although the oyster *Saccostrea* is the second most numerous taxon, it only makes up 0.2% of the shell content by MNI and weight and is only found in the lower lateritic ridge layer (Table 8.10; Figure 8.10). Another small bivalve, *Placamen*, found in the shallow, sandy/muddy substrate of protected bays and at estuary mouths, or on the muddy banks of rivers, only occurs in the upper shell layer of the deposit. The mollusc taxa *Littoraria*, *Placuna*, *Melo*, *Melina* and *Pseudonachis* with a total MNI less than three are not included in the list. The mangrove gastropods are present in relatively small numbers in this mound. Distribution of these mangrove gastropods however, which peak in the upper 20cm of the shell mound, and in the lower lateritic ridge, does not exactly coincide with that of the distribution pattern of *Anadara* shell (Figure 8.10), with which they are mixed in the deposit, but does coincide with peaks in shell fragmentation.

Other organic material

A total of 50g of very fragmented bone was recovered from the sample (Table 8.9), predominantly from the 3.2mm sieve. Bone distribution is spread throughout the deposit, with the largest peaks in density in the uppermost layers both of the shell stratum and lower ridge stratum of the mound (Figure 8.9). At least 25% by weight of the bone material was positively identified as fish and subjectively fish bone appears to be by far the dominant material. Included are the otoliths of two taxa, mainly fork-tailed catfish (*Arius* sp.). These are concentrated in the top 44cm of the upper shell layer, and the top 22cm of the lower laterite layer, while the otoliths of black jewfish (*Protonibea diacanthus*) were found in two spits, 14 and 21, at depths of 44cm and 67cm in the upper shell layer. Also present were *Arius* sp. skull fragments, vertebrae and pectoral spines, and the vertebrae of small unidentified fish.

The proximal metatarsal of a large macropod similar to that found in HI83, probably Antilopine kangaroo (*Macropus antilopine*) was recovered from spit 7, about 20cm below the surface. Fragments of possible mammal, reptile and bird bone, as well as a small quantity (9.5g) of crab exoskeleton and claws similar to those identified as mud crab (*Scylla serrata*) in HI83, and isolated pieces of mangrove worm tubes, were also identified in the excavated sample. These were recovered mainly from the lower lateritic ridge level.

Charcoal pieces (total 67g) were evident throughout the deposit. During the excavation the fine charcoal pieces and ashy sediment were observed to continuously fall through the gaps between the shells to the next spit. This explains the absence of a taphonomic pattern in the distribution of charcoal similar to that exhibited by shell and bone (Figure 8.9). The higher peak of charcoal density between the depths of 103cm and 120cm coincides with the commencement of the laterite ridge, where the denser matrix acts as a barrier through which the charcoal does not pass as easily. It would seem that this peak is due in part at least, to the excavation process, although it is also likely that there was increased cultural activity such as cooking on the ridge, as indicated by the presence of the stone hearth at this level.

Stone assemblage

Ochre, mostly red, but also one yellow piece, occurs mostly in the sample from the lower layer of the deposit. A total of 33 stone artefacts weighing 6.9g were recovered from 0.67m^3 of excavated deposit, giving an estimated average density of 41 artefacts per cubic metre, very similar to that for HI83. Quartz is the dominant raw material, comprising 88% numerically and 51% by weight. Artefact types are mainly small quartz flakes, mostly from the 3.2mm sieve residue (91%). One large (3.8g) retouched reddish/orange chert flake 30mm x 20mm and two smaller flakes, probably of the same piece, were recovered from spits 39 and 41 respectively in the lower lateritic ridge layer (Table 8.7).

Like the assemblage in HI83, average artefact weight for these artefacts is 0.1g (discounting the large flake) and they can be classed as small sized lithic debitage (cf. MacDonald 1991:12). This debitage peaks in the uppermost and lowermost layers of the deposit, with a hiatus between the depths of 47cm and 124cm, (apart from two small flakes in spit 25) (Table 8.7; Figure 8.9). Numerous lateritic rocks >6.4mm, which occur throughout the deposit, and are concentrated in the lower layer, may have been used as hearthstones (Veitch 1994). One discrete feature described above at the base of the upper shell layer is unequivocally identified as a stone hearth (Plate 8.5). Two of this cluster

Table 8.8 Total mollusc taxa identified in Section A, Site HI81

Taxon	MNI	Wt(g)	%MNI	%Wt	Wt/MNI
Anadara granosa	68592	532676	96.8%	99.3%	7.8
Juvenile *Anadara*	1685	357	2.4%	0.1%	0.2
Saccostrea echinata	162	1096	0.2%	0.2%	6.8
*Cerithium coralium**	106	4.3	0.1%	<0.1%	<0.1
Nerita sp.	66	54.3	0.1%	<0.1%	0.8
Cassidula angulata	54	14.7	0.1%	<0.1%	0.3
Placamen calophyllum	46	29	0.1%	<0.1%	0.6
Terebralia semistriata	41	250.8	0.1%	<0.1%	6.1
Telescopium telescopium	31	614.6	<0.1%	0.1%	19.8
Chicoreus capucinus	28	53.4	<0.1%	<0.1%	1.9
Xanthomelon sp.	20	20.6	<0.1%	<0.1%	1.0
Volema cochlidium	13	236.8	<0.1%	<0.1%	18.2
Geloina coaxans	10	79.9	<0.1%	<0.1%	8.0
Cerithidea sp.	10	6.1	<0.1%	<0.1%	0.6
Nassarius sp.*	9	1.1	<0.1%	<0.1%	0.1
Ellobium aurisjudae	7	2.2	<0.1%	<0.1%	0.3
Cymbiola flavicans	3	70	<0.1%	<0.1%	23.3
Littoraria sp.*	2	0.8	<0.1%	<0.1%	0.4
Placuna sp.	2	0.2	<0.1%	<0.1%	0.1
Melo amphora	1	28.2	<0.1%	<0.1%	28.2
Melina ephippium	1	0.4	<0.1%	<0.1%	0.4
Pseudonachis sp.*	1	0.3	<0.1%	<0.1%	0.3
Unidentified shell		1098.6		0.2%	
Total	70890	536695.7			

* Chenier shell

Table 8.9 Identified components of non-shell faunal remains, Section A, HI81

Type	Wt(g)	n	Av. wt
Total Bone: Mammal, Reptile, Bird, Fish (Fork-tailed catfish, Black jewfish)	50.1		
includes: Macropod metatarsal	0.8	1	
Total Otoliths:	5.5	28	
includes: Fork-tailed catfish	4.8	26	0.2
Black jewfish	0.7	2	0.4
Crab	9.5		
Mangrove worm tube	0.6		

of hearth rocks, which were all roughly the same size, were collected. One measured 60 x 50 x 25mm and weighed 98g, the other 60 x 45 x 30mm, weighing 96g.

Discussion

Data from the excavation of this mound can be discussed in the context of geomorphic data on the evolution of the morphology of the chenier plains at the southern end of Shoal Bay, to hypothesise on formation processes of the large mound complex. After the first phase of progradation began around 2300 years ago, as identified by Woodroffe and Grime (1999) for this area, mound sites such as HI81 may have formed through people harvesting molluscs from beds of *Anadara* that colonised intertidal flats fronting the ridge. Alternatively, the molluscs may have been gathered from the mudflats seaward of the mangroves, accessed with canoes via the tidal channels.

An interpretation based on the patterns of distribution of bone and stone artefacts, with peaks in density coinciding with increases in shell fragmentation rates (Figures 8.9 and 8.4), may mean that there were two main depositional periods separated in time, for this mound. This hypothesis is also supported by a correlation between peaks in distribution of mangrove gastropods and shell fragmentation in the upper layers of both the upper shell mound and the lateritic ridge, and lack of correlation with that of the distribution pattern of *Anadara* shell (Table 8.10; Figure 8.10). I interpret this to mean that gathering of mangrove gastropods increases

Table 8.10 Distribution of shell taxa with depth in Section A, Site HI81

	XU	Depth (cm)	XU Wt (kg)	*Anadara*	Juvenile *Anadara*	*Saccostrea*	*Nerita*	*Cassidula*	*Placamen*	*Terebralia*	*Telescopium*	*Chicoreus*	*Volema*	*Geloina*	*Ellobium*	*Cerithidea*	*Cymbiola*	*Cerithium**	*Nassarius**
Shell layer	1	2	18	1539	315		1	5		8	2	2							
	2	6	19	2011	152		2	8		6	1	1			1				
	3	9	17	1795	137		4	3		3	3	1	1			2	1		
	4	11	15	2155	107		2	3	5	1	1			1					
	5	15	15	2146	84		2	2	4			1						3	
	6	19	16	2398	37		1	2	5	1		1						4	
	7	22	15	1934	15		2	1	2	1								5	
	8	26	14	1762	19				3	1									
	9	28	14	1762	9			1	1									2	
	10	31	14	1606	23		1		3									3	
	11	34	13	1584	18				1			1						3	
	12	37	13	1680	31		1		2					1				3	
	13	41	16	1954	23		1	1	3			1	1					5	
	14	44	14	1710	29		2	1	2	1								3	1
	15	47	16	1859	14		1		2									2	
	16	51	15	1656	20					1			1					3	
	17	53	14	1552	8					1			1					1	
	18	57	15	1659	33		1		1	1								3	
	19	61	15	1507	23		1		1	1		1						4	
	20	64	15	1710	6				1				1					2	
	21	67	15	1639	28		1		1			1	1					3	2
	22	70	15	1838	25	1			1			1						3	
	23	73	15	2009	23	1			2									7	2
	24	77	16	2022	61		3	1	1						1			7	4
	25	81	16	2173	30		1	1	1									4	
	26	85	16	2332	13		1	1											
	27	87	16	2390	6		1	1		1									
	28	91	16	2486	6			2		1									
	29	96	16	2438	7			1											
	30	99	15	2231	3		1	1					2						
	31	103	15	1979	6		2				1							2	
	32	107	16	1690	30		1	1			2		1				1	3	
Ridge layer	33	111	20	1605	83		2				6	1		1				9	
	34	116	24	1354	81	3	2	1			3	1	2	2		2		6	
	35	120	26	1009	57	9	3				2	1		2		1		9	
	36	124	29	764	26	16	4	2		3	3		1	1	2			3	
	37	127	31	647	17	15	3			1	1	1			2	1	1		
	38	130	31	673	14	14	4	2		2	1	1	1	1	1	1			
	39	133	30	450	11	16	3	3		1	2	1				1			
	40	135	34	360	13	24	2	4			1	1						2	
	41	139	32	251	14	20	3	5		2	1			1		1		2	
	42	142	31	112	8	13	3			1	1	1							
	43	145	31	60	10	13	1	1		1									
	44	149	31	54	4	9	2									1			
	45	152	30	40	2	3	1			1									
	46	156	32	4	2	4													
	47	160	31	3	2														
			933	68592	1685	161	66	54	42	40	31	19	13	10	7	10	3	106	9

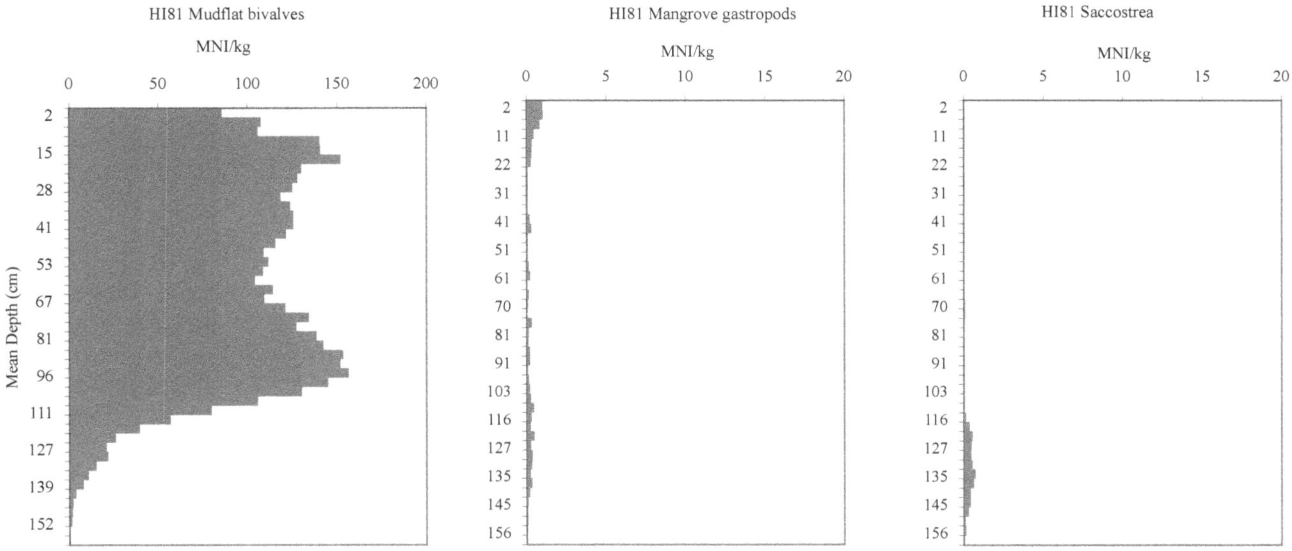

Figure 8.10 Graph showing relative densities (MNI/kg) vs depth of deposit HI81, of mudflat bivalves Anadara and Placamen and the mangrove gastropods Cassidula, Nerita, Chicoreus, Terebralia, Telescopium and Ellobium

slightly during times of more intense activity on the actual surface, at that point in time when there is a hiatus or slow down in mound formation through deposition of *Anadara* shell. The distribution pattern of the mainly *Anadara* shell component, and differences between the shell and laterite ridge layers, in the frequency of other molluscs, including *Placamen* only in the shell layer and *Saccostrea* only in the ridge, is again consistent with the notion of at least two separate depositional periods (Table 8.10; Figure 8.10).

In the first period, ca 1800 years BP, it appears that molluscs were collected and the shells cooked and deposited by people camping on the sandy lateritic gravel ridge. Exploited molluscs included mainly the roughback cockle *Anadara* from the sandy mudflats, augmented with small quantities of large oysters (*Saccostrea*), from rock platforms or mangrove roots, and the archetypal suite of gastropods (mainly *Cassidula*, *Nerita*, *Chicoreus*, *Ellobium*, *Telescopium*, and *Terebralia*) from the mangroves. It is apparent from the small numbers of mangrove gastropods in this mound that gathering of gastropods occurred at very low levels during periods of rapid processing of large quantities of *Anadara*.

The more weathered condition of shell in the ridge is consistent with a low rate of shell discard and/or high intensity of human activity but could also be due to decay processes as noted above. However, data on the distribution of the main taxa with depth (Table 8.10) support the former hypotheses. This shows that the vertical distribution of shell is thinned out in the laterite ridge, with numbers of *Anadara* for fourteen spits spread through 50cm of deposit in the ridge, less than that in four spits spread over 10cm in the shell layer. This is consistent with the shell being pushed downward through the ridge sediment by processes such as human trampling during deposition, which would also increase fragmentation (see Figure 8.4).

Fish, mainly fork-tailed catfish (*Arius* sp.), crab, mammals, reptiles and birds were some of the non-molluscan resources targeted at this site. These resources were cooked at the site during the initial period of occupation, the frequency of hearth stones in and on the ridge suggesting that cooking was a major activity. The stone hearth identified on the surface of the ridge layer (Plates 8.5 and 8.6) bears a striking resemblance to that of the typical stone oven used mostly for cooking kangaroo by the Fish Creek group of Arnhem Land, described in McCarthy and McArthur's (1960:169, Plate 3A) ethnographic food-quest study. Knapping of imported stone was also an on-site activity, of mainly white quartz, but also of a small quantity of reddish/orange chert. The concentration of these stone artefacts in the uppermost and lowermost layers of the deposit, with a hiatus between, supports the idea that these periods of activity may have been separated by some time (Figure 8.9).

During the second period around 1450 years ago, larger quantities of the *Anadara* resource were harvested. These molluscs were rapidly processed and shells deposited on site to form the upper shell mound with an estimated volume of 83m^3. Evidence supporting an interpretation of increased focus on *Anadara* is the much greater proportion of this mudflat bivalve to mangrove gastropods in the upper shell layer, of around 500:1, compared with 100:1 in the ridge layer. That this is a real phenomenon and not due to differential taphonomic processes of decay is shown by the higher densities of the relatively fragile, brittle *Saccostrea* shell and the fragile gastropods such as *Nerita* and *Cassidula*, in the lower levels (Table 8.10). Thus in the period around 1450 BP the bivalve *Anadara* continued to be the predominantly exploited resource, but harvested much more intensively, along with lesser numbers of *Placamen* from the mudflats and the commonly gathered mangrove gastropods, but no oysters as in the earlier period.

Studies have shown that *Saccostrea*, found on mangrove roots or rock platform, has a low resilience to human exploitation, with populations easily depleted due to their fixed position in the intertidal zone and favoured status as a resource (Catterall and Poiner 1987:121). Therefore Catterall and Poiner (1984:297) predict that this taxon would be found in strata accumulated early in a midden, a prediction seen to occur in this mound, with oysters concentrated in the ridge layer. As I observed, and as indicated by the Wt/MNI estimates (Tables 8.3, 8.8, 8.14), the average shell size of these oysters is much larger than those recovered from the more recent mounds HI83 and HI80, which may represent the effects of human predation and/or environmental change over time.

Fishing also appears to have been a focus of subsistence strategies apart from mollusc gathering during this period. The peak in density of bone, fish otoliths, mangrove gastropods and stone artefacts in the top 20cm of the mound is consistent with the notion that these components may be intrusive, deriving from later occupation on top of the mound after the terminal accumulation period of the staple mollusc *Anadara*. Similarly, the large macropod bone recovered from 20cm below the surface, may be the remains of cooking a large Antilopine kangaroo on the mound surface, or could be part of a lost hair ornament as discussed above.

The very low level of fragmentation in the upper shell mound suggests it accumulated through no more than two or three short-term depositional events. Such a scenario is not consistent with observations of discard practices of modern collectors like the 40-strong Anbarra group in Arnhem Land (Meehan 1982a). Rapid deposition of primarily one mollusc species in a specific site over a short time span would signify a larger group, different discard practices or more efficient harvesting technology, or a combination of these factors, than that observed ethnographically. This is so even without considering that the mound excavated is part of a much larger complex with a minimum estimated shell volume of 3,650m^3.

More efficient technology than the ethnographically observed practices of using hands and digging sticks (Meehan 1982a:97-8) could include harvesting from a boat, using a hand dredge made from material such as bamboo and plant fibre, as described in southeast Asia and Japan (Broom 1985:26-9). Canoes have been recorded historically being used to transport molluscs around the coasts of Darwin and Arnhem Land (King 1969:89; Tindale 1925:82). The presence of a high proportion of incidental bycatch that would conceivably collect through dredging, such as juvenile *Anadara* and other small shells recovered in the 3.2mm sieve, would provide support for this idea. But the relatively low proportion of juvenile *Anadara* in the rapidly accumulated large mound HI81 (Table 8.8), compared with that of the much more slowly formed smaller Middle Arm mounds MA7 and MA1 (Tables 7.4 and 7.5) is not consistent with this notion of a change in harvesting technology. It is more likely that the relatively higher proportion of juvenile *Anadara* in some middens is related to the season of collection as suggested by Richard Willan (pers. comm.).

Based on the above data, I interpret this site as a multi-component site incorporating different functions during different time periods. The original ridge was likely to have been a seasonal base camp, where a variety of resources were exploited from the surrounding ecotones of the estuarine system, swamps, patches of monsoon jungle and hinterland open forest. Periodic occupation may have occurred perhaps over hundreds of years from around 1800 BP.

The upper shell mound, which was deposited rapidly around 1450 BP, possibly represents the remains of a processing site for large quantities of *Anadara* for a specific purpose. This could include feeding a large group of people participating in ceremonial activities and/or procuring dried mollusc meat for later consumption or trade, either with inland groups or visitors from the sea. I am of the opinion that the latter option is unlikely given the recent evidence of the unsuitability of the Darwin area as a port of call for Macassan traders. The process of drying large quantities of molluscs for storage, and trading with inland groups has been recorded ethnographically for many coastal societies in North and South America, Oceania, Asia and Africa (Waselkov 1987:106-7).

Ethnographic reports of sun-drying or food storage in Australia are scarce and, apart from isolated references such as the reported drying of strips of kangaroo meat by a group in the Pilbara area of northwest Australia, refer to plant foods (Mulvaney 1975:73, 241; Withnell 1965:20). Reports of plant storage include processed loaves of cycas bread being wrapped in pandanus leaf or paperbark and stored for several months before being eaten, often said to be in connection with large gatherings of people for ceremonies (Jones and Meehan 1989:124; Thomson 1939:216, 1983:105). Similar short-term storage has been reported, of surplus cycad nuts in Arnhem Land (Peterson 1983:105) as well as sun-drying and storage of wet season green plums (*Buchanania* sp.) for weeks or months (Berndt and Berndt 1970:36; Chaloupka 1981:170; Irvine 1970:280). Thomson (1949:23-4) describes the processing of this fruit, which is dried coated with red ochre, stored in teatree bark parcels and the hardened fruit and seeds pulverised with pounding stones and mixed with water or the flesh and bones of red kangaroo into a "greatly relished" paste. On Cape York Peninsula, Thomson (1939:216) reports storage of yams (*Dioscorea sativa*) and tubers of water lily (*Nymphaea* spp.).

If ceremonies were held in the area, nearby extensive cycad stands could have provided a source of carbohydrates and calories to complement the high protein diet provided by the molluscs. Additional support for the interpretation of aggregations of people in this area comes from statistically indistinguishable radiocarbon dates of around 1450 BP for two shell mounds, HI81 and HI83, signifying that all or parts of the sites were possibly contemporary with each other. And like Site HI81, Site HI83 has a low rate of shell fragmentation and radiocarbon estimates indicating relatively rapid accumulation. It is also possible that the earth mound HI97, located a few hundred metres inland, is also contemporary, though as discussed below the

Plate 8.7 Looking seaward from Mound Complex across saltflats inundated by king tides

Plate 8.8 View of Mound Complex looking toward hinterland from saltflats Mound

radiocarbon date obtained for HI97 is on shell that may pre-date the earth mound.

Site HI80

Environmental setting

Site HI80 is a typical elongated *Anadara* mound located on the saltflats midway between the large mound complex, Site HI81, and the mangroves (Figures 8.1 and 8.11). It is one of 28 middens in this small section of saltflats and lies approximately 50m north of a small mangrove-lined tributary of a main tidal channel leading to the shallow seas of Hope Inlet. During the late dry season king tides and after big rains in the wet, the mound appears as one of many small, low islands in a flat sea of water (Plate 8.7). The large mound complex on the hinterland margin about 150m to the north-east is clearly visible from this mound (Plate 8.8).

There was little vegetation on the surface of this saltflats mound other than two small mangrove shrubs, a dead mangrove stump and short clumping grass (Plate 8.9). Auguring around the mound showed that the shell deposit extends under the surrounding saltflats surface, approximately another three metres in length from each of the north and south edges, and another 10 metres in width, eight metres from the eastern edge and two metres from the western edge. Surface measurements of the mound, of 5.5m north-south by 10m east-west and depth of 0.5m would give an estimated volume of approximately 11m^3, as opposed to 73m^3 when subsurface measurements are added, bringing it to 11.5m north-south by 20m east-west with a depth of 0.8m.

Thus topography of the mound can be likened to an iceberg, with only the tip showing, and the largest volume of material being under the saltflats surface, indicating that ongoing sedimentary infill has partially buried the mound since its formation. It is likely that other small middens in this section, many of which appear as flat circles of shell on the saltflats surface, are in fact much larger than their visible parts. Some middens may be completely buried within the plains, as Woodroffe *et al.* (1988) found for the coastal and estuarine plains of the South Alligator River.

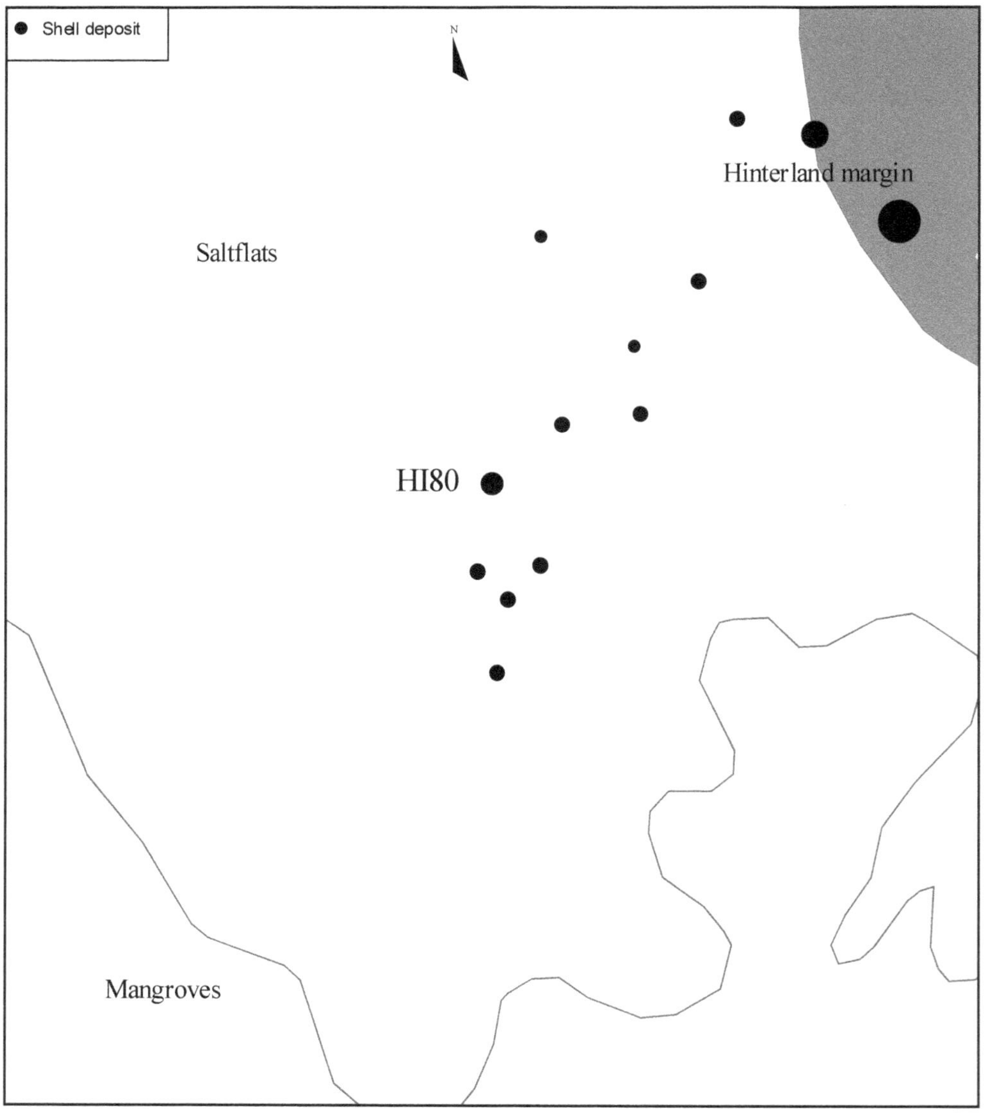

Figure 8.11 Location plan of shell mound site HI80

Stratigraphy, shell fragmentation and chronology

Four stratigraphic layers were observed during the excavation of a one metre square slightly northwest of the mound centre (Figure 8.12):

1. An upper layer of predominantly *Anadara* shells in a dark greyish black sticky muddy matrix, extending from the surface to a depth of around 40cm.

2. A thin layer (5-10cm) comprising a large proportion of *Marcia* as well as *Anadara* shells in a light-medium brown sandy matrix.

3. A medium brown sandy clay layer from around 50cm to 80cm depth, with the shell content of mainly *Anadara* decreasing with depth. The water table was reached at around 70cm depth, and water covered the base of the deposit for the remainder of the excavation.

4. A blue-grey mangrove mud layer with a few *Anadara* shells, extending to the base of the excavation at a depth of 1.2m. A large mangrove buttress root was observed insitu in this layer at the base of the west wall of the pit (Plate 8.10).

Shell and charcoal pair samples for radiocarbon dating were collected during the excavation from 3cm below the mound surface in spit 3, and from spits 18 and 20, at depths of 40cm and 48cm, immediately above and below the sandy *Marcia* layer. The surface samples returned an estimate for the shell of 980±75, which calibrated to 549 (457-678) calBP, and 590±110 for the charcoal, calibrated to 251 (81-517) calBP. The basal samples, at 40cm depth returned estimates for the shell of 1210±90, calibrated to 733 (595-936) calBP, and 860±75 for the charcoal, calibrated to 483 (316-572) calBP, and at 48cm depth returned estimates for the shell of 1090±90, calibrated to 647 (505-826) calBP, and 1010±90 for the charcoal, calibrated to 570 (464-713) calBP (Table 8.11).

These radiocarbon estimates are interpreted as indicating that deposition of mound HI80 had begun by around 750 years BP and ceased around 550 years BP, spanning some 200 years. However, this must be qualified by the fact that all the shell dates are statistically the same at the 95% level, and the mound could have formed over a shorter period. The small difference between these shell and charcoal pair dates provide preliminary support for suggestions that a smaller marine reservoir correction factor than the accepted 450±35yrs years may be appropriate for this region (Woodroffe and Mulrennan 1993:40-1; Woodroffe *et al.* 1988:98).

The proportion of broken shell throughout this more recently accumulated deposit is relatively high compared with the other two excavated shell mounds, fluctuating

Table 8.11 Radiocarbon estimates, Site HI80

Material	Depth below surface(cm)	Lab Code	dC13	Date (BP)	Calibrated Date (calBP)
Anadara shell	3	OZC956	0#	980±75	549 (457-678)
Charcoal	3	OZC957	-25#	590±110	251 (81-517)
Anadara shell	40	OZC958	0#	1210±90	733 (595-936)
Charcoal	40	OZC959	-25#	860±75	483 (316-572)
Anadara shell	48	OZC960	0#	1090±90	647 (505-826)
Charcoal	48	OZC961	-25#	1010±90	570 (464-713)

* AMS analysis # Assumed dC13 values

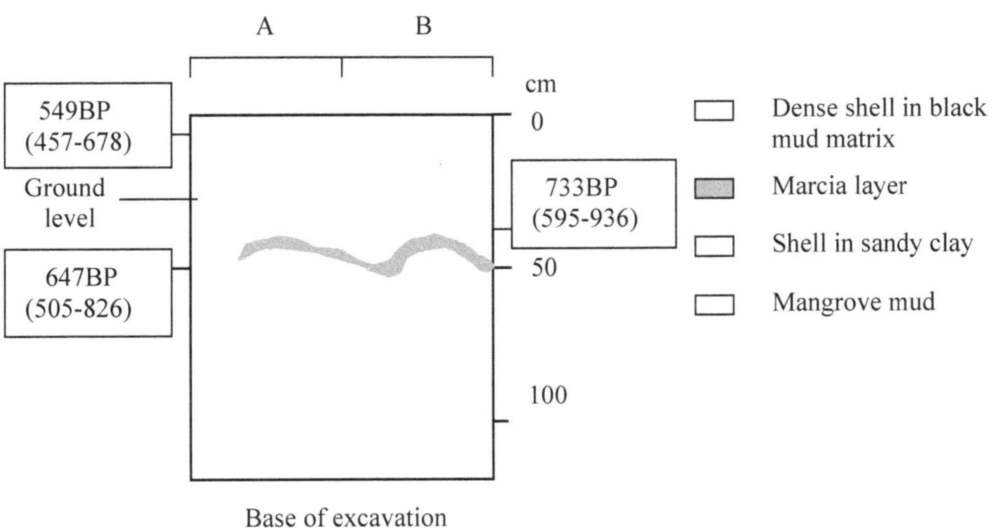

Figure 8.12 Stratigraphic profile, Site HI80

Plate 8.9 Excavated *Anadara* mound HI80 on saltflats

Plate 8.10 West wall, excavation pit, mound HI80, showing mangrove buttress root at base

between 20% and 40% throughout the deposit (Figure 8.4). Considering the mound's location on the inhospitable saltflats, where shade and vegetation that attract animals are lacking, and given that the fragmentation occurs throughout the deposit and is not significantly higher on the surface, recent animal trampling is unlikely as an explanation for the high degree of fragmentation. Periodic inundation by seawater and subsequent weakening of shell strength through chemical processes is a possible explanation for increased fragmentation. However, this is considered unlikely given the presence of well preserved bone and shell in this mound, some whole *Nerita* even retaining their colour patterns, though this is apparently common in Neritid fossils (Claassen 1998:70). The good preservation is probably due to the anaerobic mud matrix of the mound, which would actually slow decomposition processes (Stein 1992d:201). Alkalinity is high with the pH levels varying slightly between 8.5 and 9.5 throughout the deposit.

That this shell deposit with a cohesive mud matrix is partly buried, salt-encrusted, and solidly compacted, with no evident post-depositional disturbance, suggests stratigraphic integrity. Although there appear to be reversals on the two basal dates for this mound (Table 8.11), which would seem to indicate stratigraphic mixing, the samples were taken from opposing sides of the excavation pit, and within the level of precision of radiocarbon dating, statistically the dates are the same age (cf. Erlandson and Rockwell 1987:61). Taking all these factors into consideration, the relatively high rate of fragmentation throughout the mound in this case is interpreted as a reflection of a high degree of intensity of human activity during mound formation (Muckle 1985:75), rather than due to post-depositional mixing by non-human agents.

Overall, the distribution of shell, bone and charcoal in this mound follows the characteristic taphonomic pattern seen in other middens (Table 8.12, Figure 8.13). However, there are many small peaks in distribution of these organic components, which, when considered with the many peaks in rate of shell fragmentation, may represent discrete depositional episodes. One explanation for the distribution pattern of higher density of organic material in the lower levels of this mound is that it reflects increased subsistence activities such as fishing and cooking in the earlier period of accumulation of this mound.

However, the concentration of material in the lower layers is also consistent with downwards vertical movement of small particles such as fragmented bone and stone artefacts due to high intensity of human activity. It is also possible that the taphonomic pattern is related to the effect of overburden due to the muddy sedimentation processes that have partially buried this mound. This explanation is supported by the dearth of rocks >6.4mm in the top 30cm of deposit (Table 8.12). Taking all these factors into account, the stratigraphy and fluctuating patterns of fragmentation and distribution of cultural matter is consistent with accumulation of this mound over perhaps a century or two, or less, through a number of separate depositional events.

Shell

Relative to the other two sampled shell mounds, many more shell taxa - a total of 27 - were identified in the excavated material from the saltflats mound HI80. These mollusc taxa and the non-molluscan components are listed in Tables 8.13 and 8.14 sequentially. However, only 18 of the shell taxa are macro-molluscs of a size suitable for human consumption and one of these is the ubiquitous land snail, which only adds a couple of taxa to the range of molluscs targeted in this mound compared with the older shell mounds. Both these added taxa, *Anadara inaequivalvis* and *Arca* sp., are found in intertidal sandy areas. The remainder consists of very small chenier shells recovered from the 3.2mm sieve residue, and derive mainly from the sandy clay and mangrove mud layers at the base of the deposit. These small chenier shells make up around 3% numerically but still less than 0.1% by weight (or volume) of the shell content. The sandy clay layer is interpreted as being the previous sandy chenier plains surface on which the cultural shell was first deposited, and this is supported by the distribution of the small chenier shells, which as shown in Table 8.15 and Figure 8.14, are most numerous at the base of the deposit

In this sample the proportion of the dominant *Anadara* shell is 72% by MNI and 86% by weight of the total shell content, a reduction of around 20% overall compared with frequency of *Anadara* in the older mounds HI83 and HI81. Frequency of juvenile *Anadara* is slightly higher, comprising 7% numerically and 0.3% by weight of the total shell component. The mangrove gastropod *Cassidula* is the next most numeric taxon, consisting of 8% by MNI, though only 1% by weight, due to its fragility and the fact that generally only the flange survives, and the bivalve *Marcia* comprises around 2.5% by MNI and weight (Table 8.13).

The bivalve shell *Marcia* is found only between the depths of 25cm and 95cm, peaking in density between 40-50cm, and possibly represents one depositional episode (Table 8.15, Figure 8.14). Another species of *Anadara*, *A. inaequivalvis* (Brugiere), found only in the *Marcia* layer, also lives in the sandy intertidal areas inhabited by *Marcia* (Lim 1966, cited in Broom 1985:5), rather than muddy silt substrates as *A. granosa* does. Intra-midden analysis shows no progressive decrease over time in relative proportion of *Anadara*, which comprises around 80-85% of shell both below and above the *Marcia* layer (Figure 8.14). As is shown in Figure 8.14, the mangrove gastropods exhibit a few peaks in density, once immediately after deposition of the *Marcia* layer, when there is a temporary increase in their numbers, without any proportionate decrease in *Anadara*. Evidence of decreases in density of *Anadara* correlating with increases in density of other molluscs such as mangrove gastropods occurs at intervals throughout the deposit, noticeable in particular with the replacement of a

Table 8.12 Quantitative data, Section A, Site HI80

XU	Mean depth	XU wt	Shell	Char.	Bone	Otoliths		Crab	Barn.	Stone artefacts			Ochre	Rocks >6mm
	(cm)	(kg)	wt(g)	wt(g)	wt(g)	no.	wt(g)	wt(g)	wt(g)	no.	wt(g)	mat.	wt(g)	wt(g)
1	2	14	9047	1.0					4.2					
2	4	14	8184	2.0		1	0.6		9.8					4
3	5	14	7206	3.3					22.4					9
4	7	14	7197	9.5					16.8					3.4
5	9	13	6628	10.7	0.4				5.7					5.1
6	12	16	8184	4.6	0.4			0.2	4.7					0
7	14	16	8765	3.4	0.2				0.7					2.4
8	16	16	8681	3.8	0.1				0.2					0
9	18	15	7906	4.7	0.6				1				1.0	0
10	21	16	7641	4.9	0.1				0.5					0
11	23	23	10451	9.7	0.2				1					0
12	25	24	10700	12.0	0.6	2	0.3	0.7	0.7					22.5
13	29	24	11357	8.2	0.2	1	0.2		1					71.9
14	31	25	11550	5.5	0.4				0.4					76.7
15	36	29	14400	4.4	0.7	1	0.3		1.5	1	0.4	q		169.7
16	39	22	10560	0.8	2.6			0.3	1					75.4
17	41	23	8490	1.1	2.2	1	0.4	0.5	1.2					163.4
18	43	23	8087	13.1	0.3				0.9					891.5
19	46	25	10323	15.8	0.8			0.1	2.5				0.1	202
20	49	25	10963	9.3	2.8	1	0.2		3.3					254
21	52	26	11275	2.8	2.6	3	1.8	0.2	2					352.2
22	55	26	11862	4.3	2.7	4	2.0		1.2					278.5
23	59	29	11974	7.2	12.5	10	3.3	0.4	1.6	1	0.2			215
24	62	32	12706	10.9	5.8	1	0.6	0.6	1.6	1	0.2	qe		366
25	66	37	15104	8.1	1.2	1	1.0	1.1	4.2					583.4
26	69	32	12954	1.8	2.1	2	1.6	0.6	28					255.8
27	71	30	11916	3.9	6.1			7.4	22.3	2	5.7	q, d		479.8
28	76	33	11527	4.7	1.9	1	0.3	0.1	5					66.3
29	79	31	10587	17.7	1.6			0.8	8	1	0.1	q		256
30	82	33	10000	15.9	4.2	2	0.9	0.8	3	2	63.9	qe		750.6
31	86	34	10768	13.0	12.4	5	2.8	2.3	1					1351
32	90	36	11394	8.2	11.2	9	4.7	2.8	0.9	1	0.1	d		750.9
33	95	32	6856	11.4	11.5	2	1.2	0.4	0.7					177.8
34	98	34	6590	10.5	5.9			0.6	0.2					15.7
35	103	34	5137	28.7	12.6	5	1.6	1.2	0					18.6
36	107	32	2603	34.8	6.5	2	0.6	0.7	0.4					9.7
37	111	31	1208	31.3	4.0	2	1.6	0.1	0					3
38	114	27	478	27.0	2.2	1	0.7		1					6.5
39	119	28	281	16.2	2.0			0.2	0.4					8.4
Total		986	351541	386.2	121.6	57	26.7	22.1	161	9	70.6		1.1	7896.2

q = quartz Char. = charcoal qe = quartzite Barn. = barnacles d = dolerite mat. = material

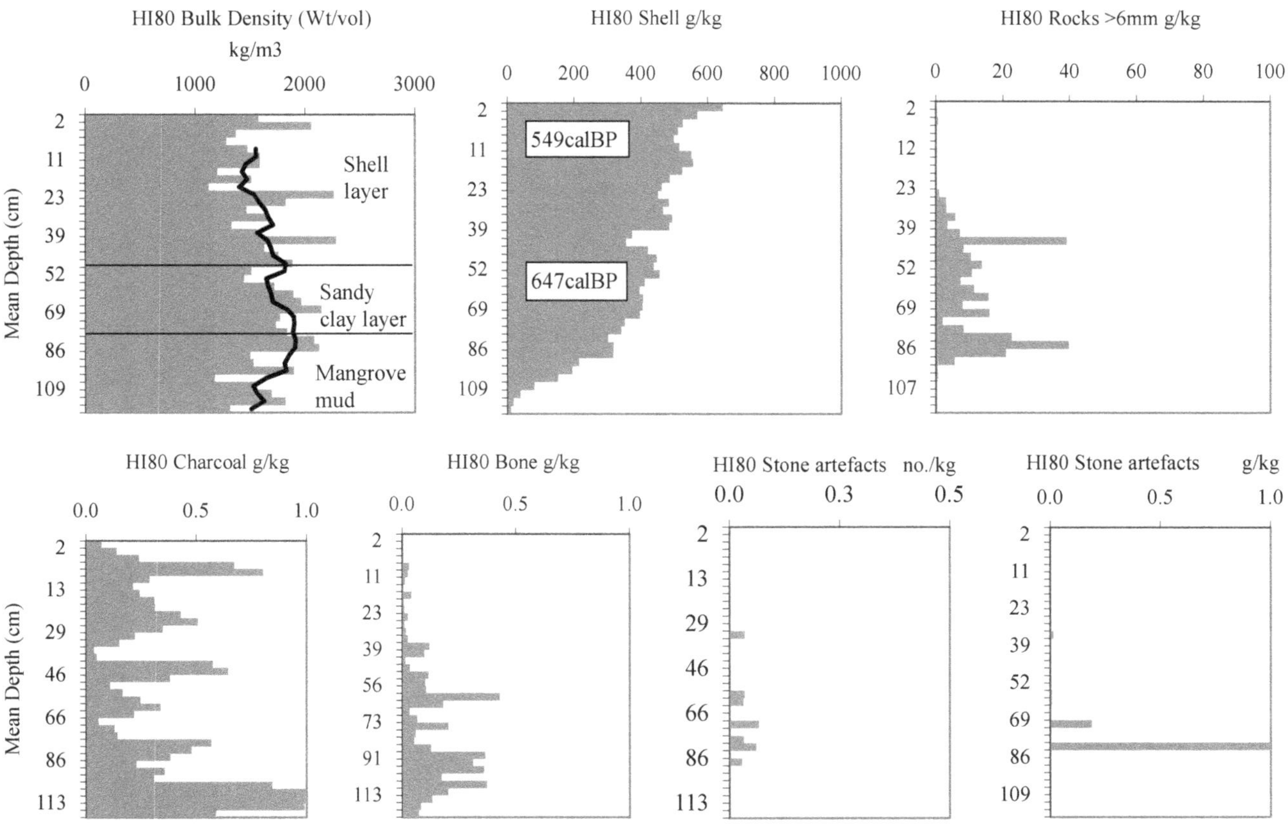

Figure 8.13 Distribution of material

large proportion of *Anadara* with *Marcia* in the stratum identified as the *Marcia* layer. The distribution pattern of mangrove gastropods in this mound is once more consistent with collecting small numbers of these molluscs whilst harvesting mainly *Anadara*, with continued collecting of gastropods when gathering of the dominant bivalve slows.

Other organic material

A total of 137g of bone was recovered from the $0.6m^3$ sample, mostly (77%) from the 3.2mm sieve and mainly from the lower levels of this mound. A relatively high quantity of crustacean remains was also recovered from this level, comprising crab and mud lobster (22.1g) and barnacles (161g) (Tables 8.12 and 8.15). The crustacean claws in this sample differ morphologically from those identified as mud crab in HI83 and HI81, being less curved and more robust, and are identified as probably mud lobster (*Thalassina*), common in mangroves in the area and throughout Southeast Asia (Holthuis 1991; Macintosh 1982:18; Murray and Hanley 1986). Distribution of bone generally follows the characteristic taphonomic pattern, but with higher concentrations in the lower layers. Large quantities of charcoal (386g), of variable density in distribution were observed throughout the deposit, also with higher concentrations in the lower layers and a number of small peaks at various depths, which possibly represent cooking episodes (Figure 8.13).

Much of the bone is burnt, but contains better-preserved larger pieces than in the other deposits, as well as the numerous unidentifiable fragments. At least 30% of the bone was positively identified as fish, and the bone assemblage appears to be predominantly fish. Five fish were identified to level of taxon (Helen Larson pers. comm.). This includes the otoliths of the ubiquitous fork-tailed catfish (*Arius sp.*) and black jewfish (*Protonibea diacanthus*) (Table 8.14). The former, other than one large specimen in spit 2 just below the mound surface, are concentrated from below 25cm depth to the base of the deposit, while the latter are concentrated in the lower layers below the Marcia layer. Other fish bone includes *Arius* sp. skull fragments, vertebrae and pectoral spines, and vertebrae of small unidentified fish. Larger pieces such as the premaxilla and other jaw fragments of bream (*Acanthopagrus sp.*), an hyperostotic bone of threadfin salmon (*Polynemus sheridani*), and barramundi vertebrae (*Lates calcarifer*) were also observed in Section B in the lower levels.

A few small fragments of reptile vertebrae, mammal and bird bone, as well as teeth and a 13mm long incisor broken at the tip from a medium to small Macropodidae, were recovered from the lower levels of the analysed Section A. In addition, a few large macropod vertebra were noted in spits 8 and 19 of Section B, a reminder of the heterogeneous nature of middens, in that rare or clustered features or artefacts may not be recovered from one column sample.

Along with the bone, fish otoliths crab exoskeleton and claws, other material including mangrove worm tubes and sea urchin were recovered concentrated primarily (93%) in the lower layers. Also identified in small quantities in the

Table 8.13 Total mollusc taxa identified in Section A, Site HI80

Taxon	MNI	Wt(g)	%MNI	%Wt	Wt/MNI
Anadara granosa	31464	303652	71.8%	86.4%	9.7
Cassidula angulata	3346	3205.8	7.6%	0.9%	1.0
Juvenile *Anadara*	3217	944.3	7.3%	0.3%	0.3
Marcia hiantina	1132	8325.8	2.6%	2.4%	7.4
Nerita sp.	1013	3912.5	2.3%	1.1%	3.9
Terebralia semistriata	858	5505.4	2.0%	1.6%	6.4
Ellobium aurisjudae	619	2166.8	1.4%	0.6%	3.5
Juvenile *Ellobium**	404	17.6	0.9%	<0.1	<0.1
Cerithidea sp.	382	106.8	0.9%	<0.1	0.3
Juvenile *Cassidula**	302	5.9	0.7%	<0.1	<0.1
Iravadia sp.*	266	2.9	0.6%	<0.1	<0.1
Corbula sp.*	133	9.1	0.3%	<0.1	0.1
Telescopium telescopium	109	653.4	0.2%	0.2%	6.0
*Cerithium coralium**	85	3.1	0.2%	<0.1	<0.1
Juvenile *Telescopium**	64	118.3	0.1%	<0.1	1.8
Littoraria sp.*	60	5.3	0.1%	<0.1	0.1
Placamen calophyllum	60	16.7	0.1%	<0.1	0.3
Salinator sp.*	42	0.9	0.1%	<0.1	<0.1
Nassarius sp.*	40	6.8	0.1%	<0.1	0.2
Notospisula sp.*	35	1.7	0.1%	<0.1	<0.1
Volema cochlidium	39	465	0.1%	0.1%	11.9
Chicoreus capucinus	29	157.9	0.1%	<0.1	5.4
Saccostrea echinata	25	40.9	0.1%	<0.1	1.6
A. inequivalvia	20	51.5	<0.1	<0.1	2.6
Xanthomelon sp.	20	27	<0.1	<0.1	1.4
Geloina coaxans	17	85.9	<0.1	<0.1	5.1
Arca sp.	16	20.8	<0.1	<0.1	1.3
Placuna sp.	13	14.4	<0.1	<0.1	1.1
Melo amphora	6	52.9	<0.1	<0.1	8.8
Pseudonachis sp.*	3	0.7	<0.1	<0.1	0.2
Dentalium sp.*	3	0.3	<0.1	<0.1	0.1
Unidentified shell		21963.5		6.2%	
Total	43822	351541.9			

* Chenier shell

Table 8.14 Identified components of non-shell faunal remains, Section A, HI80

Type	Wt(g)	n	Av. wt
Total Bone: Mammal, Reptile, Bird, Fish (Fork-tailed catfish, Black jewfish, Threadfin salmon, Barramundi, Bream)	137		
includes: Macropod incisor	0.5	1	
Macropod teeth	0.1	3	
Total Otoliths:	26.7	57	
includes: Fork-tailed catfish	15.6	37	0.4
Black jewfish	11.1	20	0.6
Crab and Mud Lobster	22.1		
Barnacles	161		
Annelid worm tubes	7		
Mangrove worm tubes	5.2		
Internal invertebrate casts	1.6	6	
Coral/Pumice	13.4		
Sepia cuttlefish	0.5		

Table 8.15 Distribution of shell taxa with depth in Section A, Site HI80

Mean Depth (cm)	Spit	XU Wt (kg)	*Anadara*	Juv. *Anadara*	*Cassidula*	*Marcia*	*Nerita*	*Terebralia*	*Ellobium*	*Cerithidea*	*Telescopium*	*Placamen*	*Volema*	*Chicoreus*	*Saccostrea*	*A.inequivalvis*	*Geloina*	*Arca*	*Melo*	*Placuna*	*Littoraria**	*Corbula**	*Cerithium**	*Iravadia**	Juv. *Cassidula**	Juv. *Ellobium**	Juv. *Telescopium**	*Nassarius**	*Salinator**	*Notospisula**
2	1	14	963	149	29		9	20	1	3	3							1												
3	2	14	1109	92	32		8	5	1	2	3		1																	
5	3	14	976	85	33		7	11	1	3	4	4	1												2	6				
8	4	14	926	78	56		7	7	2		1	8	1										1	2	2	5	3	5		
9	5	13	865	94	50		17	9	3	1		2							1								2			
11	6	16	827	93	40		17	9	2	1	1	3	1					1	1	1			1			8	1	1		
13	7	16	770	85	56		8	4	2			2	1	1					1								1			
16	8	16	754	63	104		8	12	12	1			1	1				1									1			
18	9	15	687	64	130		11	19	21	1			1	2				1	1								1			
21	10	16	740	84	97		15	22	21	2			1	1	1			1									2			
23	11	23	1219	52	70		20	27	7	7		3	2	1	1		1	1				3	3			11	2	2		
25	12	24	1260	159	225	3	36	15	10	1	4	5	3	1			2	1			1		9		3	16		4		
29	13	24	1115	15	294	1	150	19	34	7		1	2		1		1	2				1	6			12	1	1		3
32	14	25	1249	100	358	9	104	28	52	2			2		1	1	1	2	1								3			
36	15	29	1344	99	90	46	50	25	9	4		1	1	1	2	3				1		2	4			11	11	2		4
39	16	22	1041	49	45	17	10	13	9	5		1	1		1				1	1			5			12	6	3		
41	17	23	781	71	63	73	21	26	5	3	10	1				4				1			4		4	10				3
44	18	23	528	52	76	112	26	61	17	15	5	1	1		1	5	1			1			6			10	1	1		
46	19	25	680	82	92	282	50	22	6	11		7	1		3	4				1		2	17		4	11	7	8		9
49	20	25	623	115	198	327	42	18	11	5	4	2	1		2	1		1				1	11			7	6	2		5
52	21	26	879	118	163	187	32	20	24	14	6	4	1			1						1	3			4	8	6		3
56	22	26	1142	38	162	32	31	13	14		27	1	3		1	1						1			4		3			
59	23	29	1178	49	138	4	27	24	23	13		3	1	3	1			1		1		1	4	3		4	3	2		5
63	24	32	917	186	196	6	36	59	53	5	3		1	3	1			2												
66	25	37	1115	190	136	4	45	63	64	14	5	5	3	10	1		1													
69	26	32	914	168	44	2	21	47	51	21	7		2	2	2						10	3	7		3	9		2		2
73	27	30	989	195	74	4	11	75	21	5	7		1							1		3								
77	28	33	901	119	36	3	13	16	9	8			1							1	1	3	3		3	12	2	1		
80	29	31	844	71	12	3	7	22	5	9	2		1		2						2		1	8	12	12			2	
83	30	33	829	71	21	2	6	24	6	6		3			1						2				7	23			2	
86	31	34	792	85	54	3	27	24	10	22	4	1	1		2			1		1	7	5		13	30	7				
91	32	36	693	86	50	5	35	43	32	33			1							1	9			22	9	11			7	
95	33	32	520	57	39	6	42	32	24		4	1	1	2			1					4		20	20	12			9	
99	34	34	552	31	16		11	11	10	13	1				1							4		6	3	4				
105	35	34	404	28	19		10	3	8	23	2						2					22		36	22	30			4	
109	36	32	208	20	18		15	6	12	12	2	1					3					11		34	34	24				
113	37	31	81	13	10		16	3	9	30							2				7	30		42	48	28			7	
116	38	27	30	6	10		6	1	15	39	2						1				11	28		42	48	69			4	
120	39	28	19	5	10	6	6		4	22	7	1				5	1			6	4	8	22	38	44	36	36		7	

* Chenier shell

lower levels were cuttlefish, tubes of annelid worms, tusk shells (*Dentalium* sp.), coral and pumice and shell grit. Like the small shells described above, these are all found in natural chenier deposits (Burns 1994:59-60; O'Connor and Sullivan 1994b:23) and are probably constituents of the sandy base on which the cultural shell deposit accumulated. High densities of annelids have been recorded from the mangrove mud of creek banks and mud lobster mounds around Darwin (Healy and Coates 1997). Coral has been observed to enter a midden incidentally as dead coral provides a base for attachment by oyster or mussels (Alfredson 1984:58, cited in Rowland 1994b:122). In addition to a mangrove buttress root at the base of the pit (Plate 8.10), unidentified macroscopic vegetation was also recovered from the lower levels of this mound.

Stone assemblage

Nine stone artefacts weighing 70.6g were recovered, mainly from the lower levels of the 0.6m^3 of excavated deposit (Table 8.12). This gives an estimated density of 15 artefacts per cubic metre, a lower density than in the landward mounds HI81 and HI83. Two small dolerite manuports (0.2g), two large blue/grey quartzite manuports (63.9g) and a small quartzite flake in addition to two white quartz flakes were recovered from the deposit between 60cm and 90cm depth. This assemblage

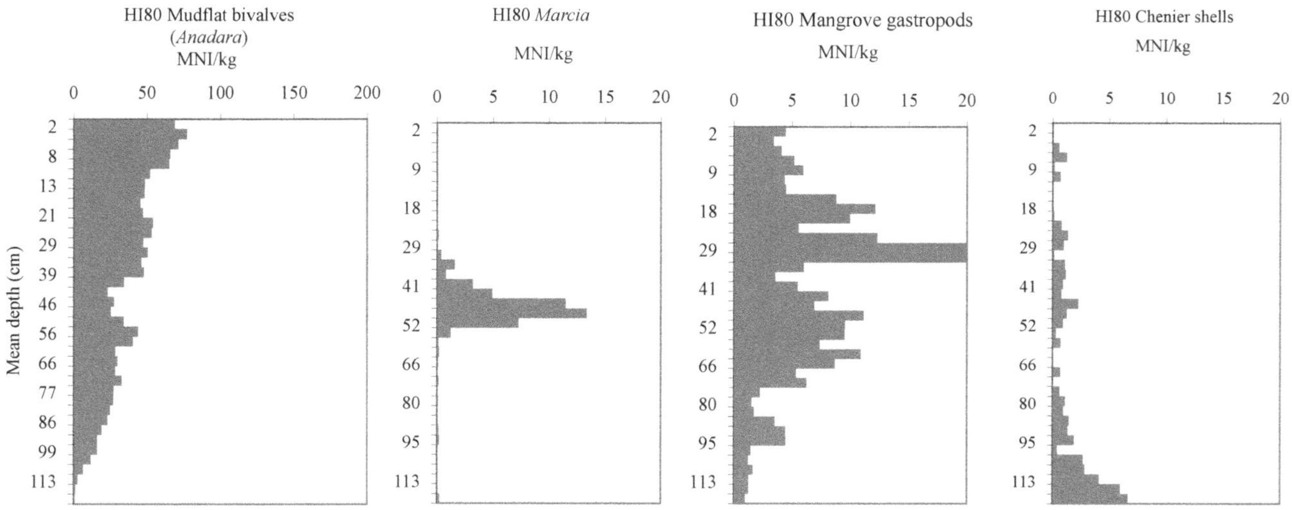

Figure 8.14 Graph showing relative densities (MNI/kg) vs depth of deposit HI80, of mudflat bivalves Anadara, Marcia and the mangrove gastropods Cassidula, Nerita, Chicoreus, Terebralia, Telescopium and Ellobium

also differs from the other sites, with five artefacts retained in the 6.4mm sieve and four in the 3.2mm sieve, and an average artefact size of 1g, excluding the large quartzite manuports, and if the medium-sized quartz flake is also excluded, 0.2g for the remaining six artefacts. Neither is quartz the dominant raw material in this site, comprising only 33% numerically and 9% by weight of the assemblage. Only two small pieces of red ochre occur in this sample. Over 1700 laterite rocks >6.4mm, totalling 7.9kg were recorded, concentrated in the lower layers, and these are interpreted as hearthstones (cf. Veitch 1994). Burnt fractured fine siliceous hearth material was also noted in Section B in the lower spits.

Discussion

Formation of mound HI80 can be placed in the context of geomorphological data from nearby Camerons Beach at Shoal Bay (Woodroffe and Grime 1999). The level at which the mangrove buttress root underlying this mound was found corresponds with the levels of mangrove wood from transects through the chenier plain at Cameron's Beach about 7km southwest of this section of Shoal Bay, radiocarbon dated to 6130±100 BP (Woodroffe and Grime 1999:309). Woodroffe and Grime (1999:309) interpret that date as indicating the mangrove-fringed shoreline around 6000 years ago, when sea level reached its present levels (Woodroffe *et al.* 1987). It is difficult to know if the wood found in the blue-grey mangrove mud in this pit also represents the position of mangroves on the shoreline here at that time.

The sandy clay layer 50cm above the mangrove wood, which forms the base of the shell mound, probably represents the past chenier plains surface resulting from a phase of build-out around 900-1100 years ago at Camerons Beach at the southern end of Shoal Bay, identified by Woodroffe and Grime (1999) (see Chapter Two). The HI80 shell deposit extends approximately 50cm below the level of the surrounding saltflats surface, indicating that sedimentary infilling of these chenier plains has continued after deposition of the mound some 700 to 550 years ago.

Mound HI80 provides a record of short-term changes in relative frequencies of molluscs, from *Anadara*, that prefers finer silty muddy substrates, to a relative increase in *Marcia* that thrives in coarser more sandy substrates, and then a return to marked dominance of *Anadara*. A reasonable explanation for the concentration only between certain depths in this deposit, of particular species associated with sandy substrates such as *Marcia hiantina* and also *Anadara inaequivalvis*, would be local micro-environmental change over time. Changes in the relative quantities of sand, mud, river and sea currents and the location of sand banks, as observed by Meehan (1982a:62) for the Arnhem Land coast, may have provided a more favourable habitat for *Marcia* at Hope Inlet for a short period around 650 to 700 years ago. Thus this is possible evidence of short-term local environmental change in the Darwin region around 650 to 700 years BP, involving a temporary increase in availability of a culturally favoured resource, *Marcia* (*Tapes*) *hiantina* (Meehan 1982a:70, 78, 142).

However, it is possible that the environmental change may have been on a regional rather than local micro-scale confined to the Hope Inlet area. On Middle Arm peninsula, mound MA7, (see above) for which a radiocarbon date of around 700 years BP was obtained, also contains a relatively large *Marcia* component (20%), compared with less than 1% for the excavated mounds MA1 and HI83 dated to around 2000 to 1300 BP and 1450 to 1550 BP respectively (Burns 1994:80; Hiscock 1997:446).

The relatively high shell fragmentation rate and relatively irregular distribution pattern of organic matter and stone artefacts is interpreted as indicating a high level of activity during the period of accumulation of this mound (Figures 8.4 and 8.13). If this period was around 200 years as suggested by the dates, this is a reasonably long time in terms of the extent of human activity that could have occurred. However, the low density and relatively large average size of stone artefacts in this mound suggests that relatively little of one type of activity - knapping - was carried out on this site compared with the hinterland mounds. A low density of artefacts can be explained by the dearth of locally available stone and long distance from raw material sources. Average artefact size is larger than that of the other mounds, but is based on very small numbers, which precludes detailed analysis other than showing that some stone reduction occurred, particularly of the dolerite and white quartz imported from outside the Hope Inlet area.

All stone must have been carried to this saltflats mound from a long distance away, up to many kilometres from inland in the case of the dolerite found in the lower levels. There is a relatively close source of the type of blue/grey quartzite artefacts in outcropping rock (Site HR2) on the saltflats 7km south of this site. Raw material for the white quartz stone artefacts may have been brought in from around Darwin Harbour 25km to the southwest. The transport of laterite rocks from the hinterland for use as hearth stones suggests that cooking was an important function of this site.

Faunal remains, charcoal and imported laterite hearth stones indicate that activities included cooking, of a variety of resources. The range of molluscs, fish, crabs, mud lobsters, barnacles and mangrove worms could all have been obtained through fishing and crabbing in the nearby mangroves, estuarine creeks and shallow nearshore waters, including the inter-tidal mudflats and flooded supra-tidal saltflats. Mammals, reptiles and birds could have been hunted in the adjacent coastal woodlands, monsoon forest patches and swamps of the transitional hinterland margins. Small macropods captured on the hinterland must have been transported out to the saltflats mound for cooking.

Interesting features of this deposit are the presence of mud lobster and relatively large quantities of barnacles recovered (Tables 8.12 and 8.15). Uses of the mud lobster (*Thalassina*) by Indigenous groups in Southeast Asia include food, medicine against asthma and as bait. It is reported to have:

> little or almost no meat, as the entire body and the tail are full of a greenish mud, and only in the pincers there is a small amount of white, crumbly meat, without a particular taste. The natives of Celebes eat the meat of the pincers, but when I myself tried this, the meat did not agree with me...(Rumphius 1705:5, cited in Holthuis 1991).

Such reports by Rumphius and others (Holthuis 1991) on the small amount of meat supplied suggest that its use is more likely to have been medicinal or for bait.

The quantity of barnacles recovered from this deposit (161g) is much higher than in all other deposits excavated for this study, where barnacles were either absent or present in minute quantities only. Barnacles attach to things like mangrove roots and crabs (Hutchings and Saenger 1987:184), and their presence may be incidental, related to the relatively high frequency of mangrove worm tubes and crab remains in this saltflats mound, which is situated within 50m of the mangroves. It is also possible that these barnacles were eaten, as they are assumed to be in some NSW middens (eg. Colley 1997). Although the barnacles are small and mostly (70%) recovered from the 3.2mm sieve, these are all broken pieces, comprising the "plates" of this crustacean (Underwood 1977:4). A few whole specimens retained in the 6.4mm sieve range in size between 12m diameter x 12mm high (0.6g) to 13x16mm base x 16mm high (0.9g), with some plates up to 20mm high, suggesting that these barnacles are large enough to be considered food (cf. Rowland 1994b). However, at artificially cultured shell bed areas in Malaysia, the barnacle *Balanus amphitrite* is also found attached to the posterior end of *Anadara* molluscs which protrude from the mud (Broom 1985:11), and this suggests another possibility.

Contemporary studies on mollusc exploitation by coastal peoples (Hockey and Brosman 1986; Moreno *et al.* 1984) found that the relative abundance of algae and sessile inedible invertebrates such as barnacles, was greater at sites that were the focus of exploitation. They also noted that the modal and mean size of exploited taxa was smaller, and that the number of other species increases with selective exploitation of one species (Hockey and Brosman 1986; Siegfried *et al.* 1985:306). Analysis of this sample does show a decrease in shell size of the main resource, *Anadara,* over time and a higher frequency of other taxa, with a marked decrease in relative proportions of the mudflat bivalves to mangrove gastropods, compared with the two mounds HI83 and HI81 accumulated hundreds of years earlier (see Chapter Ten).

If this mound, with an estimated volume of 73m^3, actually accumulated over 200 years as suggested by the dates, then deposition, assuming a constant rate, would be quite low, at slightly less than 0.4m^3 per year. This is much less than Meehan's estimated quantity of 8m^3 of shells discarded per year by a 40-strong group of modern Anbarra collectors in Arnhem Land. However, it is consistent with the ethnographic model where molluscan remains were discarded widely over the landscape, and deposits that survive the elements represent only a fraction of discarded shells (see Meehan 1982a:165-7). Such deposits, which contain some depth, a feature that adds to their likelihood of survival, are formed through the process of scraping and piling up of discarded material to clean "Home base camps" that are habitually reoccupied over long periods of time (Meehan 1982a:114). It is also possible that the deposit represents the remains of a smaller group and/or a number of seasonal occupation periods separated in time.

The above features, and diverse food remains, are consistent with an interpretation of this site as part of a home base camp that was occupied repeatedly over some years, decades, or perhaps generations. To determine the size of the groups occupying the area during this period would require further research, such as the dating of the other 27 middens, which also appear partially buried, located in this half a square kilometre area of saltflats. A radiocarbon estimate of around 700 BP for the sub-surface midden HI66 described below, located 4km

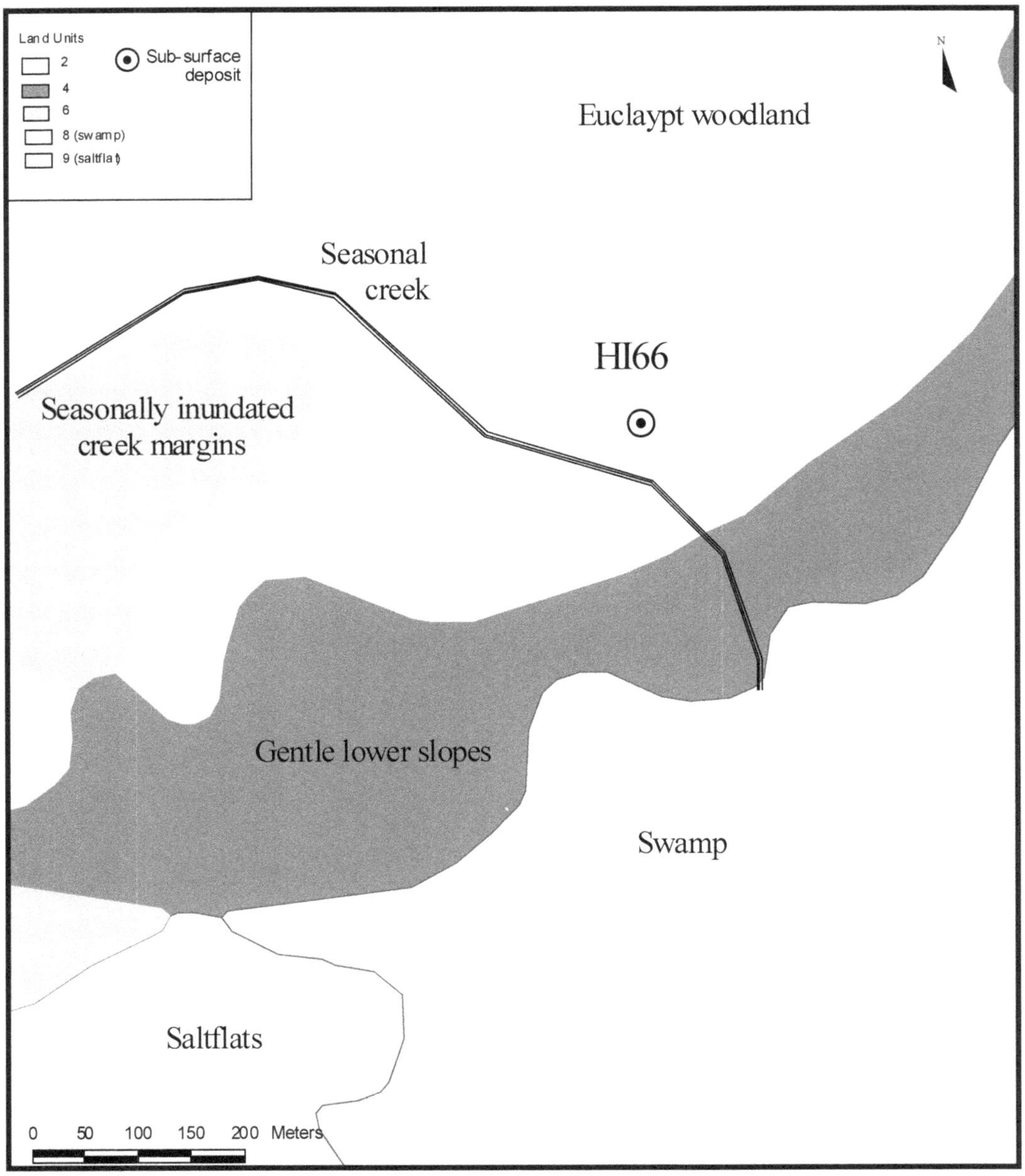

Figure 8.15 Plan of Site HI66

to the east, which is statistically the same at the 95% level as estimates for HI80, means it is possible that accumulation of these two sites at least, may have been contemporary.

Site HI66

Environmental setting

Site HI66 is a subsurface midden located 4km east of the other excavated sites (Figure 8.1). The midden is situated on the lower gentle slopes of the Northern Plains plateau some hundred metres from a large freshwater swamp on the edge of the saltflats (Figure 8.15). Seasonal creeks flow into the swamp within a kilometre on either side of the site. The site appears as a flat circular clearing of approximately 30m diameter in open eucalypt woodland, with a lateritic gravel cover and patches of midden material dug up around the edges of goanna holes. The dimensions of the subsurface midden itself were indeterminable without excavating the entire area. Historical European artefacts such as buffalo bones, tin and a silver spoon were also present on the surface. A one metre square was excavated toward the eastern edge of the clearing (Plate 8.11).

Stratigraphy, shell fragmentation and chronology

Two main strata were evident in the deposit excavated to a depth of 60cm: an upper shell layer (Stratum I) and lower clay layer (Stratum II) (Figure 8.16). Stratum I consists of two layers: a compacted light brown gravelly laterite soil a couple of centimetres deep, which caps an upper shell layer approximately 30-35cm in depth. The upper shell layer comprises a dark greyish brown fine silt with a sparse shell content of mainly *Anadara.* Stratum II, the lower layer, consists of lateritic lighter brown clay with numerous rocks. Patches of the upper shell layer project down into the lower stratigraphic layer (Plate 8.12). Measurements showed the pH level changed slightly with depth from 9.5 just below the

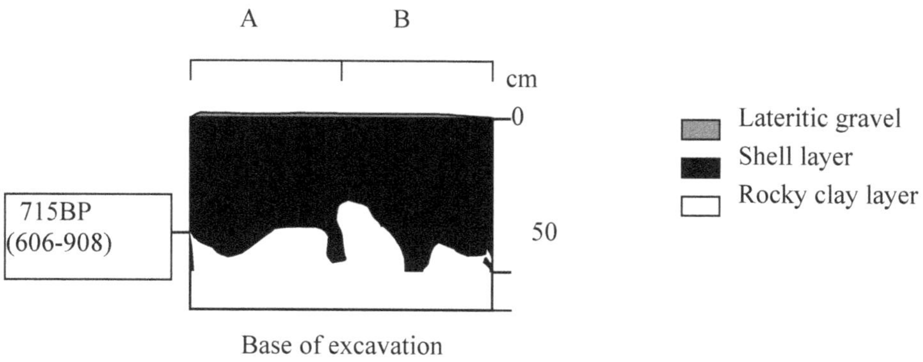

Figure 8.16 Stratigraphic profile, Site HI66

Table 8.16 Radiocarbon estimates, Site HI66

Material	Depth below surface(cm)	Lab Code	dC13	Date (BP)	Calibrated Date (calBP)
Anadara shell	35	Wk6528	-3.1	1190±75	715 (606-908)

surface to 8.5 toward the base of the deposit. Termite nest material was noted at 20cm depth and goanna holes were also evident in the lower half of the deposit. Termite nest is known from ethnographic reports to have been used as hearth material in cooking (Meehan 1988:2; Peterson 1973:177).

The proportion of broken shell is relatively high at around 60% throughout the deposit (Figure 8.17), with increased fragmentation and more weathered condition of shell at the base and surface, consistent with high levels of activity and taphonomic processes of decay. Although the shell layer in this deposit is not as dense as in the shell mounds, the pH levels are still quite alkaline, which acts to preserve the shell, and the shell in this deposit is less weathered than in HI97, which has lower pH levels.

However, this sub-surface deposit has a high soil content and is highly disturbed by bioturbation, as evidenced by the number of goanna holes present. The high degree of fragmentation throughout the sub-surface midden could be a reflection of occupational disturbance during deposition (Muckle 1985:75), or of post-depositional taphonomic processes resulting in shell breakage, or a combination of both. The stratigraphy is consistent with either relatively slow deposition of the shallow 30cm deep shell layer, or deposition of shell with termite nest material for use as an earth oven. The shell patches projecting down into the lower layer are only five to ten centimetres wide and are more likely to be goanna burrows than depressions resulting from the construction of cooking pits (cf. Meehan 1982a:110, 114). Features which Williams (1988:88, 155) identified as hearth pits in and near earth mounds in

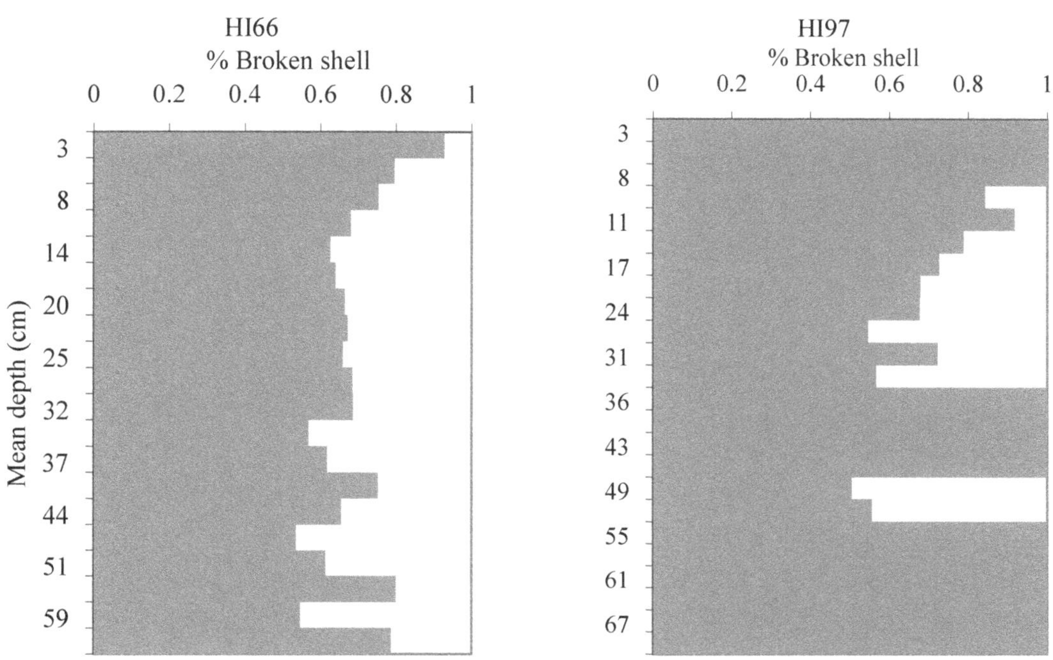

Figure 8.17 Rate of shell fragmentation in HI66 and HI97

Table 8.17 Quantitative data, Section A, Site HI66

XU	Mean depth	XU wt	Shell	Charcoal	Bone	Otoliths		Crab	Stone artefacts			Ochre	Rocks >6mm
	(cm)	(kg)	wt(g)	wt(g)	wt(g)	n	wt(g)	wt(g)	n	wt(g)	material	wt(g)	wt(g)
1	3	22.9	709	0.1	0.1				5	7.1	q		2397
2	5	21.5	1675	3	0.7						q		1346
3	8	20.1	2110	2.6	0.1	1	0.3		3	0.2	q		1205
4	11	21.6	2668	4.2	1.2			0.3					1371
5	14	20.8	3018	9.2	8.9				5	9.4	q t		1229
6	17	21.8	2797	11.7	1.2				1	0.1	q	0.2	1318
7	20	21.9	1915	9.3	0.7	1	0.1		2	2.1	q		1436
8	22	19.9	1265	7.6	0.7				1	0.1	q		1255
9	25	19.8	1116	9.9	0.4				1	0.4			1514
10	29	21.4	808	13.2	0.5	1	0.3		1	0.1	c		1543
11	32	21.4	643	13.3	0.6				4	1.9	q		1389
12	35	21.6	517	8.9	0.2								1725
13	37	21.5	228	7.3	0.1								2073
14	40	24.3	125	8.6	0.3								2637
15	44	25	107	4.8	1.1								2911
16	46	25.4	71	4.1	0.1				1	0.1		2.2	3145
17	51	25.1	65	2.7	0.3								3597
18	55	24.7	40	1.1								0.1	5426
19	59	23.8	75	0.5	0.1				1	0.1	q		6531
Total		424.5	19950	122.1	17.3	3	0.7	0.3	25	21.6		2.5	44048

q = quartz t = tuff c = chert

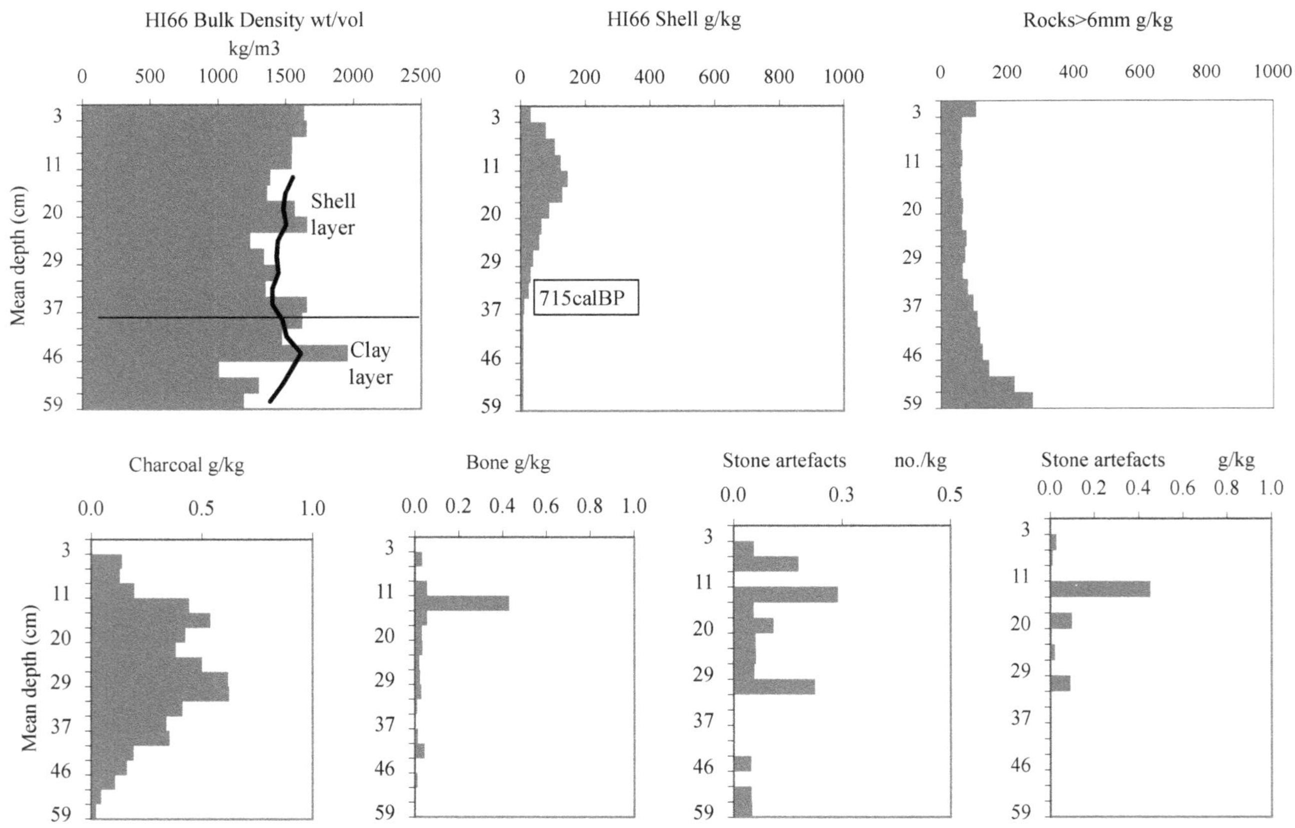

Figure 8.18 Distribution of shell, charcoal, bone and stone artefacts in sites HI66

Table 8.18 Total mollusc taxa MNI and Weight, identified in Section A, HI66

Taxon	MNI	Wt(g)	%MNI	%Wt	Wt/MNI
Anadara granosa	1186	17350	54.6%	87.0%	14.6
Juvenile *Anadara*	533	103.3	24.6%	0.5%	0.2
Nerita sp.	124	123.8	5.7%	0.6%	1.0
Terebralia semistriata	104	586	4.8%	2.9%	5.6
Cassidula angulata	89	34.4	4.1%	0.2%	0.4
Chicoreus capucinus	43	109.8	2.0%	0.6%	2.6
Ellobium aurisjudae	27	14.3	1.2%	0.1%	0.5
Xanthomelon sp.	25	46.1	1.2%	0.2%	1.8
Cerithidea sp.	13	2.2	0.6%	<0.1%	0.2
Geloina coaxans	11	22.7	0.5%	0.1%	2.1
Volema cochlidium	5	105.7	0.2%	0.5%	21.1
Saccostrea echinata	5	1.6	0.2%	<0.1%	0.3
Marcia hiantina	4	3	0.2%	<0.1%	0.8
Telescopium telescopium	2	14.4	0.1%	0.1%	7.2
Melo amphora	1	4.5	<0.1%	<0.1%	4.5
Melina ephippium	1	0.1	<0.1%	<0.1%	0.1
Unidentified shell		1426		7%	
Total	2171	19947.9			

Table 8.19 Identified components of non-molluscan faunal remains, Section A, HI66

Type	Wt(g)	n	Av. wt
Total Bone: Rodent, Fish (Fork-tailed catfish)	9		
Total Otoliths:			
Fork-tailed catfish	0.7	3	0.2
Crab	0.3		
Internal invertebrate casts	0.2	2	

southwestern Victoria are much wider, between 0.5m and 1m in diameter and 5cm to 20cm in depth.

Shell was collected for dating during the excavation from spit 19, the base of the shell layer at a depth of 35cm. Radiocarbon dating returned an estimate of 1190±75, which calibrated to 715 (606-908) calBP, suggesting that deposition of shell in Site HI66 began around 700 years BP. (Table 8.16).

Shell

The distribution of shell in this deposit exhibits the characteristic skewed taphonomic pattern seen in the shell mounds (Table 8.17, Figure 8.18). A total of 15 macro-mollusc taxa, including land snail, were identified in the excavated material from HI66, similar to the range found in the excavated shell mounds. These and the non-molluscan components are listed in Tables 8.18 and 8.19 respectively. *Anadara* dominates the material, comprising 87% by weight but only 55% by MNI of the total shell content. This MNI figure is depressed by the high number of juvenile *Anadara,* which makes up 25% of the shell component and 0.6% by weight. The relatively higher proportion of juvenile *Anadara* in this midden is possibly related to the season of collection (Richard Willan pers. comm.). The gastropods *Nerita, Terebralia* and *Cassidula* together make up another 15% by MNI and 4% by weight. An isolated specimen of *Corbula* was also seen, and is probably an incidental addition. Each of the shell taxon, including land snail, exhibits the typical skewed distribution through the deposit (Table 8.20).

Other organic material

A small quantity (9g) of very fragmented bone was recovered from throughout the deposit (Table 8.19). Only fish otoliths (*Arius* sp.) and rodent bone, and a few crab fragments and internal casts of small invertebrates were recognisable. The density of charcoal in this sub-surface deposit is relatively high (total 124g) and both charcoal and bone roughly exhibit the characteristic taphonomic distribution pattern exhibited by organic material in the shell mounds (Figure 8.18). Small peaks in density of charcoal and stone artefacts below the level of the highest concentration of shell may indicate other activity prior to accumulation of the shell midden. The high pH levels in this deposit would be expected to enhance preservation of bone and charcoal.

Plate 8.11 Excavation of sub-surface deposit HI66. Larrakia field assistants are Alfie and Jon May

Plate 8.12 East wall, excavation pit, Site HI66, showing shell patches extending down into lower stratigraphic layer

Table 8.20 MNI per excavation unit for mollsc taxa identified inSection A,Site I6

XU	Mean depth (cm)	*Anadara*	Juvenile *Anadara*	*Nerita*	*Terebralia*	*Cassidula*	*Chicoreus*	*Ellobium*	*Xanthomelon*	*Cerithidea*	*Geloina*	*Volema*	*Saccostrea*	*Marcia*	*Telescopium*
1	3	14	11	2	3	2	2		1						
2	5	58	50	6	9	6	5	1	1		2				
3	8	85	48	8	8	5	2	2	3		1				1
4	11	176	68	13	8	7	2		3		2	1	1	1	
5	14	139	48	10	13	11	7	5	6			2	1		
6	17	211	68	14	13	11	5	6	1	2				1	
7	20	131	59	12	7	10	3	1	2	1	2		1	1	
8	22	88	40	16	6	7	2		2				1		
9	25	74	30	7	6	6	2	2					1	1	
10	29	68	34	8	6	3	2	1	1						
11	32	38	26	6	5	7	2		1	4					
12	35	43	23	5	4	5	2	1	1	2	1				
13	37	18	8	3	4	2	2	2	1		1				
14	40	6	3	3	2	1	1	1	1		1	1			1
15	44	9	4	2	2	2	1	2		3		1			
16	46	4	3	2	2	1	1	3	1	1	1				
17	51	4	1	2		1	1								
18	55	4	2	2	1	0									
19	59	7	3	0	2	1									
20	66	9	4	3	3	1	1								

Stone assemblage

Stone artefacts are concentrated in the shell layer of this subsurface midden. Quartz is the dominant stone component (Table 8.17), accounting for over 80% of the number and 51% of the weight of the assemblage. A total of 25 stone artefacts weighing 22g were recovered from $0.3m^3$ of excavated deposit, giving an estimated density of 83 artefacts per cubic metre. Artefacts comprise over 80% small quartz flakes (average weight 0.1g) recovered from the 3.2mm sieve residue. Five larger flakes made from quartz were retained in the 6.4mm sieve. One flake manufactured from volcanic tuff (8.8g) was recovered from 10cm below the surface during the excavation. Red ochre was also recovered from a few spits and one piece of coral from spit 4 may have been an incidental addition transported from the sea some 7km to the west. Laterite rocks retained in the 6.4mm sieve weighing a total 50kg, occur throughout the deposit, concentrated in the lower layers (Table 8.17).

Discussion

The range of macro-mollusc taxa comparable to that found in the shell mounds (Tables 8.18 and 8.20), supports the interpretation of this shallow sub-surface deposit as *in situ* primary shell midden material deposited over a short-term period some 700 years ago (see Attenbrow 1992). Non-shell faunal remains, together with the presence of termite material and high charcoal content, indicate the site may also have been used as an oven to cook resources such as molluscs catfish and crab, and possibly small rodents. However the relatively low density of shell and bone remains, suggests this use may have been short-term, since it would be expected that any bone remains should be relatively well-preserved in this alkaline deposit.

Alternatively, the remains of this spread-out, non-mounded site may have been destroyed through repeated occupation episodes and intensive activity by humans, and climatic degradation, as has been documented in an ethnoarchaeological study in the Bloomfield River area, southeast Cape York Peninsula in the late 1970s - early 1980s (Anderson and Robins 1988). The high level of shell fragmentation exhibited by this deposit is consistent with this proposition. Peaks in density of charcoal and stone artefacts, and higher quantities of laterite rocks that may have been used as hearth material, below the level of the highest concentration of shell, also suggest activities such as cooking and stone artefact maintenance and manufacture, prior to accumulation of the shell midden. Subsequent to its deposition, the shell midden layer was covered by a thin veneer a few centimetres thick of colluvial sand and silt winnowed from upland areas and deposited on these lower slopes by sheetwash action (Pietsch 1985:16).

Plate 8.13 Water lilies in swamp at Hope Inlet

Plate 8.14 Magpie geese on swamp near HI66

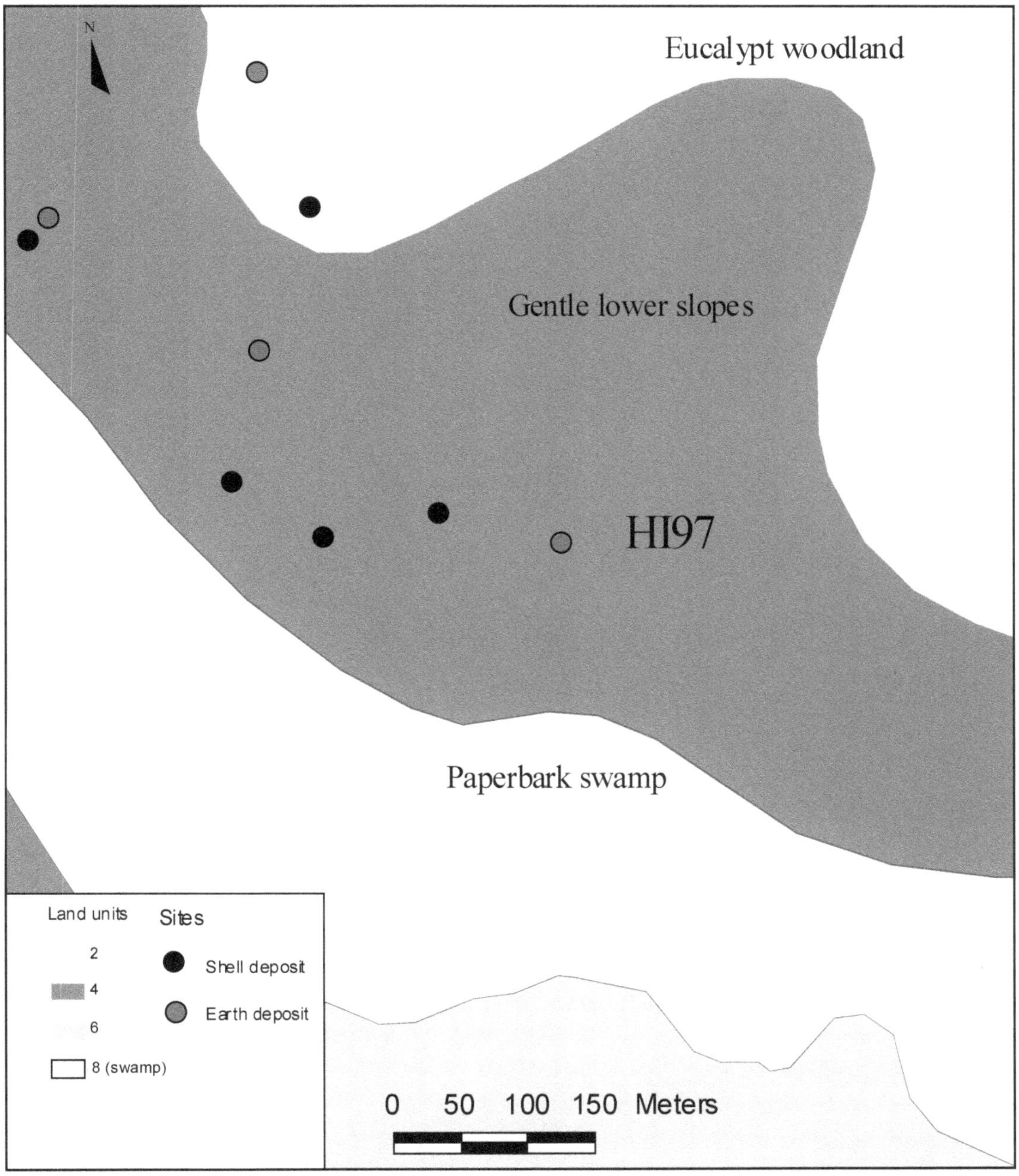

Figure 8.19 Plan of Site HI97

Stone artefacts in the upper layers of the deposit are interpreted as post-depositional and intrusive into the midden, perhaps the result of artefact manufacture and maintenance activities during later reoccupation of the site. The large exotic tuff flake in the surface layers of this deposit may have been transported from sources such as outcrops of Gerowrie tuff and dolerite over 50km kilometres south of Darwin in the Adelaide River region inland (Kim Akerman pers. comm.) Also imported to this site from an unknown inland source was red ochre. Density of stone artefacts is more than twice that in the excavated shell mounds, and this, with the small average size of 0.1g, indicating that these artefacts can be classed as small sized lithic debitage (cf. MacDonald 1991:12) suggests a relatively high intensity of knapping activity, of mainly quartz at this site.

The site would have been a favourable place to camp in the mid-late dry, situated as it is close to a persistent freshwater swamp with abundant resources available at this time, in particular water, water lilies and magpie geese (Plates 8.13 and 8.14), as well as the adjacent estuarine resources of molluscs, fish and crabs. Historical artefacts and buffalo bones on the surface indicate the site was favoured in recent times as a boning site by buffalo shooters. This could have been one of the seasonal shooting camps for buffalo for hides and pet meat that were established on the coastal plains and environs from the Adelaide River area to the East Alligator region from the late 19th century to the mid-20th century (Hockey 1981:34; Tulloch 1981:6-7). Cattle and buffalo shooting activities took place on Koolpinyah station, which extended into the Shoal Bay area (Chapman 1956:117, 121). The bone remains at this site indicates these animals were likely

Plate 8.15 Earth mound HI97 in eucalypt woodland, showing small mortar found on the surface on right hand edge of photo. Nearby shell mound can be seen in centre of photo between trees in the background

Plate 8.16 Close-up of small broken sandstone mortar found on surface of earth Mound

Table 8.21 Radiocarbon estimates, Site HI97

Material	Depth below surface(cm)	Lab Code	dC13	Date (BP)	Calibrated Date (calBP)
Anadara shell	46	Wk6525	-4.9	1830±100	1351 (1171-1595)

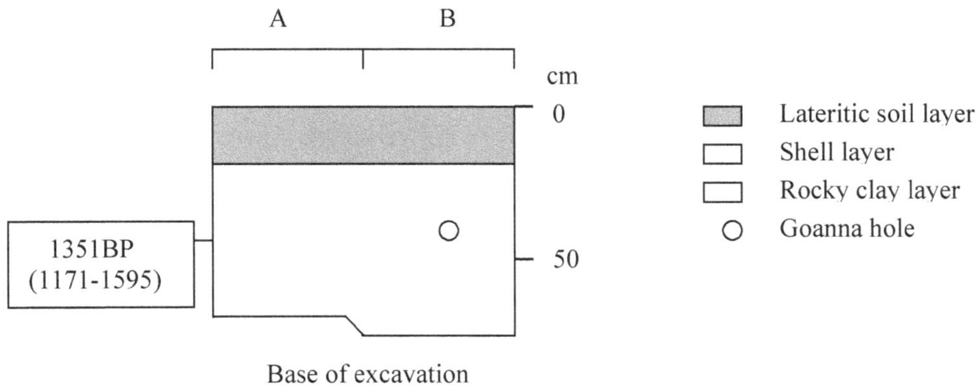

Figure 8.20 Stratigraphic profile, Site HI97

to have been shot for pet food or hides, as the carcases of animals shot for human consumption were removed whole after being bled, to be cut up at abattoirs. (Rideout 1982:36). These shooting camps usually consisted of both European and Aboriginal men and Aboriginal women (Clinch 1990:20).

Site HI97

Environmental setting

Site HI97 is located approximately 1km north-east of HI83, in eucalypt woodland adjacent to a seasonally inundated low-lying paperbark forest stand, and 50m south-east of another large *Anadara* mound (Figure 8.19). This discrete low circular earth mound, 12m in diameter and about 0.4m above the surrounding ground surface, was at the time of excavation covered in lateritic gravel and devoid of vegetation (Plate 8.15). A small broken sandstone mortar was observed on the surface (Plate 8.16). The site is one of many earth mounds, shell mounds and shell rings situated around a small inlet off the saltflats on the footslopes of the Northern Plains plateau. It lies adjacent to an isthmus connecting the main body of land to the narrow promontory on which the excavated shell mounds are situated.

Less than a kilometre to the north of this mound is a small seasonal creek, the site area is backed on the landward side by stands of cycad palms in open eucalypt woodland. A number of earth mounds with no discernible archaeological material on the surface and therefore of equivocal anthropogenic status, dot the area north of the creek. Test pits are required to determine whether these mounds were built by humans or by the scrubfowl (*Megapodius reinwardt*) which inhabit this area (see Discussion below).

Stratigraphy, shell fragmentation and chronology

Fragments of *Anadara* shell on mound HI97, and larger quantities of midden material spread around the edges of goanna holes suggested a higher shell density than revealed by excavation of a pit in the mound centre to a depth of 70-75cm. Three stratigraphic layers were observed in this 23m^3 deposit (Figure 8.20). The upper layer, Stratum I, is compacted light brown lateritic gravelly soil 15cm in depth. The second, more friable layer, Stratum II, consists of sparse *Anadara* shell in a dark greyish brown silt matrix concentrated between the depths of around 15cm to 40-45cm, and extending across half of the excavation square, with a small shell patch 10-20cm wide projecting down into the lower layer. Stratum III, the lower layer, consists of light brown clay with gravel and rocks. A pH reading of 6 was recorded in the upper layer, and between 7 and 7.5 toward the base of the deposit. Termite nest material appeared in patches between 10cm and 44cm depth, and a goanna hole was observed in the east wall at around 45cm depth.

Shell was collected for dating during the excavation from the base of the shell layer at spit 17, a depth of 46cm from the surface. Radiocarbon dating returned an estimate of 1830±100, which calibrated to 1351 (1171-1595) calBP, suggesting that deposition of shell in Site HI97 began around 1350 years BP. (Table 8.21). However, given the evidence that this is a secondary shell deposit (see below), the cooking activities that formed the earth mound would post-date the age of the shell. Further dating of the mound using charcoal, which unfortunately my funding did not extend to, could test whether the time period by which the earth mound post-dates the shell is of any significant length.

The distribution of shell, bone and charcoal in this deposit exhibits the characteristic skewed taphonomic pattern seen in the excavated shell mounds (Table 8.22, Figure 8.21). Fragmentation of shell deposited in this earth mound is higher than for any other excavated deposit, at between 60%

Table 8.22 Quantitative data, Section A, Site HI97

XU	Mean depth	XU wt	Shell	Charcoal	Bone	Crab	Stone artefacts			Ochre	Quartz stone	Rocks >6mm
	(cm)	(kg)	wt(g)	wt(g)	wt(g)	wt(g)	n	wt(g)	material	wt(g)	wt(g)	wt (kg)
1	3	22.9	5.2	0.8	0.4		1	0.3	q		0.7	10
2	5	25.3	7.8	0.3	0.8		2	0.1	q		0.5	13
3	9	25	18.7		2.5		4	0.4	q			11
4	11	23.8	27.3	0.1	4.5		4	0.4	q		0.4	12
5	14	25	42.1	0.1	3.9	0.3	5	0.4	q		0.3	11
6	17	24.5	69.8	0.7	7.8	0.8	5	0.5	q		0.6	11
7	20	25.3	74.9	0.2	6.7	1.3	2	0.1	q			10
8	23	25.2	126.7	0.4	7.1	1.1	1	0.1	q			11
9	26	26	95.6	0.7	6.7	1.3	6	0.5	q d			12
10	29	26	78.4	0.4	5.5	1.3	1	0.1	d		0.2	10
11	33	26	47.3	0.1	4.8	1.1	4	0.3	q			10
12	35	25.8	36	0.6	5.2	0.7	1	1.1	q		0.3	10
13	38	25.7	17.8	0.8	7		3	0.9	q		0.2	11
14	42	26.5	26.2	1	4.7						0.3	10
15	44	26.8	17.5	1.7	4.5	1.1					0.2	12
16	47	25.7	30.5	1.2	3.7	0.4				0.5		13
17	51	26	11.3	0.6	2							13
18	54	25.7	9.9	0.6	2.5		1	0.1	q		0.6	13
19	57	25.5	10.9	0.7	2		2	0.1	q		0.3	13
20	60	24.6	47	0.5	0.9						0.4	13
21	63	24.9	1.1	0.7	0.9		1	0.3	fs		0.3	14
22	65	24.7	2.7	1.6	1.2						0.2	13
23	68	25.2	1.6	0.4	0.6						0.3	14
Total		582.1	806.3	14.2	85.9	9.4	43	5.7		0.5	5.8	270

q = quartz d = dolerite fs = fine-grained sedimentary (pink)

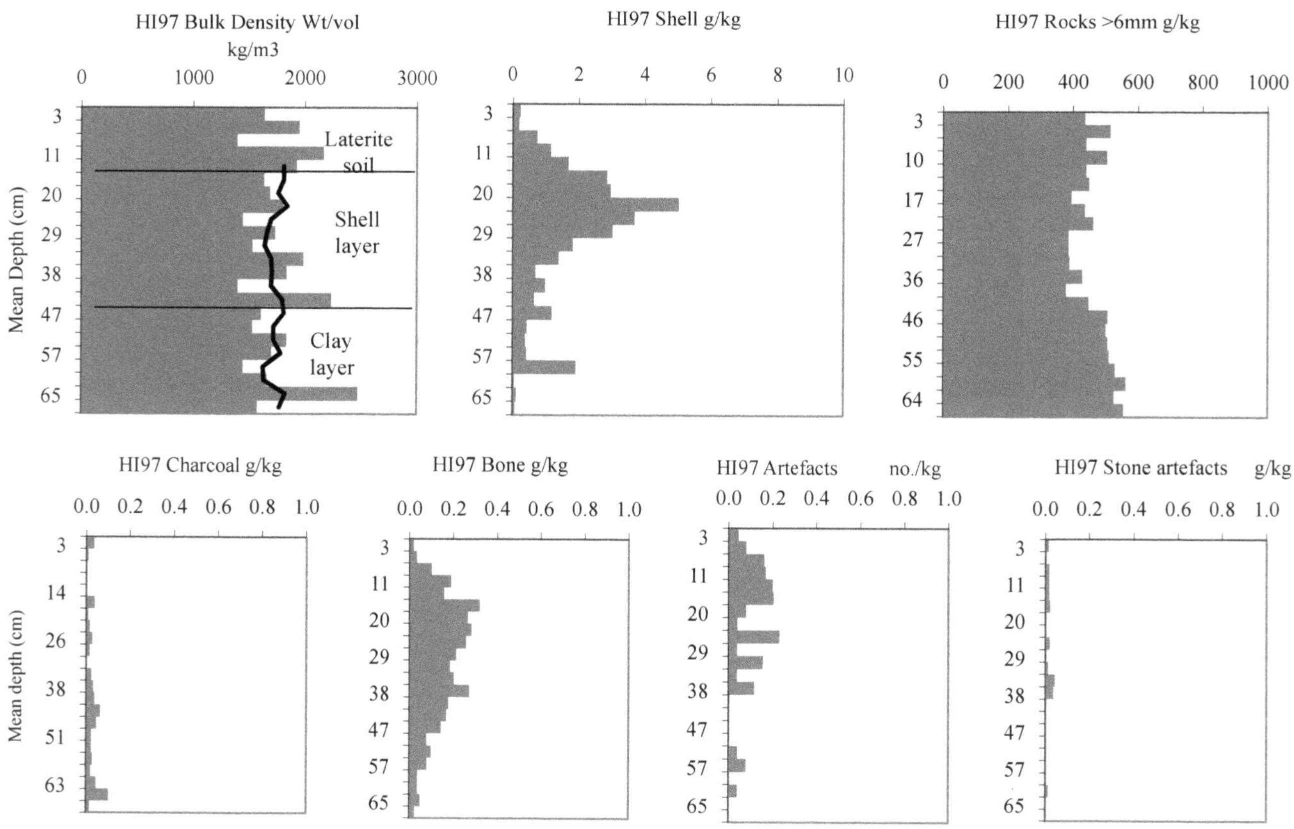

Figure 8.21 Distribution of material

and 100% (Figure 8.17). Very few shells are present below a depth of 50cm, and the condition of shell is weathered throughout, consistent with the low preservation properties of a deposit with relatively low pH levels. The presence of a number of goanna holes indicates that this earth mound is highly disturbed by bioturbation, which may have contributed to shell fragmentation. The shell patch projecting into the lower layer of this deposit may be part of a hearth pit or the result of goanna burrowing (Figure 8.20).

Shell

Five shell taxa were identified in Section A of the excavated material from HI97, and these and the non-molluscan components are listed in Tables 8.23 and 8.24. Including an isolated specimen of *Chicoreus* observed in Section B, the range of macro-mollusc taxa in this deposit comprises only one third of that identified in the other excavated deposits. *Anadara* dominates the shell material, comprising 83% by weight and 81% by MNI of the total shell content. *Terebralia* makes up 14% by MNI and 9% by weight. Table 8.25 shows the paucity of taxa other than mainly *Anadara*, and *Terebralia* in this mound, which nonetheless exhibit the characteristic taphonomic pattern in density distribution. Juvenile *Anadara* were not recovered from this mound.

Other organic material

Bone (86g) was recovered from throughout the deposit, distributed in the characteristic taphonomic pattern (Figure 8.21). A relatively large number of macropod teeth as well as snake vertebrae, rodent and bird bone fragments were identified from spits throughout the mound (Table 8.24). Only a small quantity of fish bone (7% of total bone) was identified, as fork-tailed catfish (*Arius sp.*), and a wrasse tooth (*Wrasse* sp.). Unidentified vertebrae of very small fish, and crab exoskeleton and claw fragments (16.4g), were also recovered. No fish otoliths were recovered from this earth mound, unlike the shell mounds and sub-surface midden excavated at Hope Inlet, but like other earth mounds in the region in which fish bone was found (Sally Brockwell pers. comm.). A total of 14.2g of charcoal was recovered from this sample, concentrated in the lower level of the deposit (Figure 8.21).

Stone assemblage

Once more quartz is the dominant raw material used for stone artefact manufacture in this earth mound, accounting for 89% numerically and 81% by weight of the assemblage (Table 8.23). A total of 43 stone artefacts weighing 5.7g, comprising mostly (93%) small flakes recovered from the 3.2mm sieve residue were identified. Density for 0.35m^3 of excavated deposit is estimated at 123 artefacts per cubic metre, the highest for the excavated deposits at Hope Inlet. Average artefact weight for this assemblage is 0.1g and can be classed as small sized lithic debitage (cf. MacDonald 1991:12). In addition to flakes made from the ubiquitous white quartz, one rose quartz flake and three flakes manufactured from dolerite were recovered from the shell layer, and one piece of pink fine-sedimentary stone from the lower clay layer. Non-diagnostic quartz stone was also recovered (Table 8.22).

Distribution of artefacts generally follows the characteristic taphonomic pattern seen for organic matter in this earth mound. Whilst the small artefacts are concentrated in the upper and shell layers, a few quartz flakes retained by the 6.4mm sieve came from the lower half of the shell layer and the lower clay layer, as did one piece of red ochre. Numerous laterite rocks occur throughout the deposit. Laterite rocks >6.4mm that may have been used as hearth stones exhibit a density two to four times that of any of the other excavated deposits (Figure 8.21).

Discussion: Is HI97 an earth oven?

Excavation and analysis of earth mound HI97 supports the hypothesis that it is of human origin, contra Stone's (1989; 1992:158) theory of a megapode or natural origin for earth and shell mounds in northern Australia. Significant differences in composition are shown, between earth mound HI97 and two nearby earth mounds of Scrubfowl origin, in the Leanyer swamp area (see Figure 2.2), just as they were for the Middle Arm *Anadara* mound MA1 and a nearby Channel Island scrubfowl in my Honours thesis (Burns 1994). Excavations of the two earth mounds of Scrubfowl origin, in savannah adjacent to Leanyer rainforest, by Bowman *et al.* (1999), found a composition of sandy loam, with small quantities of charcoal fragments and gravel. There was no evidence of strata, stone artefacts, hearths, or remains of molluscs or animal bone as in HI97. Radiocarbon dates obtained from the Leanyer earth mounds by Bowman *et al.* (1999:30-31) showed a lack of temporal stratification, consistent with bioturbation associated with the scrubfowl, *Megapodius reinwardt* working and reworking material in their nests (see Crome and Brown 1979; Stone 1989). Bowman *et al.* (1999:30) conclude that mounds similar to these in savannah adjacent to monsoon rainforests in the Northern Territory are all abandoned scrubfowl nests.

However, given the difference revealed through excavation of HI97, an earth mound of similar superficial morphology to scrubfowl mounds, I am of the opinion that testing through excavation would be necessary to differentiate between mounds of Scrubfowl or human origin. This applies particularly to a number of earth mounds located not far from rainforest patches, in the vicinity of the cluster of HI97 and shell mounds. It may be significant in determining origin that these Hope Inlet earth mounds, like earth mounds in south-eastern Australia (Balme and Beck 1996:42; Williams 1988:214) are also located on poorly drained clay soil adjacent to the nearby, better drained gently sloping Koolpinyah surface (Wood *et al.* 1985).

A number of features suggest that mound HI97 may have been used as an earth oven. The stratigraphic profile reveals a concentrated hearth-like feature containing shell, bone, crab and a relatively high density of macropod teeth remains (Figure 8.20; Table 8.24). Shell content is sparse in this 0.5m wide, 25cm deep shell layer within the earth mound, which is similar in size and shape to features identified by Williams (1988:88; 155) and Coutts *et al*. (1976:24) as hearth pits in and near earth mounds in southeastern Australia. The shells in this deposit fit the description provided by Spennemann and Colley (1989:56), of shells exposed to high temperatures, being chalky white, crumbly and fragile, and heating is considered to be the most likely cause of this weathered condition, and of the high degree of fragmentation (see above, Robins and Stock 1990).

There is a paucity of taxa other than *Anadara* and a few *Terebralia* in this mound, which is consistent with the notion the shell feature in this earth mound represents secondary redeposition of "dead" midden shells for use as hearth material, much as Meehan (1982a:89, 114) observed in her 1970's ethnographic study of the Anbarra in Arnhem Land. Re-use of old midden shells is also consistent with the absence of any small, juvenile *Anadara* shells in this mound, which comprise 2-25% of shell in all other excavated deposits.

The dearth of otoliths in earth mounds such as this one, when fish bone is present, is consistent with their use as ovens, where the otoliths are exposed to repeated heating over prolonged periods, and do not survive as well as shell and bone, due to their composition. In Spennemann and Colley's (1989) experimental study on the effects of burning on faunal remains, fish otoliths did not survive, even though some parts of the fish bones did, as well as relatively more robust bones. Otoliths, found in the inner ear of fish, are composed primarily of calcium carbonate and organic matter in the crystal form of aragonite (Casteel 1976:17-20; Degens *et al*. 1969, cited in Weisler 1993:132). When calcium carbonate is heated to temperatures between 600-700°C, an approximate cooking range for earth ovens (Spennemann and Colley 1989:55; Robins and Stock 1990:86), it forms white calcium oxide (CaO) with carbon dioxide (CO_2) given off as a gas. Aragonite also breaks down more readily than calcite (Robins and Stock 1990:87), which is a component of the more complex structures of molluscan shells and bone (Claassen 1998:22; Schiffer 1987:182-5). Shell and bone go through processes such as carbonisation and calcination as they are exposed to higher and prolonged heating, and may permineralize and strengthen through chemical conversion (David 1990:74; Hare 1980:217-8; Lima *et al*. 1986:87, cited in Claassen 1998:66; Shipman *et al*. 1984). As others have noted, chemical changes that strengthen bone structure when it is burnt may increase bone survival rate in archaeological deposits (Aplin 1981:399; Walshe 1998:203), but this may be offset by associated increased brittleness, which increases the risk of fragmentation through trampling (Mountain 1990:212).

I contend that while otoliths preserve well in shell mounds, where they may be subjected to heating but in a more alkaline environment, their preservation is much less likely in the more acidic soils of earth mounds, particularly if these are used repeatedly as ovens (see Behrensmeyer 1975:482; Colley 1990:214; cf. Kefous 1977). Differential discard of fish body parts, as an explanation for the lack of otoliths in this mound, is countered by the presence of skull fragments of fork-tailed catfish and a wrasse tooth. I interpret the faunal remains, of highly fragmented and burnt shell and bone, and high density of macropod teeth, as accumulating through repeated use of this oven using old midden remains as hearth material. The presence of termite nest material and high density of laterite rocks suitable for use as hearth stones as described in ethnographic accounts (eg. Foelsche 1882:13-14), is also consistent with use of the mound as an earth oven. Based on ethnographic reports of the use of ant nest material for earth ovens by 20th century Arnhem Land Aboriginal peoples (Peterson 1973:177; Meehan 1988:2), Meehan *et al*. (1985:147) assert that the use of termite nest material in cooking ovens contributed to the accumulation of the Kiina mound. Similarly in this case termite nest nodules may have been deposited with shell for use as hearth material in an earth oven. The shell layer feature in mound HI97 is consistent with a depression resulting from the construction of cooking ovens as observed ethnographically (Meehan 1982a:89, 110, 114).

The relatively low density of charcoal in this earth mound compared with the other excavated sites may not directly reflect the extent of cooking that occurred. It may be that the relatively low pH levels (between 6 and 7.5) in this deposit affect preservation of organic charcoal. Repeated use of the mound for cooking would further reduce remaining hearth material through burning, including old charcoal, to ash. In her analysis of Kimberley rockshelter deposits O'Connor (1990:110) noted a dearth of charcoal even though bone and shell material showed evidence of burning. Analysis of lower level sediments from the rockshelters found that they "had been subjected to levels of firing intensity sufficient to alter the clay mineralogy", "despite the near absence of charcoal" (my emphasis, Singh *et al*. 1991; cited in O'Connor 1999b:29). A type of cooking in earth ovens described in ethnographic accounts, which involved cooking in a pit with stone or termite hearth material removed from the original fire, would conceivably result in little or no charcoal (Clark and Barbetti 1982:149; see also Foelsche 1882:13-14).

The data, of diverse faunal remains together with stone artefacts, suggests there may have been a combination of uses for HI97, as posited for many earth mounds in northern Australia (Brockwell 1996c) and elsewhere. Functions proposed for earth mounds in south-eastern Australia on the basis of ethnographic and archaeological data, include earth ovens, shelter foundations, burials and camping places (Williams 1988:18-21), and more speculatively, gardens (Balme and Beck 1996). I argue that the small sized lithic debitage concentrated in the upper layers may be intrusive in this deposit, representing artefact manufacture or maintenance by people camped on the mound after it had formed. Peaks in density of stone artefacts in the upper and lower levels separated by a hiatus (Table 8.22; Figure 8.21), suggest depositional episodes separated in time, consistent with successive use of the mound as a camping place.

Summary

Analysis of excavated material from 1m x 0.5m sections of three Hope Inlet shell mounds - HI83, HI81 and HI80 - confirmed surface appearance that these deposits are broadly similar in composition. Fifteen, seventeen and eighteen macro-mollusc shell taxa were identified in the samples from mounds HI83, HI81 and HI80 respectively. A comparable range of fifteen shell taxa in the sub-surface midden HI66, supports the interpretation of this deposit as *in situ* primary shell midden material. Only around one-third of the range of shell taxa in those deposits was recovered from the earth mound HI97, consistent with the interpretation that the shell layer in this deposit represents secondary use of old midden shells as hearth material.

In each deposit the dominant shell constituent is the bivalve *Anadara granosa*, with smaller numbers of other taxa including *Cassidula*, *Chicoreus*, *Nerita*, *Telescopium*, *Terebralia*, *Ellobium*, *Volema*, *Saccostrea*, *Geloina*, *Placamen*, *Melo* and *Xanthomelon* (land snail). These thirteen macro-mollusc taxa are common to each deposit and make up over 95% of the total shell numbers in each sample (Tables 8.3, 8.8, and 8.13, and see Chapter Nine, Table 9.2). The vertical distribution of shell in each deposit exhibits a typical taphonomic pattern, which is also seen in the distribution of bone, charcoal and stone artefacts through each deposit and is discussed below.

Despite their similarities, examination of excavated material revealed significant inter and intra-midden variability in these deposits, in vertical distribution of organic matter and stone artefacts, relative frequencies of shell taxa and shell fragmentation. Mollusc variability within and between sites, specifically the relative frequencies of mudflat bivalves to mangrove gastropods, is investigated further in Chapter Ten. Shell fragmentation rates is an additional variable measured in these deposits, following Muckle's (1985:75) experimental studies on the effects of trampling on shell deposits that found degree of shell fragmentation generally reflects intensity of human activity (see Chapter Six). Shell fragmentation rates are comparatively low at 5-10% and less than 5% through much of the deposit for mounds HI83 and HI81 respectively, but 20-40% for HI80 (Figure 8.4), similar to that of 30-40% for the Middle Arm deposits (Figure 7.6). Relatively high shell fragmentation of over 60% for the sub-surface deposit HI66 and earth mound HI97 are consistent with high intensity of human activity and/or post-depositional taphonomic processes. Low rates of shell fragmentation for HI83 is consistent with relatively rapid shell deposition as indicated by the statistically indistinguishable radiocarbon dates. Higher shell fragmentation rates for HI80 and the Middle Arm mounds are consistent with similar relatively high levels of intensity of activity and/or slower deposition rates over hundreds of years as indicated by their radiocarbon dates. Although radiocarbon dates for HI81 also span hundreds of years, the low shell fragmentation and replicated taphonomic pattern of this mound, in peaks of organic material and stone artefacts separated by a hiatus, supports the interpretation of two major depositional periods separated in time.

Two assumptions made in the interpretation of these deposits are that the samples are generally representative of overall mound content, and that the deposits are cultural rather than natural formations, such as scrubfowl mounds (Stone 1989; 1991; 1992; cf. Bailey 1991; Burns 1994; Cribb 1991; Mitchell 1993; Roberts 1994; Veitch 1994). With regard to the former, it is recognised, as discussed earlier, that interpretations based on this data are limited by the proviso as to how representative these samples are, given that shell mounds may be heterogenous, and contain a range of spatially discrete remains from a variety of occupation episodes (O'Neil 1993; Waselkov 1987:143). However, there are two factors that mitigate against this limitation. The first is that the replicated vertical patterns of material distribution and shell fragmentation seen in each of these mounds, when the test pits are dug in various locations with respect to the centre of the mound, implies a certain consistency to the internal structure of these mounds. In other words lateral heterogeneity is less likely since there is some lateral variation in where the pits were dug. The second factor, related to the first, is what Veitch (1999:165) refers to as the general low level of variation between these sites and other excavated *Anadara* shell mounds in northern Australia, which were dug with a variety of excavation techniques (eg. Allen 1969; Bailey 1977; Bailey *et al.* 1994; Beaton 1985; McCarthy and Setzler 1960; O'Connor and Sullivan 1994b; Veitch 1996).

The latter assumption is based on the employment of criteria developed by archaeologists and geomorphologists (eg Attenbrow 1992; Bailey *et al.* 1994; Coutts 1966; Gill 1954; Hughes and Sullivan 1974; Mitchell 1993; O'Connor and Sullivan 1994b) to differentiate between natural and cultural formations as discussed in the next chapter.

Discussion: Taphonomic patterns and implications for interpretations of archaeological deposits

> Soils (and shell deposits) are...dynamic, open systems in which numerous processes operate to pedoturbate profiles, and to move objects vertically and horizontally within them. Through these processes cultural materials... may sink into the soil,...be thrust to the surface, or...moved horizontally on a plane or downslope (Wood and Johnson 1982:593).

Shell in each of these deposits exhibits a taphonomic pattern resembling a skewed normal distribution. This is characterised by first an increase in shell density relative to surface density, then decreasing shell density with depth, and occurs for all shell taxa as discussed above. As discussed in Chapter Six, the decrease in density with depth is more marked for thin walled shells than for the more robust thick-walled shells in the assemblage (Figure 7.9). This pattern can generally be seen for other organic material such as charcoal and bone, which supports the hypothesis that the patterns represent differential preservation due to taphonomic processes of decay.

The characteristic taphonomic pattern exhibited by organic material in deposits also appears for inorganic material such as stone artefacts (eg. Figures 7.5, 8.5 and 8.22). A uniform pattern of vertical dispersal of stone artefacts, which correlates with peaks in fragmentation rates, is consistent with the idea that they are intrusive objects deposited on the surface of a mound during periods of a slow down or hiatus in deposition of the dominant material, shell. This pattern is consistent with the argument that considerable vertical, horizontal and differential sub-surface movement of artefacts occurs in middens after deposition, catalysed by human or animal trampling on the surface, biogenic activity and cyclical wetting and drying.

Deviations from the typical taphonomic pattern, when considered in conjunction with data on stratigraphy and rate of shell fragmentation, and interpreted within the context of the history and structure of each individual deposit, allows the nature and chronology of cultural activity to be teased out from the natural taphonomic processes. Thus in Site HI81, the distribution pattern is replicated in the upper levels of the upper shell mound and the lower lateritic layer, with roughly correlating peaks in bone, charcoal and stone artefacts suggesting depositional episodes separated in time.

Hiscock (1984:134) has argued that if human behaviour is responsible for an increased rate of discard, then when the amount of occupation increases but nature of occupation remains the same, the rate of deposition of other material, such as bone and shell, should parallel stone artefact discard rates. Following this line of argument, a decrease in rate of shell discard during increases in density of bone, charcoal and stone, would suggest that the nature of occupation has changed. Densities of non-shell material, shown in the graphs, are really proportionate to shell density, since shell is by far the dominant material. A pattern exhibiting peaks in density of one component could arise when deposition of the dominant component –shell- ceases or slows so that the other components, -bone, charcoal and stone artefacts- increase markedly in their manifestation. Thus a decrease in shell discard, rather than increase in discard of bone, charcoal and stone, could result in the pattern seen. In other words, people are discarding bone, charcoal and stone at a relatively constant low level during accumulation of (mainly) shell deposits. At times there is a change in the nature of occupation, in that there is a hiatus or slowing of shell deposition, whilst other activity on the site continues, plausibly by the same group who were previously discarding shell.

Archaeologists traditionally obtain information on the past by excavating a site such as a midden or deposit in a rockshelter, recording cultural matter such as shells, bone, charcoal and stone tools, and dating material (usually charcoal) in the deposit, then dividing the deposit into units or levels according to stratigraphy, and/or the dates obtained, or some other arbitrary fashion, and interpreting the finds. Numerous instances of interpretations based on vertical segmentation of the deposit occur in the archaeological literature, where the deposit is divided by stratigraphy (based on mainly colour), and bounded by dates obtained from charcoal at those depths. Many researchers have interpreted higher concentrations of shell, bone, charcoal and stone artefacts between depths delimited by dates they obtain, as increased rates of cultural discard indicative of greater intensity of site use for that period. However, there are problems, identified by Frankel (1988), with explanations for cultural change being derived from the construction and definition of temporal analytical units (excavation units and their aggregates), because of the often untested or clearly faulty assumptions of an equivalent time span for each unit and a constant rate of accumulation of deposits.

Another often-untested assumption, prevalent in Australian archaeology today, is that a change in quantity or proportions of material distributed vertically through a deposit reflects changes in intensity or the nature of human activity over time. But as argued in this study, the pattern of vertical distribution of material in a deposit is more likely to be due to post-depositional taphonomic decay or movement. Support for this argument comes from a recent paper by Spriggs (1999), who discusses the skewed "normal curve" distribution patterns similar to those seen in this study, of concentrations for pig bones, and for pottery sherds in Southeast Asian sites, where high concentrations in the upper layers of a deposit fall off rapidly with depth, to a "tail" of distribution. This "tail" as represented by one or a few bones or sherds is then dated by association on charcoal or on stylistic similarity in the case of pottery, and argued to reflect the time of introduction of the pig or pottery styles. Due to the high improbability of such phenomena as the presence of pig at 10,000 BP or stylistic continuity in pottery for up to 5000 years in New Guinea, Spriggs (1999) is able to argue convincingly that the distribution patterns seen are the result of post-depositional processes.

In the same way an argument for early introduction of the "Small Tool Tradition" characterised by the appearance of points (eg. Jones and Johnson 1985:181-5, 194-6, 200-6; cf. Bowdler and O'Connor 1991; Hiscock 1993a; White and O'Connell 1982) are based on a few stone points represented as a tail in distribution of the cultural material and dated by stratigraphic association (usually on charcoal). As Hiscock says in his 1986 paper on technological change in the Hunter River valley and the interpretation of late Holocene change in Australia: "Translating amounts of archaeological debris into estimates of quantities of prehistoric activities, be it population size or complexity of social relations, is a difficult task" (Hiscock 1986:47). Nonetheless, it is with this aim that the meaning of the mounds is discussed in the following chapters.

9 Mounds of meaning: The nature of the archaeological record and past occupation of the Darwin coast

Chronology of archaeological sites on the Darwin coast: Contemporaneity and continuity of occupation

Radiocarbon dating of shell and charcoal samples from 13 shell and earth deposits in the Darwin region shows they accumulated for some 1500 14C years, between around 2000 and 500 BP (Table 9.1). Based on the 14C age estimates returned for the shell mounds (including Beta-72152-5 and Beta-55464-7 from Hiscock 1997), which are clustered between 1500 and 500 years BP (Figure 9.1), the major period of mound building spanned is quite similar to that reported for the *Anadara* mounds of Cape York Peninsula (Bailey *et al.* 1994; Beaton 1985, and unpublished dates cited in Stone 1992:103). Radiocarbon dates indicate there is a faster rate of accumulation for the larger Hope Inlet mounds than those on Middle Arm Peninsula, Darwin Harbour. For example, HI80 and HI83 with volumes of $73m^3$ and $57m^3$ accumulate over 100-200 years and within 100 years respectively, while MA1 and MA7 with volumes of $14m^3$ and $12m^3$ accumulate over 700 and 650 years respectively. Bailey *et al.* (1994) noted similar variability in the accumulation periods of the Weipa mounds, with dated samples suggesting that some large and relatively small mounds accumulated rapidly while others formed over several hundred years.

Whilst the mounds dated thus far around Darwin Harbour are not contemporary with each other, at Hope Inlet the pooled mean age estimates for the upper large shell mound HI81 and shell mound HI83 are statistically the same at the 95% level. Thus all or part of the accumulation of these two sites, and possibly the earth mound HI97, may have been contemporary. Moreover, accumulation of these three deposits takes place within the period during which MA1, a smaller mound on Middle Arm peninsula accumulates, between ca. 2000 and 1300 years ago. The pooled mean age estimates for basal levels of shell mound HI80 and sub-surface midden HI66 four kilometres away, are not statistically different either, making accumulation of these sites too broadly contemporary with each other, within the limits of conventional radiocarbon dating (see Head 1990; Ward 1994).

The data from this study suggest not only differing function of different site types, but also different functions for the site type shell mounds, which appear superficially similar. Evidence from the excavations supports the hypothesis that the large composite shell mounds HI81 and smaller mounds such as HI83 at Hope Inlet are specialised processing sites accumulated over very short periods of time around 1450 BP, and are possibly related to ceremonial activity. These are quite different to the "processing sites" described by Meehan (1982a:117), where discarded shells, usually of small quantities of one species, are cooked at the site and the meat removed and carried back to the home base. That type of processing site leaves small remains that would be easily destroyed by natural processes. Some deposits, such as the Middle Arm mounds, that accumulated over much longer periods, prior to, during and after accumulation of the large Hope Inlet mounds, are consistent in nature with the remains of "home base" camps (Meehan 1982a:114-6), accumulated gradually, through repeated scraping up of material left over from previous episodes of use.

One of the questions that needs to be addressed in determining settlement strategies is that of continuity of occupation of a site. Remembering that it would only take a year for a group of 40 people to accumulate a shell deposit of around $8m^3$, it is clear that shell deposition at MA1 ($14m^3$) and MA7 ($12m^3$), spanning some 750 and 650 years respectively, was not continuous at these sites. Even accumulation of HI80 ($73m^3$) over a much shorter period of 200 years would not be continuous. A relatively high shell fragmentation rate of 40-60% for these Middle Arm mounds and 20-40% for the Hope Inlet mound HI80 provides corroboration for the notion of intermittent periods of shell depositional hiatus, and/or high intensity of human activity during their accumulation.

Yet a hiatus in deposition of shell does not necessarily equate to a hiatus in use of the sites. Continuity of occupation during periods of slow down or hiatus in shell deposition are attested by the peaks in bone, charcoal and stone artefacts correlating with peaks in shell fragmentation rate, as seen in the Hope Inlet mounds. Occupation of the area, if not of particular sites, was likely to have been ongoing and seasonal. The excavated shell and earth mounds at Hope Inlet are a small sample of around 50 sites clustered within a kilometre radius on the same section of headland and associated saltflats, and an even smaller sample of the total cluster of over 200 sites in less than $100km^2$. Other headlands in the Hope Inlet area may have been occupied during intervening years, and as already noted, some mounds on Middle Arm peninsula also span the period of use.

There is also the possibility that the recorded sites may represent only a partial sample of an originally larger number in the past. In Chapter Four, I have already commented on the ongoing disturbance processes of sedimentation, aeolian deposition and argilliturbation, which result in burial of some sites, and particularly affect visibility of stone artefact scatters. In addition, it is highly likely that some sites have been disturbed or destroyed, if not by storms, cyclones or

Table 9.1 Radiocarbon age estimates from excavated sites in the Darwin region. The two-sigma calibrated ages were determined using the marine calibration model with a ΔR correction value of -5±35 in the CALIB computer program (Stuiver and Braziunas 1993; Stuiver and Reimer 1993)

Site	Material	Depth (cm)	Lab Code	dC13	Date (BP)	Calibrated Date (calBP)
MA1	*Anadara* Shell	5	Beta-76830	-2.6	1780±60	1305 (1197-1469)
MA1	*Anadara* Shell	25	Beta-76831	1#	2430±90	2056 (1829-2311)
MA7	*Anadara* Shell	5-15	Beta-95257	1#	1870±70	1395 (1264-1572)
MA7	Charcoal	5-15	Beta-95256	-24	850±80	480 (296-609)
MA7	*Anadara* Shell	30-40	Beta-87872	0#	1220±60	742 (642-910)
MA7	Charcoal	30-40	Beta-87873	-25#	1070±80	637 (504-771)
MA10	*Anadara* Shell	4	Beta-94184	1#	750±90	407 (230-529)
HI80*	*Anadara* Shell	3	OZC956	0#	980±75	549 (457-678)
HI80*	Charcoal	3	OZC957	-25#	590±110	251 (81-517)
HI80*	*Anadara* Shell	40	OZC958	0#	1210±90	733 (595-936)
HI80*	Charcoal	40	OZC959	-25#	860±75	483 (316-572)
HI80*	*Anadara* Shell	48	OZC960	0#	1090±90	647 (505-826)
HI80*	Charcoal	48	OZC961	-25#	1010±90	570 (464-713)
HI81	*Anadara* Shell	5	Wk6524	-1.6	1900±70	1427 (1281-1612)
HI81	*Anadara* Shell	140	Wk6523	-2.4	2220±70	1814 (1614-1986)
HI83	*Anadara* Shell	67	Wk6526	-2.3	1910±70	1448 (1287-1623)
HI83	Charcoal	67	Wk6527	-25.3	1850±70	1369 (1259-1530)
HI83	*Anadara* Shell	16	Wk8252	-3.2±0.2	2020±90	1561 (1345-1810)
HI66	*Anadara* Shell	35	Wk6528	-3.1	1190±75	715 (606-908)
HI97	*Anadara* Shell	46	Wk6525	-4.9	1830±100	1351 (1171-1595)

* AMS analysis # Assumed dC13 values
The stable isotope values for the *Anadara* shell indicate that they came from an estuarine environment. More negative dC13 values indicate less saline environment (Head 1991). All samples were whole *Anadara* shells or charcoal pieces collected during the excavation

tidal surges common in this area, as documented for coastal sites elsewhere in Australia (eg. Bird 1992; O'Connor 1989; Rowland 1989), then by human activity as discussed above (and see Ceci 1984). The known factors of visibility, with some deposits possibly buried under the prograding chenier plains, and differential rate of destruction of smaller, more diffuse sites, through either intensive human occupation or climatic hazards (Anderson and Robins 1988; O'Connor 1989), must bear on interpretations in which mounds form the primary database. Furthermore, in the context of estuarine development (Woodroffe *et al.* 1988) and association of shell middens with cheniers (O'Connor and Sullivan 1994b), while these are useful in determining relative chronology of occupation, it is necessary to discriminate between natural and cultural accumulations of shell, and this is dealt with below.

Nature of the deposits

Field evidence on the *Anadara* shell mounds of the Darwin region, which are typical of the *Anadara* mounds of northern Australia (Roberts 1994:185; Bailey 1993:112-4), shows unequivocally that these are cultural deposits. They meet criteria considered useful for differentiating between shell mounds deposited by natural or human agencies, which have been developed by archaeologists and geomorphologists (eg Attenbrow 1992; Bailey *et al.* 1994; Coutts 1966; Gill 1954; Hughes and Sullivan 1974; Mitchell 1993; O'Connor and Sullivan 1994b) in Australia over many years of research. These criteria include the range, number and relative proportion of mollusc taxa present, shell size and range of ages, condition of shells, presence or absence of materials other than shell, including other non-molluscan fauna, stone artefacts, charcoal, pumice, shell grit, and coral.

Mollusc fauna

The condition of the mollusc shells from all the shell mounds excavated at Hope Inlet, although variably weathered, generally according to their depth in the deposit, are not bleached or waterworn like shell debris found on beaches and in chenier ridges. They consist of a mixture of mainly intact single valves, a small proportion of un-opened bivalve shells, and shell fragments, as predicted for tropical cultural mounds by Bailey (1977:138). Most of the shell from each sample was retained by the 6.4mm sieve, with less than 10% of *Anadara* (mostly broken fragments) passing through to the 3.2mm sieve. Nonetheless, the importance of analysing the 3.2mm sieve was reinforced by the fact that a quite inaccurate representation of the relative proportions of mollusc taxa other than the dominant *Anadara* would have been gained if this were not undertaken. For example, in each of the shell mounds a large percentage of *Cassidula* and *Nerita* was consistently recovered from the 3.2mm sieve residue (around 60% and 30% respectively), due to the relatively high breakage rate of these more fragile shells and survival of only the flange, which was small enough to pass through the 6.4mm sieve.

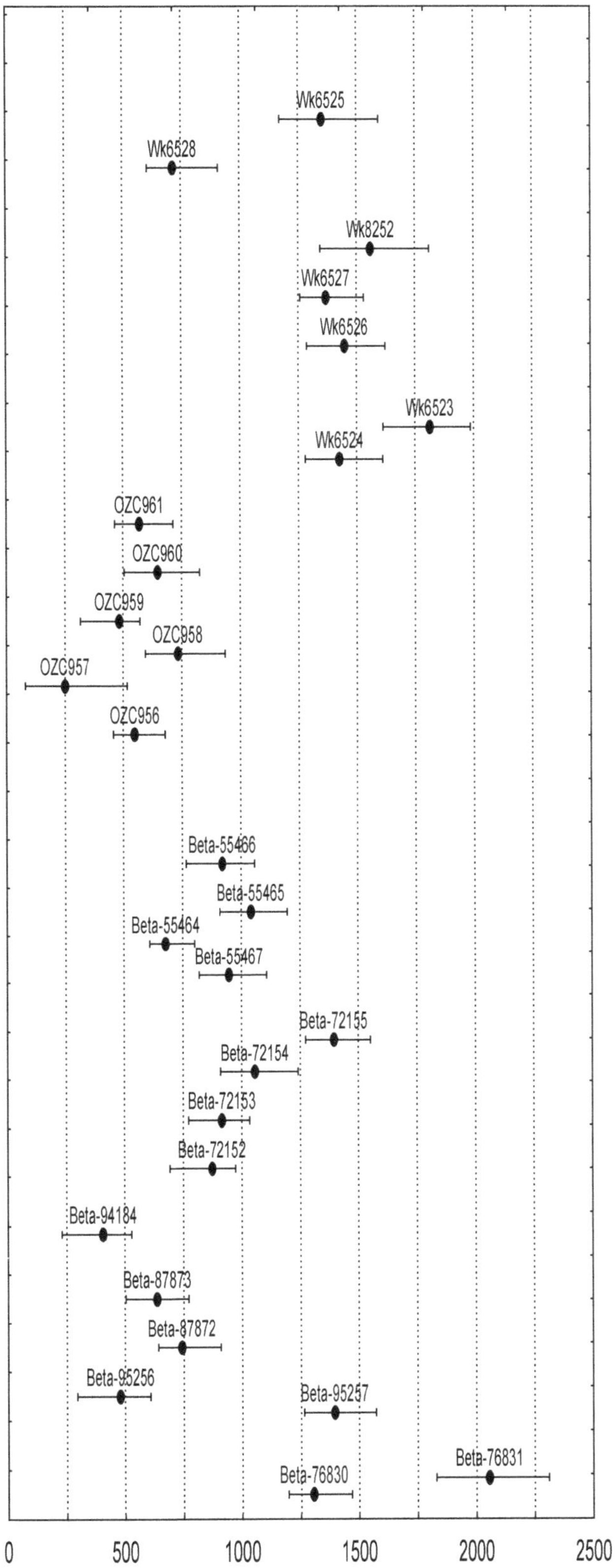

Figure 9.1 Graph showing calibrated and corrected shell dates clustered between 500 and 1500 BP

A summary of the total mollusc taxa identified in excavated sites, eleven of which occur regularly in midden deposits in the Darwin region, is given in Table 9.2. All of the shells recovered from each deposit are of edible mollusc taxa and most are much larger than 6.4mm - quite large enough to provide a reasonable amount of flesh for human consumption, though size isn't everything in considering whether a mollusc was eaten by Aboriginal people (Rowland 1994b). It is these large edible molluscs that I refer to as macro-molluscs. Analysis of shell size of these mounds in Chapter Ten below demonstrates that *Anadara* shell length is tightly clustered around mean values of between 27-36mm throughout in each of the mounds. A comparable mean value of 25mm was obtained for size of *Anadara* shell in the Darwin Harbour shell mound MA1 (Burns 1994:91). These values are significantly larger than mean values of between 5mm and 10mm obtained by Burns (1994:91-2) for size of *Anadara* shell from chenier ridges and scrubfowl mounds on Middle Arm peninsula.

A minute quantity of what I term chenier shells were recovered from some deposits, as well as juvenile *Anadara* shells from most deposits (Table 9.2). Chenier shells are small whole shells <15mm in length collected in the 3.2mm sieve residue, of both adult individuals and juvenile specimens of macro-molluscs, which are water worn and bleached, typical of shell taxa commonly found in chenier deposits (Burns 1994), and are therefore considered non-dietary. The small shells make up less than 10% by MNI and 0.3% by weight of total shell in all sites except HI66, in which juvenile *Anadara* makes up 25% by MNI and 0.6% by weight. These proportions are still less than 1% by volume and within the range of incidental bycatch one would expect to find in a cultural deposit (cf. Attenbrow 1992:16; Hughes and Sullivan 1974:6). The presence of a relatively higher proportion of chenier shells in the lower levels of the saltflats mound HI80 is consistent with its location on a sandy clay base that would have been the surface of the chenier plains when mound accumulation began.

An assessment of the past mollusc community in the Darwin region is possible from an analysis of samples from chenier ridges that accumulated in the last 2000 years (Stone 1987:133-7) and a scrub fowl mound built from this chenier material on Channel Island, Darwin Harbour (Burns 1994:55-7; 88-92). Analysis found some 65 mollusc taxa in the chenier material, mostly small adult individuals between 2mm and 5mm in size from sandy substrate or intertidal rocky platform habitats, dominated by the diminutive gastropod *Cerithium coralium* and relatively small oyster (*Saccostrea echinata*) shells.

Most of the macro-molluscs found in the cultural shell mounds are also found in this chenier material (Burns 1994:89), if in small numbers or as broken or juvenile specimens, including lesser numbers of relatively small specimens of *Anadara granosa*. The chenier ridges at Camerons Beach near Hope Inlet, dated to the last 2500 years, are reported as comprising medium to coarse sand with shell fragments and whole shells of *Anadara granosa* and *Telescopium telescopium*, as well as *Terebralia palustris* and insitu beds of bivalves *Austriella sordida*, *Glauconome cf. rugosa* and *Meretrix meretrix* from within the chenier plain (Hickey 1981; Woodroffe and Grime 1999).

Thus the shells in the mounds are limited in the number and range of taxa present in relation to available mollusc taxa in the past local environment (Burns 1994), indicating a "degree of selection" which does not occur in natural shell deposits (Attenbrow 1992:9). They represent a small selection of macro-molluscs from the estuarine environment of mangroves and mudflats in the Darwin region (Table 2.1), known to be exploited by north Australian coastal forager groups (eg. Meehan 1982a). As noted in other analyses of north Australian shell mounds (eg. O'Connor and Sullivan 1994b:22; Veitch 1996:80) the molluscs consistently exhibit breakage consistent with ethnographic methods of extraction of flesh. The gastropods *Chicoreus, Telescopium, Terebralia* and *Volema* are generally broken, with the posterior end removed if the flesh cannot be removed easily from the shell after cooking (Meehan 1982a:86-9, 105-9).

Mound environment

Also considered are stratigraphy and the location of the shell deposit in the landscape, in particular whether the mound content reflects the surrounding surface, which the Middle Arm mounds do not, as shown in Burns (1994:86-96). The Hope Inlet shell mounds on a terrestrial surface, HI83 and HI81, exhibit a sparse, fine, dark brown grey ashy matrix encompassing the shell, that does not reflect the sediment of the surrounding ground surface of brown clay and dark brown red gravelly soil of the lateritic ridge respectively. In the case of HI80 on the saltflats, the dark greyish black sticky muddy matrix, and brown sandy matrix of the upper and *Marcia* layers respectively, is distinct from the surrounding surface light brown sandy silt clay and sub-surface bluish-gray sandy mud. Like the *Anadara* mounds at Weipa (Bailey 1994:77), these mounds exhibit clear differences in content and structure, to beach and chenier ridges and scrubfowl mounds, not explained by Stone's (1989; 1992) hypothesis that both types of deposit are formed through natural processes such as wave-action, or megapode activities (Burns 1994).

Other organic material

These mounds met other criteria for the identification of shell deposits as human, such as the presence of other organic material including charcoal, animal bone and teeth and fish otoliths. Non-molluscan faunas identified in excavated sites are given in Table 9.3. Along with some bone which appears unburnt, the full progressive range of burnt bone colours, from brown through black, grey, bluish white to white, occurs in each of these deposits, consistent with anthropogenic burning (David 1990; Nicholson 1993:415; Shipman *et al.* 1984; Spennemann and Colley 1989). However, some colours may be due to diagenetic processes associated with weathering or fossilisation of buried bone, noted by researchers to also cause colour changes similar to those effected by fire (Stiner and Kuhn 1995). Most of the bone and crab remains were recovered from the 3.2mm sieve (Table 9.4), and only comprised a very small percentage by weight of the excavated material. The relative proportions of bone in the 3.2mm sieve gives an indication of the level of fragmentation, and by inference, the preservation of bone in each site, and are consistent with the site chronology and interpretations above. As for shell, the quantities of bone and fish otoliths would have been grossly underestimated without analysis of the 3.2mm sieve (Table 9.4).

Marine/estuarine non-molluscan fauna

High fragmentation made identification of bone even to order level difficult, but fish bone appeared by volume the most common non-molluscan faunal material in these coastal sites. Fish bone and crab is found in all excavated sites except the shallow midden MA10. Fish bone makes up a significant proportion of the faunal remains in mound HI81 on the hinterland margin and the more seaward saltflats mound HI80 (at least 25% and 30% respectively), but only a small proportion (around 7%) in the more inland earth mound HI97. The most common species identified, mostly on the basis of otoliths and skull fragments, was the Fork-tailed catfish (*Arius* sp.). Most of the crab remains were fragmented and unidentifiable, but on the basis of claw morphology the mud crab (*Scylla serrata*) was identified in HI81 and HI83, and mangrove lobster (*Thalassina* sp.) in HI80.

Terrestrial non-molluscan fauna

Of the terrestrial non-molluscan fauna recovered from these deposits, Macropodidae were the most identifiable, mainly on the basis of teeth. Macropod remains were recorded in all sites except the sub-surface midden HI66 at Hope Inlet and the oldest mound MA1 on Middle Arm Peninsula, from which only a minute quantity of bone overall was recovered (Tables 9.3 and 9.4). In mound HI81, part of the large mound complex, macropod teeth were not recovered, only mammal bones and the proximal metatarsal of a large macropod. A wide range of teeth size indicating small through to medium and large macropods were found in the more inland sites HI97 and HI83, and the saltflats mound HI80. Dental specimens identified as from large-sized macropods are probably Antilopine kangaroo (*Macropus antilopine*) found in coastal woodlands. Small to medium teeth are most likely to be the common agile wallaby (*Macropus agilis*), found in abundance in the sub-coastal lowlands and seen daily during fieldwork in the Darwin area (Goodfellow 1993:29, 31; Williams and Newsome 1991).

Some possum teeth recovered from a few of the deposits may be the nocturnal Northern Brushtail Possum (*Trichosaurus arnhemensis*), which is common in the Darwin area and a favoured Aboriginal food (Goodfellow 1993:27). The remains of rodents recovered from all the Hope Inlet sites but not the Middle Arm deposits (Table 9.3), could be from a variety of species that commonly inhabit the Top End of Australia. A likely candidate is the Dusky Rat (*Rattus colleti*), which is abundant on floodplains, living in burrows and feeding on sedge corms

Table 9.2 Mollusc taxa identified in excavated sites in the Darwin region

Mollusc taxa	MA1	MA7	MA10	HI81	HI83	HI80	HI66	HI97
Anadara granosa	*	*	*	*	*	*	*	*
Juv. *Anadara*	*	*	*	*	*	*	*	
Chicoreus capucinus	*	*	*	*	*	*	*	*
Ellobium aurisjudae	*	*	*	*	*	*	*	
Nerita sp.	*	*	*	*	*	*	*	*
Telescopium telescopium	*	*	*	*	*	*	*	
Terebralia semistriata	*	*	*	*	*	*	*	*
Volema cochlidium	*	*	*	*	*	*	*	*
Cassidula angulata	*	*		*	*	*	*	
Marcia hiantina	*	*	*		*	*	*	
Saccostrea echinata	*	*		*	*	*	*	
Xanthomelon sp.	*	*	*	*	*	*	*	*
Geloina coaxans				*	*	*	*	
Melo amphora				*	*	*	*	
Cerithidea sp.		*		*		*	*	
Melina ephippium		*		*		*		
Cymbiola flavicans				*	*			
Placamen calophyllum		*		*	*	*		
Placuna sp.				*		*		
Syrinx aruanus		*	*					
Anadara inequivalvis						*		
Arca sp.						*		
Cerithium coralium #		*		*	*	*		
Nassarius sp. #				*	*	*		
Littoraria sp. #				*		*		
Juv. *Cassidula* #						*		
Juv. *Ellobium* #						*		
Juv. *Telescopium* #						*		
Corbula sp. #						*		
Fragum hemicardium #		*						
Iravadia sp. #						*		
Notospisula sp. #						*		
Pseudonachis sp. #						*		
Salinator sp. #						*		
Turbo cinereus #	*							
Total	13	18	11	21	18	31	15	6

# chenier shells = small water worn and bleached whole shells <15mm in length from the 3mm sieve residue, typical of shell taxa commonly found in chenier deposits (Burns 1994)
MA = Middle Arm
HI = Hope Inlet

and seeds (Goodfellow 1993:36-40; see also Madsen and Shine 1996). Reptile remains identified from vertebrae in three sites include snake in the earth mound HI97, possibly python (*Boidae*), different species of which are common around Darwin (Goodfellow 1993:69-72). Bird remains occur in most sites, though confined to a few fragments not identifiable as any of the wide variety of species, including the abundant magpie geese on the Hope Inlet swamps (Plate 8.14), that can be seen in the Darwin region wetlands today (see Chapter Two).

Stone assemblages

Although formal stone artefact types are lacking, there is generally a higher number of stone artefacts in the assemblages from within these deposits (Table 9.5) than that usually seen in tropical shell mounds (eg. Bailey 1993:9; Mitchell 1993:188; O'Connor and Sullivan 1994b:21; Veitch 1996:82). In all mounds except HI80, the largest numerical proportion of artefacts recovered from the excavated samples comprises small flakes and flaked pieces from the 3.2mm sieve size class (Table 9.5). The lower proportion of artefacts

Table 9.3 Non-molluscan fauna identified in excavated sites in the Darwin region

Fauna	**MA1**	**MA7**	**MA10**	**HI81**	**HI83**	**HI80**	**HI66**	**HI97**
Marine/Estuarine								
Arius sp.(Fork-tailed catfish)	*	*		*	*	*	*	*
Protonibea tiacanthus (Black jewfish)				*	*	*		
Polynemus sheridani (Threadfin salmon)						*		
Lates calcarifer (Barramundi)						*		
Wrasse sp.					*			*
Acanthopagrus (Bream)						*		
Small unidentified fish (vertebra)		*		*	*	*	*	*
Crab	*	*		*	*	*	*	*
Mud lobster						*		
Barnacles		*			*	*		
Mangrove worm				*		*		
Annelid worm						*		
Sepia cuttlefish						*		
Terrestrial								
Macropod		*	*	*	*	*		*
Possum		*			*	*		
Rodent				*	*	*	*	*
Reptile		*				*		*
Bird		*		*	*	*		*

MA = Middle Arm
HI = Hope Inlet

Table 9.4 Total bone and crab and fish otoliths in each sample, estimated relative densities and proportion recovered from the 3.2mm sieve

	MA1	**MA7**	**MA10**	**HI81**	**HI83**	**HI80**	**HI66**	**HI97**
Bone and crab (g)	0.3	29.4	1.5	59.6	107	143.7	17.6	95.3
Density (g/m3)	3	235	24	75	213	240	59	272
3.2mm Proportion (%)	100	83	87	80	94	77	88	96
Otoliths (n)	4	3	0	28	7	57	3	0
Density (n/ m3)	36	24	0	35	14	95	10	0
3.2mm Proportion (%)	75	100		89	86	74	100	

Table 9.5 Total number, weight and average density of stone artefacts in each sample, proportion recovered from the 3.2mm sieve and average artefact weight (Anomalously large artefacts are excluded from average weight estimates)

	MA1	**MA7**	**MA10**	**HI81**	**HI83**	**HI80**	**HI66**	**HI97**
Total no.	56	107	69	33	19	9	25	43
Total wt(g)	60.6	27.5	68.3	6.9	15.9	70.6	21.6	5.7
Av. density (n/m3)	498	856	1104	41	38	15	83	123
3.2mm Proportion (%)	72	89	87	91	83	44	70	92
Av. wt (g)								
	0.2	0.2	0.3	0.1	0.1	1.0	0.1	0.1

retained in the 3.2mm sieve in HI80 consists of four out of a total of nine artefacts in 0.6m^3 of excavated deposit and estimates of average weight for this assemblage must be considered in the context of this very low density (15/m^3).

Low artefact density in most north Australian mounds is often attributed to lack of stone sources in an area (Mitchell 1993:188; O'Connor and Sullivan 1994b:24) or seen as reflecting a limited range of activities at a site (Veitch 1996:82). The low number of artefacts in shell mounds has also been used as evidence by Stone (1989:60; 1992:20) to argue against the validity of using the presence of stone artefacts as one of the criteria to differentiate between cultural and natural origins for shell and earth deposits (Bailey 1993:9). According to Stone (1989:60; 1992:20), such few artefacts could have been scraped up by scrubfowl from nearby Aboriginal debris, or deposited by Aboriginal people camping on top of scrubfowl mounds. However, the consistently higher number of artefacts recovered from these samples suggests that relatively low artefact density in mounds may also be related to sampling methods, in that most of the artefacts are small sized lithic debitage that would not be recovered without analysis of the 3.2mm sieve material.

The relatively small average size (0.1-0.2g weight and 2-10mm length) of (mostly quartz) artefacts for these assemblages is in contrast with stone artefacts recorded, mostly during consultancy surveys, in surface scatters in the Darwin area, which are predominantly larger than 10mm in both width and length (eg. Navin and Officer 1995). In one analysis of some 7,300 artefacts salvaged from six sites in Berry Springs (Figure 5.1), Guse (1998:54-5) found that average artefact weight of all artefacts was between 2-8g in four out of the six sites, and average weight of quartz artefacts was between 1-2.5g for all sites.

The predominance of stone artefacts larger than 10mm recorded in surface scatters is probably a bias of archaeological collection techniques. The small sized lithic debitage (cf. MacDonald 1991:12) within these excavated samples, which have been subjected to detailed examination, is a real phenomenon but probably represents just those artefacts small enough to escape re-use and pass into the sub-surface deposit. Their presence indicates manufacture and/or maintenance of stone artefacts was carried out on these sites, reducing larger artefacts of mainly quartz material, but also exotic stone that was transported to the mounds. In the case of Hope Inlet, an area lacking in local sources of any stone other than porcellanite and one outcrop of grey/blue quartzite (see Chapter Five), the wide range of exotic stone such as white quartz, dolerite, chert, and tuff found within the deposits, must have been transported from sources over 60km inland (Kim Akerman pers. comm.).

Hearth stones

There are a number of indicators that laterite rocks recorded throughout each of the samples (Tables 7.1-7.3, 8.2, 8.7, 8.12, 8.17 and 8.22) from these stratified deposits may have been used as hearth stones. Many of the rocks appear to be heat shattered and/or charred on the surface as Veitch (1994) recorded for hearth stones in mounds on the Mitchell Plateau, similar to stones seen on the surface associated with camp fires. The resemblance is striking between the lateritic stone hearth revealed in stratified context during excavation of HI81 (Plates 8.5 and 8.6) and that of a typical stone oven, described in McCarthy and McArthur's (1960:169, Plate 3A) mid-20thC ethnographic study, as used by an Arnhem Land group for cooking kangaroo. In HI80 over 1,700 laterite rocks weighing 7.9kg recovered from the 6.4mm sieve, from 0.6m^3 of excavated sample gives an estimated average density of 2,833 rocks per cubic metre for this saltflats mound. Extrapolating these figures to the whole mound would amount to 206,809 rocks (1,034 per year) being transported to the mound at least 150 metres across the saltflats from the hinterland. While it is recognised that these figures are conjecture given the heterogenous nature of shell mounds, the high frequency of laterite rocks in this saltflats mound far from any source argues for their classification as manuports and against the role of scrubfowl as agents in their accumulation (contra Stone 1991:26).

The nature of past occupation

Implications of the faunal remains

while the data supports a human origin for the faunal remains in these deposits, and it is clear that the mollusc assemblage is dominated by the bivalve *Anadara granosa*, it is much more difficult to estimate from the data the relative contribution of molluscs and other non-molluscan resources. Despite their substantial bulk, mollusc shell may represent a minor, if regular and reliable, portion of the diet (see Meehan 1982:57-80; 141-59). The non-molluscan faunal remains from these sites, recovered mainly from the 3.2mm sieve, show that a wide range of other resources that are much less visible archaeologically, such as fish, crabs, macropods, possums, reptiles and birds, were clearly exploited.

Most researchers acknowledge that fishing was an important part of subsistence activities of Indigenous Australian coastal economies in prehistory, despite poor preservation of remains. In his analysis of the Ballina oyster mounds of southeastern Australia, Bailey (1975b:55) concluded from his calculations that fish, rather than oysters, played a dominant role in the annual economy. The proportion of bone, particularly when small, light and fragile such as fish or bird bone, in archaeological sites is generally accepted to be an under-representation of this aspect of past economies (see Spennemann and Colley 1989). This is due to processing/heating and discard practices and/or preservation factors, including the excavation process itself, that result in total destruction or high fragmentation of bone (Barker 1999:124; Casteel 1976; Nolan 1986; Walters 1979). In view of the self-preserving nature of alkaline shell deposits, other researchers (eg. McNiven 1989:46; Walters *et al.* 1987:92) have suggested that a lack of vertebrate remains in shell deposits can not be explained by taphonomic factors alone, but is probably a reflection of a limited range of subsistence activities at a particular site.

However, as discussed above, decay of bone does occur in shell deposits with removal of the mineral or organic components. Given this, is fairly safe to assume that the bone content recovered in excavated deposits represents an underestimate of the resources consumed on the site. All of the fish and animal resources represented by the bone remains identified in the archaeological sites are present in the local environment today, and some of their use and methods of procurement by Indigenous people recorded at the time of European settlement are given in Chapter Three. According to the historical reports, fishing gave high returns and was apparently just as predictable and abundant a resource in the tropics as reported for temperate zones (Barker 1999:124), given the large quantities of fish exchanged almost daily for damaged flour from the Europeans.

Implications on the nature of occupation may be drawn from the inter-site distribution of the range of faunal remains in relation to the micro-environmental location of the sites. Of interest is that macropods were transported from the hinterland to a mound on the saltflats HI80, where mollusc gathering and fishing appear to have been major activities. Fish and crab also occur, albeit in small quantities relative to the saltflats mound, in the more inland shell mound HI83 together with terrestrial game, and in earth mound HI97, where cooking of macropod and other terrestrial game appears to predominate. Clearly the data does not support the notion that these deposits are merely temporary camps for the sole function of gathering and cooking molluscs. As can be seen in the location plans for each of these sites (Figures 7.1, 7.4, 8.2, 8.7, and 8.11), they are generally located at the junction of a number of environmental land units, in optimal positions to access this range of units in a non-linear landscape (Rowland 1994a:155), including the tidal flats, estuarine creeks and rivers, swamps and monsoon jungle and mixed forest patches, and more open eucalypt woodland. The range of faunal remains in each site suggests a home-base aspect for these sites, with variation in relative frequency of marine/estuarine resources and terrestrial game according to the micro-environmental location of each site (cf. Bailey 1975a:VII:40-1 and Meehan 1982a:114-6).

However, there is some difference apparent in the nature of site HI81, part of the large mound complex. This site exhibits a relatively low density and variety of faunal remains, including molluscs other than the dominant *Anadara*, compared with all other excavated sites. A low density of faunal remains in other deposits, such as the small, older mound MA1, the shallow midden MA10 and the sub-surface midden HI66 (Table 9.4), may be easily explained as due to post-depositional disturbance and taphonomic processes of decay (see Chapters Six, Seven and Eight). However, the high alkalinity and bulk of the large shell mound HI81 with its "humus-zone" cap would be expected to provide a favourable environment for preservation of faunal remains. There is also a much lower rate of shell fragmentation in HI81 than the other deposits (Figures 8.4 and 8.10, Tables 8.8 and 9.4). Significantly, mound HI81 exhibits even lower rates of shell fragmentation than mound HI83, for which radiocarbon estimates indicate relatively rapid accumulation. These data suggest that there may be another explanation for the low density and variety of remains other than *Anadara* in HI81, namely that the mound represents a change in people's behaviour, with a temporary increased focus on rapid harvesting of larger than usual quantities of the *Anadara* mollusc.

Stone artefact discard patterns at Hope Inlet versus Darwin Harbour

There is a wider range, be it in very small quantities, in the variety of raw materials present in the excavated sites at Hope Inlet, compared with the excavated Darwin Harbour sites (Table 9.6). One chert flake and two small flaked pieces are present in the lower levels of HI81, dated to 1800 BP, and minute pieces of flaked tuff in the lower levels of HI83 dated to around 1450 BP. Two large tuff artefacts, one flake 10cm below the surface in HI66 at Hope Inlet, and one bifacial point less than 6cm below the surface in MA1 at Darwin Harbour dated to 700 BP and 1300 BP respectively, are considered likely to be intrusive items discarded on the surface (see below). Small flaked pieces of dolerite, similar to the raw material of the edge-ground axes in the surface scatters on the saltflats, are found in the upper levels of HI83 and HI97, 10cm and 30cm below the surface respectively, dated to around 1450 BP, but are also considered likely to be intrusive items discarded on the surface of the mounds. These and the dolerite pieces found in the lower levels 70 to 90cm below the surface of HI80, which commenced within the last 1000 years, are possibly related to the edge ground axes in the surface scatters.

Estimates of artefact density for excavated sites (Table 9.6) demonstrate a much higher density in the Darwin Harbour sites, where numerous outcrops of quartz stone suitable as raw material occurs, compared to the Hope Inlet sites. Average artefact weight is also higher in the Darwin Harbour sites, consistent with a greater reduction of raw material in the Hope Inlet sites, where most raw material must have been imported. Quartz is the dominant stone material numerically in both areas, excepting site HI80, where quartzite in the form of a few large artefacts is the dominant material. The source of these is probably an outcrop of quartzite (Site HR2) on the saltflats 7km to the south.

Calculation of rates of discard based on the density of stone artefacts in the samples and time spanned by each deposit gives variable results, with no apparent relationship to chronology or to location (Table 9.7). Considering the artefacts as post-depositional intrusive objects, resulting from relatively short-term activities on the then mound surface, gives the more parsimonious explanation of discard

Table 9.6 Comparison of total weight, numbers and density for different raw materials in excavated sites at Hope Inlet and Darwin Harbour

Site	Quartz		Quartzite		Dolerite		Chert		Tuff		Other		Vol. of sample
HOPE INLET	wt(g)	n	wt(g)	n	wt(g)	n	wt(g)	n	wt(g)	n	wt(g)	n	m3
HI83	5.8	15	8.7	1	1.1	1			0.3	2			0.5
HI81	3.5	29					3.8	4					0.8
HI80	6.1	3	64.1	3	0.2	2					0.2	1	0.6
HI66	11.9	20					0.1	1	8.8	1	0.9	4	0.3
HI97	4.7	39			0.7	3					0.3	1	0.35
Total	32	106	72.8	4	2	6	3.9	5	9.1	3	1.4	6	2.55
Total Density g/m^3, n/m^3	13	42	28.5	1.6	0.8	2.4	1.5	2	3.6	1	0.5	2.4	
DARWIN HARBOUR													
MA1	46.7	54							13.7	1	0.2	1	0.1125
MA7	25.7	105									1.8	2	0.125
MA10	68.3	69											0.0625
Total	140.7	228							13.7	1	2	3	0.3
Total Density g/m^3, n/m^3	469	760							*		6.7	10	

* Calculation of density not considered applicable for isolated large artefacts

Table 9.7 Rates of stone artefact discard for sites in the Darwin region

Site (approx.age)	Time span	Wt of sample	Vol. of sample	Total Stone artefacts		Density stone artefacts				Kg per 100 years	No. per 100 years
		kg	m3	n	wt	n/m^3	g/m^3	n/kg	g/kg		
MA1 (2050-1300BP)	750	131	0.1125	56	60.6	498	539	0.43	0.46	17	7
MA7 (1400-750BP)	650	100	0.125	107	27.5	856	220	1.07	0.28	15	16
MA10 (400BP)		91	0.0625	69	68.3	1104	1093	0.76	0.75		
HI81 (1800-1450BP)	350	933	0.8	33	6.9	41	9	0.04	0.01	267	9
HI83 (1550-1450BP)	100	588	0.5	19	15.9	38	32	0.03	0.03	588	19
HI80 (750-550BP)	200	986	0.6	9	70.6	15	118	0.01	0.07	493	5
HI97 (1350BP)		582	0.35	43	5.7	123	16	0.07	0.01		
HI66 (700BP)		425	0.3	27	22.3	90	74	0.06	0.05		

patterns of behaviour. As seen in Chapter Eight, each of the mounds exhibits a taphonomic pattern consistent with this explanation. Thus in most sites the bulk of the artefacts, which are concentrated in the upper layers, post-date the main period of accumulation of the dominant shell. In HI81 artefact density peaks both in the upper layers of the shell mound and the upper layers of the lower laterite ridge, consistent with two depositional episodes separated in time. The distribution of stone artefacts and other organic material, along with other evidence such as shell fragmentation, suggests artefact discard on the surface of the ridge, followed by rapid deposition of mainly *Anadara* shell, then discard of other material including stone artefacts once more on the surface. In this explanation the distribution of artefacts integrates with the data on raw material and estimates of density and average size of artefacts, which shows a clear relationship to location of the sites and availability of raw material.

The dearth of quartzite, dolerite or chert in the excavated Darwin Harbour sites augments conclusions from consultancy surveys that locally available rocks were predominantly exploited as sources of raw material for stone artefact manufacture (Hiscock and Mitchell 1992; Guse and Gregory 1994). A wider variety of exotic raw materials within stratified deposits at Hope Inlet compared with those at Middle Arm is consistent with the alternative hypothesis proposed in Chapter Five, that variability in stone artefact types and raw materials in surface scatters between the two areas reflects different activities being carried out. As the distance from the source of exotic raw materials would not be more for Darwin Harbour than Hope Inlet, there is some support for the idea of different behaviours, inherent in the act of transporting material to Hope Inlet but not Darwin Harbour sites.

Stone artefact discard and theories on exchange and mobility

Examination of the stone artefact assemblages within midden and earth deposits in the Darwin region found them to be characterised by amorphous unretouched and retouched flakes and flaked pieces, predominantly of the ubiquitous north Australian white quartz. As can be seen in Tables 9.5, 9.6 and 9.7, changes in discard rates, average size of artefacts and raw materials show little relationship to chronology of these sites, which span the period roughly 2000 years BP to 500 years BP in both areas. Rather, these changes are clearly related to the different locations, with Darwin Harbour deposits exhibiting relatively higher density, larger average size of artefacts and local instead of exotic raw materials, as at the Hope Inlet deposits.

In studies elsewhere in north Australia increased discard rates, changes in raw material and flake size over time in excavated deposits have been associated with chronological changes in technology and the introduction of points in the late Holocene (eg. Allen and Barton n.d.:43; Jones and Johnson 1985:203; O'Connor 1996:38). On the Mitchell Plateau Veitch (1996, 1999) links the appearance of *Anadara* mounds with point manufacture and a decrease in exotic raw material and increase in locally available stone around 3000 to 3500 BP. An emphasis on local stone for artefact manufacture in late Holocene sites has been interpreted as an indicator of decreased mobility (Hiscock 1994b). Hiscock (1994b) proposes a correlation between highly reduced local stone, increased bipolar knapping and decrease in points and backed blades in assemblages in the late Holocene, with low residential mobility. He argues that there was less need for technology-based risk reduction strategies (involving the use of points) as people became more familiar with their landscape (and less mobile) after the rapid environmental changes of the mid-Holocene.

Some authors have linked an emphasis on local stone with possible fissioning of growing populations into new groups, which are more circumscribed both territorially and socially (McNiven 1999; Williams 1988:110; Veitch 1999:375-7). Regionalisation in rock art styles have also been posited by some researchers to reflect social fissioning processes (eg. David 1991; Morwood and Hobbs 1995). Thus alternative explanations for a greater emphasis on locally available flaked stone include low levels of trade/exchange, low residential mobility and regionalisation (Hiscock 1994b; Veitch 1999:375). A greater emphasis on exotic stone is explained as one of indices of regional elaboration of social networks between groups (Lourandos 1983; and see McBryde 1978).

This study is at odds with such interpretations. Here in the period spanning 2000 to 550 BP a wide range of exotic material appears in *Anadara* mounds in one location, Hope Inlet. But in the neighbouring location around Darwin Harbour there appears to be little change within the deposits over this period in emphasis on the local raw material, quartz. Quartz also occurs in surface context around Darwin Harbour, with the addition of exotic materials such as dolerite and tuff in the more inland scatters within the vicinity of the upper reaches of the Blackmore River at Berry Springs (Figure 5.1; Table 5.3), and the sole Gerowrie tuff bifacial point in the surface layer of the Middle Arm mound MA1. Imported stone at Hope Inlet, which includes the quartz found locally around Darwin Harbour, also appears predominantly in a surface context. In an area deficient in stone sources suitable for artefact manufacture, exotic stone occurs in small quantities in mounds spanning the period 2000 to 550 BP, then appears to increase in quantity with the cessation of mound building, as manifest in the stone artefact scatters on the saltflats surface.

Contradictory interpretations may result by direct translation of types of archaeological material into types of past human behaviour. For example, if changes in raw material were associated with levels of mobility, the data here would suggest low mobility and a localised focus on stone procurement around Darwin Harbour, with very little evidence of trade/exchange until after 500 BP, but with the emphasis on local stone continuing. At Hope Inlet low mobility is suggested not by local stone but by large mound sites, and there is evidence from these for ceremonial activity around 1450 BP, with low levels of trade/exchange, increasing markedly within the last 1000 years.

If, in a second example, Binford's (1980) concepts of forager subsistence-settlement strategies based on the types of archaeological remains, were applied to this study, then groups in Darwin region increased their mobility in the last 500 years. According to a simplified reading of Binford's (1980) theory, foragers, who move consumers to goods with frequent residential moves would be represented as low visibility remains scattered over large areas, and collectors, who move goods to consumers with fewer residential moves would leave large highly visible sites. Thus the archaeological remains would suggest a major change in this region around 700 to 500 years ago, from a collector strategy, leaving large shell mound remains, to a forager strategy, leaving much less visible smaller middens and scatters of shell and stone artefacts.

One hypothesis that could be used to explain the difference between these neighbouring areas is McNiven's (1991a, 1999) model of group fissioning and increasing regionalisation in response to population pressure. McNiven (1999:612) proposes that in the third stage of this fissioning process separate coastal groups would develop, each with highly localised settlement-subsistence strategies and stone artefact raw material distributions. There are problems with the applicability of this hypothesis in this instance. These include the data showing that the Darwin Harbour mounds represent a low level of mollusc procurement by the inhabitants over some 1500 years, not "a dramatic increase in shellfish gathering resulting from a localised intensification of foraging activities associated with increased residency" (McNiven 1999:163).

While the large shell mounds at Hope Inlet might be seen to better reflect intensification, a small group could have easily collected the total volume of shell in these sites over the time period (see Chapter Eleven). An explanation more in line with the data is that the large Hope Inlet mounds reflect groups from around the region taking advantage of a limited, localised and seasonal resource surplus that was available for a time in this particular location. Meanwhile the smaller middens, like those at Darwin Harbour, represent continued low levels of mollusc procurement, representing a low intensity coastal foraging pattern something akin to that proposed by Zobel *et al*. (1984:22-3) for the coast of Victoria (Freslov and Frankel 1999:251).

Comparing ethnohistoric records with the archaeological record

Introduction

As stated in Chapter Three, ethnographic and historic observation of contemporary practices can be used as ideas on possible past human behaviour to be compared with the types of archaeological material present. Since information for the Darwin region is scant and 20th century ethnographic observations indicate similarities between foraging strategies of coastal groups along the north Australian coastline (eg. Davis 1985; Meehan 1982a), data from those ethnographies will be drawn on to compare with the archaeological record. Ethnographic studies have been used to show a link between the extent of forager mobility and the seasonal nature of their resource base (Kelly 1983; Pickering 1994:155; Rowley-Conwy 1983) and such studies provide information on the parameters of seasonal resource availability within which resource procurement strategies must operate.

Similarities between the archaeological record and ethnography

Faunal remains and subsistence

The remains of kangaroo, wallaby, snake and bird present in the excavated deposits are consistent with 19th and early 20th century ethnographic accounts of types of game procured by Indigenous people as discussed in Chapter Three. Identified species of fish and crab remains found in most of the archaeological deposits (Table 9.3) are still fished for today by Aboriginal people in the Northern Territory. Ethnographic accounts, such as Davis' (1985) study on contemporary Yolngu people on the Arnhem Land coast, describe traditional Aboriginal knowledge of seasonal fish movements. Large fish found in the deposits such as the barramundi and threadfin salmon, are said to be easy to spear when they move in large numbers to creek mouths, around the mangroves, and to shallow water on the saltpans flooded by king tides of the late dry/early wet season, documented as the peak time for fishing (Baker 1981:51-2; Davis 1985:303). Catfish and numerous small fish species are found in abundance in estuaries (Larson 1989) and are said to be "fat" in the wet (Davis 1985:304). The mud crab, *Scylla serrata* and many other crabs living in the mangrove muds and along tidal creeks banks (Hanley 1988:142) are said to be "fat" in the early dry season (Davis 1985:309).

All of the fish species found in the deposits could be procured from the estuarine system of rivers, mangroves and swamps or nearshore shallow waters and reefs of open areas as described in ethnographic accounts, through using spears, nets, poison, or wood or stone traps (Basedow 1907:23; 1925; Foelsche 1882; 1886:255; Hodgson 1997:38). Records at Escape Cliffs in 1866 indicate that the Indigenous people camped nearby caught large quantities of fish, including mullet, salmon, bream and catfish, some of which were reputedly speared in the creeks (see below). Small fish as represented by the numerous small vertebrae are consistent with procurement by nets or traps, such as one described at Milingimbi by Donald Thomson in the 1950s utilising the big tides. It consisted of logs and mangrove brushwood and grass placed across creeks, and Thomson recorded that one of these traps took a large haul of small fish (Dixon and Huxley 1985:188). The archaeological fish remains from these deposits is consistent with ethnographic quantitative reports of Arnhem Land coastal community diets, indicating that fish, taken from the freshwater and estuarine system and shallow seas, as opposed to marine deep-sea fish, contributed a relatively high proportion of the diet (Jones 1980b:116-7, 123, 125, 131-2; McCarthy and McArthur 1960; Peterson 1973:182).

Methods of processing and cooking food

Ash and charcoal occurs throughout the shell deposits, consistent with 20th century ethnographic methods of cooking large quantities of shellfish, by steaming them on a bed of old shells and coals covered with bark and green branches (Meehan 1982a:89). Also present are numerous rocks that have probably been used as hearth stones (Veitch 1994), some carried hundreds of metres to some mounds on the saltflats, which suggests that cooking was a major focus of activity. Late 19th century ethnographic accounts describe kangaroo being cooked in earth ovens in a manner consistent with the archaeological remains (Basedow 1907:27). The presence of ant nest material as well as faunal remains, particularly the relatively large quantity of macropod teeth, revealed in the earth mound HI97, is consistent with the use of ant nest material in earth ovens as reported by Meehan (1988:2) and Jones and Meehan (1989:124-5) Peterson (1973:177) in Arnhem Land and Thomson (1939:221) in Cape York, and Basedow (1907:27) for the western coastal tribes, including the Larrakia.

Ethnographic studies describe a range food processing methods employing large grinding stones or mortars and pestles such as those found at Hope Inlet. In their 1970s ethnographic studies in Arnhem Land, Jones and Meehan (1989:124-5) observed stone mortars and pestles being used to crush cycad nuts, which were then baked in earth ovens with heated ant-bed lumps. In Thompson's (1939) early 20th century reports very large cakes were made by processing cycad nuts with these implements, when food had to be sent a long way for ceremonies. Here at Hope Inlet, there are extensive stands of cycads, which would provide enough nuts to feed large gatherings of people for ceremonies in the same manner as recorded in Arnhem Land (Meehan 1982a:32,76).

Other traditionally important plant staples, including yams, spike rush or water chestnut and waterlilies, as well as wild rice available during the mid-late dry season, may have been gathered from the Hope Inlet swamps and processed using mortars and pestles in the same way as seen in ethnographic studies (Foelsche 1882:13-14; Harney 1951:245-6; Jones and Meehan 1989; Meehan 1982b, 1991; Peterson 1968; Russel-Smith *et al.* 1997). Flat mortars/grinding slabs and pestles were also used to pound and grind a variety of plant resources besides the cycad nut and waterlily seeds and roots, including wet season fruits such as green plum (*Buchanania obovata*) found in association with the Hope Inlet mounds (Peterson 1968:567). At Oenpelli Berndt and Berndt (1970:36) reported that these plums were stored for several months before being processed by soaking in water to soften them and pounding to prepare them for eating. In early 20th century Arnhem Land, Thompson (1939) observed that vegetable food was available seasonally, was the responsibility of the woman, and was an important part of the diet, occasionally comprising the staple food and to a large extent controlling the movement of the family. He reported the use of 50 or so plants, including several yam species, water lily roots and seed capsules and corm of the mat rush (*Heliocharis sphacelata*) (probably Eliocharis), which became a staple towards the end of the dry when it grew in brackish water at the edge of the saltpans (Peterson 1983:105).

Ethnographic data suggest that besides processing plant foods, the edge ground axes, stone mortars and pestles found in the saltflats surface sites could have been used for a variety of activities, including woodworking, breaking up animal bones for marrow, and grinding ochre for pigment to decorate implements or bodies for ceremonies (Hayden 1979:109; Peterson 1968). Brockwell (1989:98-111,278) describes the various uses that have been reported ethnographically for axes, grinding slabs/mortars and pestle/pounders of the type found on the Kakadu wetlands, and at Hope Inlet. Stone axes were used for fashioning goose-felling sticks, chopping open logs and branches in search of the native honey, "sugar bag", animals or eggs, or notching trees for climbing, and hollowing out canoes (Dickson 1981; Roth 1904:18; Spencer and Gillen 1904:660). Use-wear and residue analyses are two methods that could be employed to test inferences based on ethnography, on the past use of these stone artefacts (Boot 1993; Fullagar 1993; Hall *et al.* 1989; Kamminga 1982).

Stone artefacts and ceremonial life

It is apparent in the above discussion that food remains and stone artefacts are not limited to being subsistence related only, and are often linked to ceremonial activity in ethnographic accounts. With respect to stone artefacts, the presence of a wide range of exotic stone, such as found at Hope Inlet, is often associated with ceremonial exchange networks (eg. Berndt 1951; Roth 1904; Spencer and Gillen 1904; Thomson 1949; Warner 1969; and see McBryde 1984 and Chapter Five). Formal artefact types present in the Hope Inlet scatters, such as the large quartzite blades, and edge ground axes described by Mulvaney as "crucial indicators of secondary exchange" (Mulvaney 1990:92), have been specifically identified as prestige items for trade or exchange. Mulvaney (1990:92) cites Thompson's (1949:71,87) observations in Arnhem Land, where edge-ground axes were ground and polished over many hours by the craftsmen of the country where the special stone from which it is manufactured occurs. These and certain pounding stones were then passed into the ceremonial exchange system.

Red and white ochre found in the deposits is an item of material culture frequently described in these ethnographic accounts of funeral and initiation ceremonies around Darwin in the late 1800s and early 1900s (Basedow 1907:15, 1925:208; Foelsche 1882:5). Red and white ochre were used to decorate implements, weapons or message sticks (Basedow 1907:36, 46), or mixed with emu fat and used to paint youths during initiation ceremonies (Basedow 1925:249-250; Spencer 1912:20), and warriors preparing for ceremonial battles. It was also used to paint the bodies of the dead (Basedow 1925:184; 208; Foelsche 1882:11).

Seasonal foraging strategies and the ecological and ritual context of shell mounds

Similarities in foraging strategies of north Australian coastal groups can be related in part to environmental similarity and nature of the resource base across the tropical north, with seasonality identified by ethnographic studies as the most striking factor to influence Aboriginal behaviour generally (Chase and Sutton 1981; Jones 1980b; Meehan 1982a; Thompson 1939, 1949). Much of the ethnographic data used as an aid to interpreting archaeological remains, in particular shell middens and mounds, and understanding the nature of coastal occupation in the past comes from studies done in Arnhem Land. Some of these ethnographies describe discard practices of 20th century Aboriginal groups which form new, or add to pre-existing midden deposits (Meehan 1982a; Peterson 1973; Roth 1984a: Warner 1969). On Milingimbi Island, Warner (1969:462-3) observed that the *Anadara* bivalve contributed a large part to the diet and that the Murngin women were still discarding shells on the large *Anadara* mounds there in the 1950s. Meehan's ethnographic work with the Anbarra on the Arnhem Land coast in the 1970s showed an annual resource procurement strategy which included molluscs as a minor component, with a seasonal increase in harvesting of this resource, especially bivalves such as *Marcia* and *Anadara*

Ethnographic accounts describe seasonal patterns of movement and resource exploitation on the Arnhem Land coast, with gatherings of large groups of people for major ceremonies during the mid-late dry season months, at a time when there was a surplus of molluscs (Jones 1990; Meehan 1982a:89; Warner 1969:463). Meehan's (1982a:32,76) ethnographic observations show that although *Anadara granosa* was eaten year round, larger quantities of this mollusc and another bivalve, *Modiolus microperus*, were collected by the women during July and August, and contributed significantly, along with loaves of cycad bread, to the diet of (male) participants of ceremonial activities. Meehan (1982a:80) identifies three periods of the year associated with major mollusc gathering events – the October king tides, the extreme end of the dry and the onset of the north-west monsoon. A link between *Anadara* and ceremonies is also noted by Warner (1969:463), who in connection with the *Anadara* shell mounds of Milingimbi, refers to a "shell and ritual season" involving large gatherings of people over two to three months. Modern day harvesting of *Anadara* from culture beds in Asian countries is seasonal too, carried out over a period of two to three months in the year (Ling 1977:25).

Ecological data are also consistent with the ethnographic information on the seasonal nature of mollusc harvesting. Research on *Anadara* populations in Malaysia and Thailand found evidence of a two-month seasonal spawning period varying between July and December, depending on location. Prior to and during the spawning period there is an increase in the dry:wet weight ratio of flesh (Broom 1985:24-5), probably due to a build up of glycogen and carbohydrates, which Bailey (1975b:59) noted gave oysters their "quality of "fatness" and appeal to the palate". "Fatness" is a quality indicating that a food is in prime condition, which is emphasised by contemporary Indigenous people in their resource procurement strategies in Australia (Davis 1985:300; Meehan 1982a:142). High fat content of meat has been identified as important and universally and cross-culturally preferred in the diets of hunting and gathering societies (Abrams 1987; Speth and Spielmann 1983). In an analysis of seasonality studies on shells from prehistoric sites in southeastern United States, Claassen (1986) also found that mollusc gathering by humans occurred mostly during the mollusc's fast growth period, when it contained relatively high carbohydrate levels and low protein levels. Claassen (1986:34) suggests that "prehistoric people were deriving the maximum amount of carbohydrates and the minimum in protein available from these animals" (see Protein Poisoning below).

Oral accounts by Aboriginal people in the 20th century, include various explanations for north Australian shell mounds. Apart from an interpretation of the Garrki shell mounds of Milingimbi in Arnhem Land by a senior Aboriginal man as "birds nests", reported by Stone (1989:63), most of the mythical stories relate the mounds to the work of ancestral beings. These "creation myths and story cycles" (Cribb *et al.* 1988:68) have been recorded for mounds at Aurukun in anthropological work carried out by McConnell (1957) and Cribb (1986; Cribb *et al.* 1988) in Cape York, and in Arnhem Land at Milingimbi (Peterson 1973) and the Blyth River (Meehan 1982a). Cribb *et al.* (1988:69) comment on the similarity of origin myths associated with shell mounds at Aurukun with those of the Northern Territory. At Aurukun, in a carpet snake totemic story associated with shell mounds of the lower Love River, the snakes coils are said to have pushed up the shell mound (Cribb *et al.* 1988:69). At Blyth River in Arnhem Land the large Kula Kula shell mounds or Dog mounds were said to have been piled up by the first dog in the country, while others were formed by the large stingray, Yuluk, as he flapped his way along the dune (Meehan 1982a:167-8).

One interesting ethnographic example of a link between totemic beliefs and mounds of shell comes from early 20th century records on Indigenous northwest Australian groups by Elkin (1978) and Frazer (1937), who describe ceremonies or rituals performed at totemic sites for the increase or maintenance of resources of importance to a group. Mounds of "cockle" shell are one of the types of "increase centres" recorded (Elkin 1978:27-9, 76-7, Plate 1 opposite p.42; Frazer 1937:81). One ceremony described by Frazer (1937:81), performed by the men and assisted by women of the Burung-Karimba moiety of the Karadjeri "tribe" at Lagrange Bay south of Broome, entailed cleaning out a hole on the top of a single mound situated on the southern bank of a small mangrove creek. This mound of shell was said to represent the fish inside the shell and was the centre for the increase of cockles, while mounds of shell on the northern bank were said to each represent a shell.

Mundane information on resource location was reiterated and confirmed as part of this ritual. During the ceremony, the participants named places where the totem should become abundant, and "at the cockle increase ceremony a performer was corrected by his fellows for naming a part of the coast where cockles are not found" (Frazer 1937:78). The ceremony was said to be instituted in mythical times by Djui, the bower bird who came with his wife from the district to the north of Broome:

> Djui made a nest in a tree and a playground, his diet consisted solely of fish and shell-fish;...He made a small yard of stones, and one day, on looking into it, saw a number of fish. He then travelled down the coast making the present native fish traps, semi-circular rows of stones in which fish are caught as the tide recedes...
> After a while the exclusively fish diet began to disagree with Djui and his wife. They became very sick and at last died, leaving the fish traps, and telling people not to live on fish alone but to eat "nalgoo" and other fruits as well. They also gave the tradition that men caught fish (because the male Djui did so) while women gathered and cooked cockles because the female Djui performed these duties (Frazer 1937:81).

Similar rituals and increase centres featuring archaeological visible stone cairns and piles of dugong bones, and holes in the ground have been recorded for the northeast Australian coastal groups (Hale and Tindale 1934; Sharp 1939).

Such myths, while known to be subject to change, "reflect specific circumstances relating to Aboriginal traditional social living", and as Berndt (1979:28) says "are deeply rooted in the natural environment".

Site location and season of use

Mounds in northern Australia have previously been interpreted as seasonal wet season camps in locations which provide elevation above the surrounding landscape and optimal access to a variety of seasonal resources (Bailey 1975aVII:9-12, 1975b; Meehan 1988, 1991; Peterson 1973). I would argue, as Baker (1981:75) does for Chambers Bay, that the extent of elevation provided by the mounds on the saltflats is suitable only for occupation during the early and late wet season, and may only have been feasible through the wet on the relatively higher ground of the hinterland promontories. In addition, in the Hope Inlet case the location of the stone artefact scatters on the saltflats, which are inundated during the wet season and remain wet for some time into the dry, would limit the use of these site types to the mid-late dry. It is worth noting Peterson's (1968) observation in Arnhem Land that large stone artefacts such as mortars and pestles were owned by the women, and were left behind when people moved, at sites that were semi-permanent pre-wet or wet season camps. In general, the distribution of archaeological sites on raised areas along the coast, including on the mounds would be consistent with occupation during the early wet to the early dry, while sites in close proximity to sources of water is consistent with occupation during the dry.

Availability of fresh water would clearly be an important factor in forager resource management strategies in coastal wet-dry tropical areas during the long seven-month dry season, when all streams except major rivers and springs dry up for variable periods. For the Weipa area Bailey (1975a:VII:11) proposes an alternating pattern of population dispersal during the early dry when water is freely available, and aggregation as water becomes scarce. In this area water would have been available from shallow wells as ground water tables remain high, and the lower horizons of soils on the coastal plains are perennially waterlogged (Williams 1969:75). In some coastal areas fresh water can be found in depressions between sand dunes behind the mangroves, (Christian and Stewart 1953:117). Peterson (1973:187) described a number of wells dug along a seasonal watercourse close to shell mounds in Castlereagh Bay and early explorers noted wells dug by Aboriginal people as they traversed the coastal plains of Chambers Bay, north of Hope Inlet. This information supports the idea that coastal areas were occupied at times when surface water was scarce, during the mid-late dry season (Baker 1981:55-7).

Environmental constraints on site location and resource utilisation.

> The hallmark of the Aboriginal economy was diversity: the combination of resources from a wide range of terrestrial and aquatic ecosystems. The culling of these resources was orchestrated to be in harmony with their seasonal availability. Lacking any substantial means of food storage, the strategy was less concerned with maximising during seasonal glut, although this was indeed the time for most strictly "non-productive" activities such as the great ceremonies, but rather with gearing the whole economic pattern to cope with the season of stress, so that there was least disruption to the society as a whole (Jones 1980b:130).

Decades ago Jones (1977) made a case through demographic studies, that Aboriginal population levels were proportional to resources rather than to technology. He argues that increasing the yield of one resource is of no benefit to a forager group unless other foods can also be increased at the same time and at different times of the year, and there is an adequate supply of water. Ethnographic examples indicate that surplus resources, whether natural or obtained through new technologies, were not used to extract more food, but freed up more time for the maintenance and expansion of kinship relationships and the network of social and ritual activity. In this way the Indigenous Australians have survived the "vagaries of abundance and disaster" of this "capricious continent" for millennia (Jones 1977:202), because the essential requirements for survival, a balance between humans and their land, were met. While some researchers, like Jones, argue that it is the limits of maximum stress exerted by environmental cycles and resource fluctuation that is vital to subsistence and settlement activities of forager groups (Beaton 1990), others as in the example below, argue for primacy of social over environmental factors in determining what (if any) human behaviour caused perceived changes in the archaeological record.

An ethnographic example of constraints on site location

In her 1970s ethnographic study on the role of shellfish in the economy of the Anbarra people of Arnhem Land, Meehan (1982a) argues that cultural factors override spatial and seasonal environmental factors in site location. Meehan (1982a:66) says that the primacy of social factors is demonstrated by the higher frequency of mollusc gathering at a camping site called Ngalidjibama, which is over three kilometres from the main shell beds, than at Lalarr-gadjirripa, another site only one kilometre from the shell beds. Ngalidjibama is the site chosen for a dry season Kunapipi ceremony, and large quantities of molluscs are collected for the ceremonial participants from the 3km distant shell beds. Subsequently Shrire (1984:87-8) cites Meehan's observation as an example of the "cognitive aspect of human life" as opposed to elements in the physical landscape that influence where people locate their sites.

What this reading of the situation neglects to take into account is the proximity of other equally important physical resources, such as water and cycads, which may have played a part in determining the location of this ceremonial site. Earlier in her richly detailed volume, Meehan (1982a:31) notes that at the Ngalidjibama camp where the ceremony was held on the eastern side of the Blyth River:

> Water from the wells surrounding the site was clean and remained in plentiful supply until the camp was abandoned. In fact, water was abundant enough to enable a group of women to leach for use as food a large quantity of the toxic plant ngatja (*Cycas media*) in a specially dug waterhole adjacent to the camping area, thus rendering the water un-drinkable until after the wet season. Firewood was in good supply in the area.

Cycads, the important food for ceremonies, only grew on the eastern side of the Blyth river. Moreover, at the wet season camp at Lalarr-gadjirripa, located on the western side of the river closer to the shell beds, water had to be transported from wells a kilometre to the west until heavy rain had fallen (Meehan 1982a:35,38). As Baker (1981:77) declared in his examination of environmental factors that may have determined site location at Chambers Bay, fresh water would exert a greater pull on where people camp, as it is easier to carry shellfish and fish than it is to carry water.

Meehan (1982a:26,44) recorded that after water, mosquitoes were the most important determinate of campsite location, particularly during the wet season. Bailey (1983:567) suggests that in the wet season during rain people may have sheltered in the trees, and when the rain stopped they moved to open areas raised above the waterlogged ground, and exposed to breezes to escape the heat, humidity and insects. Certainly the Anbarra in the 1970s acknowledged the influence of environment on their camping strategies, describing sites with reference to their place in the annual cycle, for example, as "cold weather camps" and "rain-time camps" (Meehan 1982a:166).

Thus it would seem in this case that environmental factors, albeit a range of them, were the primary factors that influenced people's determination of location of the ceremonial site. It appears that the role of sites as landmarks in a physical landscape, and as landmarks in the minds of people using them do in fact converge. Importantly for archaeologists, this convergence would seem to allow that interpretation of prehistoric patterns of land use based on an investigation of the environmental context of a site and the accessibility of more essential resources than the visible molluscs – water and plants - provide a reasonable picture of the factors people consider in its location. This context needs to be carefully examined, and includes landforms and resources present in the site. Culturally determined aspects of the ceremonial camp structure that would not be retrievable from the archaeological record alone would be such details as where people camped in relation to each other.

> The occupation of Ngalidjibama was organised around the women's Kunapipi ground located in the central portion of the dune. The Anbarra were camped to the north-west of this feature, the Matai to the south, the Gulala to the east, each community occupying ground closest to its home estate (Meehan 1982a:31).

I would argue that it is these more fine-grained details of site structure and its workings rather than site location in the landscape, that is determined by people's views on creation and identity as regards the land and resources (cf. Schrire 1984:88). Such details are not available from archaeological remains when, as Jones (1977:201) reports, all the evidence that remained three months later from major ceremonial gatherings of 200 to 300 people at Ngalidjibama in 1972-73, was "whirling red dust over midden debris, and strips of paperbark rattling against bleached poles of collapsed hut structures".

Past human strategies for resource procurement are not only constrained by the landscape, but by innumerable short and long-term fluctuations of the physical surroundings and resources. As an element of their environment, constraints also exist on humans with regard to their physiological requirements and limitations and capacity to utilise those resources.

Constraints on resource utilisation: Plants and protein poisoning

> On an otherwise treeless plain the first indication that a mound was present was usually the characteristic clump of trees or bushes…of a type…associated… with dune systems, the calcium and silicon from decomposing shell providing an environment similar to the sand ridges (Cribb 1986:148).

As Cribb (1986:148) and Claassen (1998:227) suggest, it may well have been plants associated with shell-bearing sites, rather than just the molluscs that drew people back to reoccupy the same shell midden sites each season. Trees, an important resource in the seasonally hot and wet tropics, provide camping places with shade from the relentless heat of the sun and shelter from heavy rains. Many of the plants on shell mounds were a source of food or medicines, and material that could be used for implements (Cribb 1996:166-7).

The remains of plant foods are poorly represented in excavated archaeological material due to lack of preservation, and that applies to this study as well. Yet at a very basic level it is widely acknowledged and supported by ethnographic and a few archaeological studies (eg. Beck *et al.* 1988, 1989; Bird and Beeck 1988; Chaseling 1957:42; Clarke 1985, 1988; Thomson 1949:21), that plants were essential to survival of forager groups, providing a substantial portion of the diet and a good source of carbohydrates, calories and vitamins. For example, Clarke's (1985, 1988, 1989) analysis of archaeological deposits at Anbangbang rockshelter in Arnhem Land found evidence of regular use of the aquatic plants *Nymphae* sp. and *Triglochin procera* over the last 800 years.

A range of other wetland and terrestrial plants known to have been utilised by Aboriginal people were also present in this and other Arnhem Land deposits (Clarke 1989; Schrire 1982:58-60). The concentration of plant remains in the upper levels of archaeological deposits, generally dated to within the last few thousand years, are argued by some to represent an expansion in plant food procurement strategies in the late Holocene, as part of the process of "intensification" (Lourandos 1980; Hall *et al.* 1989). However, research such as Clarke's (1989, 1999) illustrates how the vertical and horizontal distribution of plant material in a deposit is determined by post-depositional processes of decay. Others argue for a much longer time span for intensive plant use in land management strategies (eg. Hallam 1986).

Inferences have been made on plants that contributed to past forager diets through analysis of use-wear and residue analysis of stone artefacts (eg. Boot 1993; Fullagar 1993; Hall *et al.* 1989; Kamminga 1982), and from the presence of plant species found on and around shell mound sites in northern Australia (Cribb 1996; Cribb *et al.* 1988; Hynes and Chase 1982). According to Cribb (1996), plants found in association with shell mounds on Cape York Peninsula may be the result of domiculture, with exploitation of plants and molluscs producing a kind of ecosystem manipulation. In this domiculture system, shell middens created by people provided new habitats for colonisation by certain plant communities, which are also utilised by people for food and implements.

Whether deliberate or accidental manipulation of this kind occurred is still open to debate, but the Hope Inlet shell mounds are certainly well placed to exploit plant foods just as Bailey (1993:14) noted for the Ballina mounds, particularly cycads. Table 9.8 gives a list of plants quite similar to those on Cape York, found in association with mounds at Hope Inlet, which are recorded ethnographically as utilised by Aboriginal people. These are in addition to the dense stands of cycad palms in open eucalypt woodland backing the sites on the transitional hinterland margins and nearby patches of paperbark, pandanus and monsoon vine forest.

Ethnographic data supports the idea that carbohydrates and calories, provided mainly from plants, were resources much more difficult for coastal groups to procure than sources of protein. Processing of some staple plant foods such as cycad nuts, so that they were suitable for human consumption in adequate quantities, was by all accounts a tedious, time-consuming process undertaken by the women (Beaton 1977; Meehan 1982a:149). Plants that required treatment were among the first to drop out after arrival of the Europeans, curtailed to use in ceremonial situations (Jones and Meehan 1989:123; Meehan 1982a:149). A basic source of carbohydrates was obtained from the Europeans in the form of flour and sugar, in exchange for more easily procured fish, crabs and turtle (see above).

Table 9.8 Plant species found on and around shell mound sites at Hope Inlet

Scientific name	Common name (Larrakia name)	Use
Dioscera transversa	long yam (*Mariyanggwa*)	Tuber favoured Larrakia food
Ficus opposita	sandpaper tree	Leaves used to sandpaper wooden artefacts, sap applied to sores
Planchonia careya	cocky apple	Fruit eaten
Sterculia quadrifida	bush peanut (*Duldinbar* or *Dundil*)	Seeds eaten when pods red and opened
Abrus precatorius	Crab's eye vine	Poison red and black seed –medicinal or craft use
Flacourtia territorialis		"Good" fruit eaten
Canarium australianum	Canarium	Fruit eaten, wood used for canoes
Peltophorum indicum	Yellow flame tree	
Grewia retusifolia	Emu berry	Fruit eaten, sweet, favoured by children
Flueggea virosa	White currant	Fruit eaten, sweet, favoured by children
Gardenia megasperma		Fruit eaten, gum for fixing spear heads
Terminalia ferdinandiana	Billy-goat plum	Fruit eaten, dye from bark used for crafts and medicinal wash for sores
Syzygium suborbiculare	Red bush apple	"Cheeky" fruit eaten
Premna acuminata		Wood favoured for firesticks
Erthrophleum chlorostachys	Ironwood (*Dalanggwa*)	Wood used for fighting sticks and spears, leaves used for smoke in funeral ceremonies
Celtis phillipinensis		Fruit eaten
Acacia auriculiformis	Black wattle (*Gwalmarrwa*)	Calender plant –flowering indicates stingrays fat, beginning of hunting season for saltwater turtle, shark, mullet and crab. Green pods soaped up and used as antiseptic
Buchanania obovata	Green plum	Fruit eaten
Eucalyptus clavigera		Good firewood
Breynia cernua		Fruit?

Yet pre-contact, plant foods would have been just as necessary a companion food to protein-rich and fat-poor marine fish and molluscs, as a good source of calories necessary for energy (Dwyer 1988; Noli and Avery 1988). Resources such as marine fish and molluscs, and terrestrial game such as macropods are a good source of protein but a poor source of calories. Yesner (1980) argues that in low latitude areas, as in the Darwin region, plant foods high in calories and vitamins would have been required to supplement a high protein diet. An abundant supply of high protein foods such as fish and shellfish may have enabled trade with inland groups for high calorie plant foods, as seen between the Indigenous coastal groups and Europeans at contact.

It is unlikely, as O'Connor (1999a:48) speculates, that the *Anadara* mounds represent substitution by women of non-visible plant resources for highly visible *Anadara* resources. In fact, at Hope Inlet, it appears that high protein molluscs and high calorie cycad nuts, both abundant in the late dry within a relatively small area, are ideal companion foods, particularly for people "engage(d) in time-demanding social activity" (Dwyer 1988:151) such as ceremonies. Large quantities of molluscs may be easily harvested within an area and time restricted by the lengthy preparation time required for the processing of cycads, which are nonetheless a good source of calories (energy) for large gatherings of people (Dwyer 1988:151).

Plants are generally accepted as having always been an essential component of forager diets. This is reinforced by reported deleterious and possibly lethal effects of protein poisoning for people on diets of molluscs for more than a few days at a time, if not complemented by carbohydrates or fats (Noli and Avery 1988; Speth 1990), though as Buchanan (1988:60-64) points out, there is some latitude to human protein tolerance. Carbohydrates derived from plants and, to a lesser extent, fats derived from both plants and animals are essential sources of energy (calories) for human bodily function, and are not subject to the same physiological limitations the body faces in processing protein as an energy source (Noli and Avery 1988:396).

Thus Noli and Avery (1988:399) argue that carbohydrates and fat provided the basic energy source for foraging peoples, and it would have only been safe to derive between 20% to 50% of daily energy needs from protein provided by resources such as molluscs. Consistent with this nutritional data is stable-carbon isotope analysis of human bone from late Holocene coastal populations on the east coast of Australia (Collier and Hobson 1987) which shows that marine protein contributed a maximum of 50% contribution to the diet and possibly less.

There is nothing new in the idea that molluscs contributed a relatively minor, if seasonally important part of the diet of human foragers. Nonetheless, as Deith and Shackleton (1988:49) point out, a minor food resource, which is not vital to the economy, may nevertheless be a useful way of monitoring aspects of cultural change because "whilst the environment provides a set of limiting factors, choices are made within that framework and, secondly, that a socioeconomic system works *as a system*, so that changes in one aspect will affect the working of the whole." And archaeological data has the potential to indicate population limiting factors that occur at longer time intervals than observed in short term ethnography (Beaton 1990:28). It is with this in mind the changes seen in frequency in different molluscs and shell size of the main mollusc harvested - *Anadara*, within and between sites in northern Australia will be examined in the next chapter.

Differences between the archaeological record and ethnography

Ethno-historic evidence of change in forager use of north Australian coasts and wetlands in the recent past has been discussed by Meehan (1988) and subsequently Brockwell (1996a), who found correspondence between some aspects of the archaeological and ethno-historic data, with differences related mainly to site location and use. In general for the Darwin region as for the South Alligator wetlands, the archaeological data are consistent with ethnographic and historical reports that indicate a land use pattern of cyclical aggregation and dispersal linked to seasonal resource availability and ceremonial responsibilities (Brockwell 1996a). However, the timing of this pattern of land use may have changed with European settlement. Parkhouse (1895:639) comments on the large gatherings at Port Darwin during the wet season, whereas ethnography in more remote areas suggests that prior to contact there existed a common pattern of land use by north Australian coastal groups with large gatherings on the coast for ceremonies during the mid-late dry (Meehan 1982a:89; Warner 1969:463).

Clearly the archaeological evidence of the large shell and earth mounds at Hope Inlet indicates differences exist between the archaeological record and ethnography. Here, as Meehan (1991:205) notes for the Blyth River region, the archaeological evidence of large earth mounds indicates that past use of the wetland swamps was a more significant part of Aboriginal subsistence strategy than observed ethnographically. Ethnography also suggests a change in practices of shell discard from mollusc consumption, resulting in low, horizontally spread out middens rather than the large shell mounds that accumulated before living memory. Peterson (1973:186) makes the point that large mounds were not recognised by Aboriginal people as left by themselves, but considered the work of ancestors in spite of there being little difference in composition. The difference is one of structure and in the Anbarra case, of location (Meehan 1982a:167-8 and see below for further discussion on midden structure).

Meehan (1982a:117; 168) also noted that the pattern of shell discard as observed for the contemporary Anbarra was consistent with the structure of older shell middens described as "dead men" sites, but differed from the "dreaming mounds", which are large, discrete shell deposits. She points out that the relative proportions of mollusc species also differed between contemporary and prehistoric sites. As discussed above, some of the sites in this study, namely the Middle Arm mounds, are consistent with ethnographic observations of discard practices in "home base" camps. Others, such as the large, rapidly built mounds at Hope Inlet, have no ethnographic parallel.

It is to be expected that perishable items such as the remains of plant foods, canoes, wooden spears and fibre nets would not survive in the archaeological record. However, these are other differences between the archaeological data and the enthno-historic reports that suggest changes in Indigenous subsistence activities had taken place prior to European settlement. Most notable of these are the use of shell, bone or pearl fishhooks, and the hunting of dugong and saltwater turtles from canoes (Basedow 1907:25).

The most detailed historic account of resource procurement by inhabitants in this area is from the daily journal of Clement Young, a clerk and accountant at Escape Cliffs, which the acting government resident J.T. Manton enclosed with his report on the Northern Territory to the South Australian Parliament. This journal describes how from February to August 1866, Wulna and Larrakia people, camped at the Escape Cliffs settlement near the Adelaide River mouth, provided the settlers with abundant fresh food from the estuarine wetlands, including geese eggs "buckets full at a time", procured from the swamps (Manton 1866). Every few days over the seven months they brought in to the camp large quantities of fish, crabs and turtle, exchanged for European items, particularly sought-after damaged flour and bread (Young 1866). The turtle are recorded as weighing between 5kg to 12kg, and are more than likely to be sub-adult sea turtles caught in the local coastal area than the much smaller (up to a maximum 5kg - Rod Kennett pers. comm.) freshwater northern snake-necked turtle (*Chelodina rugosa*) that inhabits the more inland floodplains (Cogger 1992; Ernst and Barbour 1989). Sub-adult specimens of four sea turtle species, the green turtle (*Chelonia mydas*), flatback turtle (*Chelonia depressa*), olive-backed turtle (*Lepidochelys olivacea*) and hawksbill turtle (*Erermochelys imbricata*) are common on the north coast in the Port Essington and Darwin mainland beach areas (Cogger and Lindner 1969; Vanderlely 1996:13; Matthew Fegan pers. comm.).

These and other historic accounts describing the hunting of dugong and turtle in the Darwin area (Brown 1906:9) and their cooking in earth ovens (Harney 1951), indicate Indigenous procurement of these resources at contact. However, no shell or bone fishhooks or dugong or turtle bone were recovered from excavations in the Darwin coastal area, despite the recovery of other faunal remains. It is possible that their absence is due to unsuitability of local environmental conditions for these animals in the Darwin region at the time the deposits formed. However, a similar situation exists on Cobourg Peninsula where turtle and dugong remains are absent from shell mounds and rare in middens dated to pre-Macassan contact, yet palaeoenvironmental change does not appear to explain their absence (Mitchell 1994:289-290). There, as in the Darwin region, ecological data suggests that turtles are more likely to be found in the same types of areas of sandy mudflats that support the mollusc species *Anadara* and *Marcia* from which the mounds were built.

Macassan contact and changes in cultural shell discard practices

It seems that archaeological evidence here is consistent with Mitchell's (1994) hypothesis that resource procurement strategies such as hunting of dugong and large saltwater turtles, and use of fish hooks for fishing were practices recently accelerated by Macassan contact (in the last few hundred years), post-dating the mound building period. From at least 1720 AD, and possibly as early as 1100 AD (Macknight 1976:153-4,158; McIntyre 1984:51) Aboriginal people along the coasts and islands north of Darwin were in regular contact with the Macassans (Macknight 1986; Mitchell 1994a:42-56). However, it is likely that changes to economic practices as observed around Darwin were the result of contact and exchange of ideas and material culture with Aboriginal groups to the north, rather than direct contact with Macassans, as the Macassan trepang fishers did not frequent the Darwin coast (see Chapter Three). The only evidence of these visitors to the Darwin area is a reported Macassan well at Gunn Point, at the northern extremity of the study area (Figure 1.1) (Conservation Commission n.d.). Nonetheless, radiocarbon estimates and the lack of Macassan remains, or of dugong and turtle bone on the surface of the large mound sites at Hope Inlet is consistent with concentrated use of this area before Macassan contact to the north (see below).

Chronological changes in extent of non-local stone, turtle and dugong remains, and size and structure of middens in pre and post-contact Arnhem Land sites, form the basis of Mitchell's hypothesis on the types of economic changes that were brought about by Macassan contact. These include a change in Aboriginal settlement patterns to larger groups resident for longer periods on the coast, close to foreign camps, with a decline in importance of molluscs, adoption of dugout canoes and subsequent increase in capture of large marine animals and an expansion of the ceremonial exchange system, as well as possible population decrease due to introduced disease (Mitchell 1994b:197).

Mitchell (1994a) argues that non-mounded middens resulting from this changed behaviour of shell discard were formed during the post-Macassan contact period. He (Mitchell 1994a:153,233-5,264-8) found an increase in average size of post-Macassan-contact to pre-contact middens and argues that if intensity of Aboriginal occupation increased across the entire coastal zone in response to Macassan contact, there should be an increase in size of middens across a range

Table 9.9 Shell MNI counts, dimensions and surface features recorded for some archaeological sites at Hope Inlet

Site	Form	*Anadara*	*Terebralia*	*Telescopium*	*Chicoreus*	*Volema*	*Ellobium*	*Cassidula*	*Nerita*	*Cerithidea*	*Geloina*	approx vol (m^3)	Environment	Surface material
HI24	M	220	6	3	1	0	0	0	0	0	0	584	open saltflats adj. to tidal channel	quartz manuport, quartzite flake, sandstone pestle, crab
HI92	M	120	5	0	1	0	0	0	0	0	0	205	hinterland margin	laterite rocks
HI25	M	55	4	3	0	1	0	0	0	0	0	126	open saltflats adj. to tidal channel	quartz, laterite rocks, dolerite edge-ground axe, ochre, crab, fish otolith
HI40	m	130	16	5	0	3	3	0	0	3	0	47	saltflats/mangrove creek network	quartzite flakes, quartz, quartzite and sandstone manuports and pestles (15)
HI45	M	140	35	2	0	1	0	0	2	0	1	26	saltflats/mangrove creek network	quartzite flakes, quartz, quartzite and sandstone manuports and pestles (5), laterite rocks
HI41	M	100	10	6	0	1	1	0	0	3	0	14	saltflats/mangrove creek network	quartzite flake and grinding stone, sandstone pestle
HI48	M	230	54	0	2	1	0	0	0	0	0	10	saltflats/mangrove creek network	quartzite flake
HI47	M	250	44	1	0	1	0	0	2	0	0	7	saltflats/mangrove creek network	quartz and siltstone flakes, laterite rocks
HI43	M	87	4	2	0	0	0	1	0	0	0	7	saltflats	quartz cores
HI46	M	185	47	1	0	1	1	0	0	0	0	4	saltflats/mangrove creek network	quartz and quartzite flakes, sandstone manuports
HI56	m	140	70	5	1	2	0	0	3	0	0	1.6	saltflats/mangrove creek network	no stone
HI55	m	90	8	1	1	1	0	0	2	0	0	1.2	saltflats/mangrove creek network	quartz, quartzite and dolerite flakes, sandstone manuport
HI52(1)	m	350	3	17	0	4	0	0	2	2	10	0.1	saltflats/mangrove creek network	quartzite flakes
HI42	s	42	21	12	0	9	0	3	1	1	4	295	saltflats/mangrove creek network	quartz, quartzite and dolerite flakes, sandstone and quartzite pestles
HI44	s	57	14	7	0	2	0	0	3	2	1	323	saltflats/mangrove creek network	quartz, quartzite and dolerite flakes, sandstone and quartzite pestles (30), Sandstone (1) and dolerite edge-ground axes (3), sandstone grinding stone (mortar)
HI52(a)	s	3	1	1	0	1	0	0	0	0	1	780	saltflats/mangrove creek network	quartz, quartzite flakes, quartzite manuport, sandstone pestles (2), WW2 green bullet casing
HI49	s	0	90	1	1	0	0	0	2	1	0	28	saltflats/mangrove creek network	no stone
HI137	s	3	20	10	0	1	0	0	0	0	0	624	saltflats/cracking clay	quartz, quartzite and dolerite flakes and cores, dolerite edge-ground axes (4), sandstone and quartzite pestles (18) and grinding stone, large orange striped quartzite flake and broken quartzite flakes (leilira blades)
HI129	s	0	60	2	0	0	0	0	0	0	1	154	saltflats/cracking clay	quartz, quartzite and dolerite flakes, dolerite edge-ground axes (3), sandstone and quartzite pestles (13) and grinding stone, laterite rocks
HI138	s	15	40	10	0	0	0	0	0	0	0	312	saltflats/cracking clay	quartz flakes and manuports, sandstone pestles (10)
HI139	s	5	22	12	0	0	0	0	0	0	0	112	saltflats/cracking clay	quartz, quartzite and tuff flakes and cores, dolerite edge-ground axes (4), sandstone and quartzite pestles (40), laterite rocks
HI128	s	0	3	0	0	0	0	0	0	0	0	780	saltflats/cracking clay	quartz flake, sandstone pestles (12), dolerite edge-ground axe
HI112	s	0	1	1	0	0	0	0	0	0	0	702	saltflats/cracking clay	quartz, quartzite, porcellanite and dolerite flakes, dolerite edge-ground axes (2), sandstone, quartz and quartzite pestles (100)

M = mound m = midden s = surface scatter

of areas. But Mitchell's (1994a:233) argument for increased midden size after Macassan contact is flawed, given that he based his estimates on area rather than volume, which naturally would bias the figures in favour of higher values for middens spread out horizontally over those built up vertically as in the mounds.

At Hope Inlet roughly 60% of the shell middens are mounded in form, and both middens and mounds are spread evenly across hinterland and saltflats. The largest mounds are concentrated on the hinterland margins and smaller middens tend to occur on the saltflats, though this observation may be biased by difficulties in estimating dimensions of sites that are partially buried by mud. Surface inspection revealed no marked differences in composition between shallow middens that are not mounded and shell mounds. Comparison of MNI surface counts and total MNI counts of excavated samples from three shell mounds found poor correspondence between these two measurements, probably related to higher fragmentation of more fragile shell on the surface of mounds in comparison with the subsurface material, as discussed in Chapter Six. Nonetheless MNI counts were undertaken on the surface of a number of sites (Table 9.9), as it is not feasible or desirable to excavate all sites.

Surface shell MNI counts in one metre squares on a number of middens and mounds (Table 9.9), show no observable trends in relative frequency of different molluscan taxa, or addition of other material such as turtle or non-local stone, which might indicate that non-mounded forms of shell deposit are of more recent age than the dated mounds. Non-local stone artefacts, including pestles and mortars, were found in association with both small mounded and non-mounded shell deposits in one section of the saltflats situated amongst a network of mangrove-lined seasonal creeks and ponds, as well as in surface scatters (Table 9.9). Some middens contain no stone. The geomorphic context of these particular mounded and non-mounded middens, which probing indicated did not extend below the saltflats surface, suggests they all post-date the most recent phase of chenier plain progradation identified by Woodroffe and Grime (1999) in the Shoal Bay area around 900-1000 BP. Unfortunately, like many archaeologists, I succumbed to focussing limited resources on dating large mounds rather than smaller middens, but further dating would give a clearer picture of the age of non-mounded middens in this area.

Analysis of a subset of Hope Inlet middens and mounds for which dimensions were measured found that non-mounded middens (n=58) ranged in area from 0.8m^2 to 343m^2, and in volume from 0.1m^3 to 69m^3, with an average area of 43m^2 and an average volume of 9m^3. Shell mounds (n=110) ranged in area from 0.8m^2 to 1170m^2, and in volume from 0.1m^3 to 1185m^3, with an average area of 130m^2 and an average volume of 85m^3 (Table 9.10). Clearly middens in this area are not larger in size than shell mounds, but are in fact much smaller, if somewhat more spread out horizontally.

The evidence suggests that whilst cultural discard behaviour may have changed, with cessation of mound building, the change would have resulted in a more horizontal form of the middens and decreased midden size rather than an increase (contra Mitchell 1994a:396). This is consistent with Mitchells' (1994a:387-8,396-9) hypothesis of a decline in intensity of exploitation of molluscs post Macassan contact, but makes problematic an inference based on this for more intensive occupation of coastal areas across the region.

At Hope Inlet stratified shell deposits occur in a range of sizes and forms, including mounded and non-mounded circular and elongated middens and rings of shell and non-local stone artefacts are found in association with both mounded and non-mounded deposits as well as shell scatters. There are no additional features on non-mounded middens at Hope Inlet such as Macassan remains, or turtle and dugong remains to suggest these are more recently formed than the mounded deposits, dated to have ceased by some 500 years ago (see Figure 9.1 and Hiscock 1997). Here as at Cobourg peninsula (Mitchell 1994a) there is archaeological evidence for continued exploitation of sandy-mudflat species by the Aboriginal inhabitants, without mound building. For example, Site MA10, one of many shallow non-mounded middens situated on rocky knolls or hillcrests around Darwin Harbour, gave a recent date of around 400 BP. Two non-mounded middens in Arnhem Land have also been dated as forming within the last 400 years (see Table 9.1 and Meehan 1983:14-15). But this site type does not always post-date the mound building period, as Mitchell's (1994a:259-260) date of around 1500 BP for a midden on Cobourg Peninsula attests. A date of 715 BP was also obtained on another site type, HI66, a shallow sub-surface midden at Hope Inlet. Non-local stone occurs in association with both small mounded and non-mounded shell deposits at Hope Inlet. The evidence suggests a change that involves cessation of particular types only (specifically shell rings and mounds) of a range of cultural shell discard behaviour.

Shell mounds and stone axes: Cultural change and ceremonial life

Chronology of occupation of mounds and surface scatters

At Hope Inlet the low density of stone artefacts within the mounded stratified shell deposits, dominated by *Anadara,* contrasts with the diversity of artefact types and wide range of exotic raw materials found in the surface scatters of shell and stone, in which the dominant shell component is *Terebralia*. It is conceivable that these two site types, which evidently reflect different activities, may not have been occupied concurrently. Excavation of the shell mound HI80 on the saltflats revealed that this deposit extends some 40-60cm below the saltflats surface, on which the artefact scatters are located. The mound therefore probably accumulated prior to the sedimentation processes that formed the present day saltflats. This would suggest that the deposit predates the surface stone artefact scatters. Geomorphological data indicating a phase of chenier plains build-out some 900 years ago is consistent with this proposition (Woodroffe and Grime 1999). However, as considered in Chapter Five, disturbance processes such as argilliturbation may result in vertical movement of artefacts through the clay substrates on which they rest, making it difficult to determine the original stratigraphic relationship of the artefacts with the present day saltflats surface.

Moreover, probing around a number of shell mounds on the saltflats indicated that not all of them extend below the saltflats surface. This includes midden and mound sites associated with large grinding and pounding implements, clustered in one small area amongst a network of mangrove-lined creeks and shallow ponds. It is likely that different shell mounds accumulated over time following progradation of the coastal plains, so that the mounds are younger from landward to seaward, as Veitch (1996:80) suggests for *Anadara* mounds on the Kimberley coast. Thus occupation of the stone artefact scatter sites appears to be subsequent to accumulation of some of the mounds, but may have been concurrent with occupation of more recent mounds.

Comparison can be made between the Hope Inlet scatters containing exotic stone and the much larger South Alligator wetland assemblages, comprising a wide range of exotic raw materials, which Brockwell (1989:211; 1996a) interpreted as evidence for regional mobility or extended contact over large distances. Brockwell (1989:214-24; 1996a) found variability between the stone artefact assemblages of wetland sites and the upper levels of rockshelter deposits dated to within the last 1000 years in the South Alligator region, that she argued reflected differences in proportion of type of artefact related to site location and resource availability. A high proportion of points at the more inland rockshelter sites is said to reflect a focus of hunting on large terrestrial game, while use-polish flakes that occurred only at the floodplain sites reflected processing of silica-rich plants from the freshwater wetlands (Brockwell 1989:223; 1996a). Within the last 1000 years, an emphasis on exotic stone in these sites is seen as reflecting a system of trade/exchange networks throughout the Kakadu region, of wetland resources for raw materials from the stone country (Brockwell 1996a:101). The timing of appearance of exotic raw materials in this area is relevant to archaeological (Mitchell 1994a, 1994b, 1995) and ethnographic (Berndt 1951; Thomson 1949) hypotheses, that systems of regional exchange in Arnhem Land intensified after Macassan contact began.

Support for the interpretation of a similarly relatively recent age for the Hope Inlet scatters comes from the presence of the large blades manufactured from an exotic quartzite in these surface scatters, a type of artefact also found in open sites on the Alligator Rivers. This type of artefact are also more common in the upper levels of inland rockshelter sites, and are considered to have been introduced relatively recently, within the last one thousand years (Allen and Barton n.d.; Brockwell 1989:74-5; 1996:97-99; Kamminga and Allen 1973:87-8; Schrire 1982:244).

On the evidence thus far, it is probably safe to say that the Hope Inlet stone scatters were deposited within the last millennium, yet are at least pre-European contact, and possibly pre-Macassan contact. Macassan sites or material are absent in this area, consistent with historic information that the Macassans did not frequent the Darwin coast, as Port Darwin was not a suitable area for harvesting trepang (Macknight 1976:49,108). European material is also absent, apart from buffalo and pig bones and pieces of tin and other scrap metal on the surface of a few sites on the hinterland. This material is interpreted as resulting from 20th century activity by buffalo shooters in the Hope Inlet area, which was part of the Koolpinyah Station pastoral lease from the late 19th century until surrendered to the Northern Territory Government in 1994 (Registrar-General's Office NT, Document No. 308909; Chapman 1956:117, 121).

The dearth of Macassan or early European material on the Hope Inlet midden sites, as found elsewhere, is consistent with abandonment of Hope Inlet as a focal area of activity by the Indigenous occupants prior to Macassan or European contact. In other parts of the wider Darwin region, including around Darwin Harbour, sites with modified early European material such as flaked glass or porcelain and worked metal indicates continued activity in an area by the Indigenous occupants (eg. Hiscock and Mitchell 1992) who, as ethnohistoric and archaeological studies report, incorporated European material into traditional lifestyles. For example, Basedow (1907:30-35; 1925:170-202), in describing a variety of spears and spear throwers used by the Larrakia, states that one of these is a type of long reed-shaft spear with a stone-head of quartzite or slate "or latterly of glass" attached by wax or gum (Basedow 1925:196). Many spears with heads of flaked glass and occasionally porcelain were collected from around Darwin in the late 19th century (Hodgson 1997:7). Reports during Goyder's surveys of the Darwin region describe Aboriginal people collecting any wire or metal pieces left behind and grinding them to use as knives and spearheads (Kerr 1971:105,133). Archaeological evidence of this practice is seen at the European settlement of Port Essington (Allen 1969) where a large assemblage of flaked glass artefacts were recorded, and on post-Macassan-contact midden sites on Cobourg Peninsula, containing glass and a range of foreign material such as ceramics and metal (Mitchell 1994a:232, 1995:46).

Macassan contact and disease, exotic stone and ceremonial exchange

According to archaeological (Mitchell 1994a, 1994b, 1995) and ethnographic (Berndt 1951; Thomson 1949) hypotheses on the manner in which Macassan contact affected Indigenous economies, an increase in foreign goods stimulated and expanded ceremonial exchange with other groups, as observed historically on Cobourg Peninsula (Mitchell 1994a:143-4). Foreign goods such as iron, cloth and tobacco obtained by coastal groups through trade with the Macassans, were brought into existing ceremonial exchange systems, which then intensified as inland peoples increased trade in exotic stone artefacts and ochre to gain access to these valued items. The archaeological expression of this intensification is argued to be a higher frequency of imported exotic stone artefacts in post-contact than pre-contact coastal sites (Mitchell 1994a:377).

As reported in Chapters Eight and Nine, very small quantities of exotic raw material, in particular small flaked pieces of dolerite, of the kind from which the edge-ground axes in the surface scatters are manufactured, are found in the upper levels of mound HI83 and lower levels of mound HI80 at Hope Inlet. If the provenancing of the dolerite pieces found in the HI83, dated to around 1450 BP (500AD) was to reveal that there was an association between this material and the edge ground axes in the surface scatters, then it could be argued that the surface stone scatters are pre-Macassan contact. However, as discussed above, it is considered likely on taphonomic grounds that the dolerite artefacts in the upper levels of HI83 may be post-depositional intrusive objects.

It is more difficult to argue on the grounds of vertical movement that the dolerite pieces found in the lower levels up to 90cm below the surface of mound HI80, dated to 750-550 BP. (1200-1400AD) are post-depositional intrusive objects. An association between the dolerite pieces within this mound and the edge ground axes in the surface scatters would still place the exotic raw material as pre-Macassan contact, if the 1720 AD date for start of regular pre-Macassan contact is accepted (Macknight 1986; Mitchell 1994a:42,399). In this situation consigning the stone scatters as post-Macassan would require acceptance of the earlier date of 1100AD for the start of Macassan contact, as suggested by some authors (McIntyre 1984:51) and some radiocarbon dates (Macknight 1976:158).

If the imported stone in surface scatters at Hope Inlet is contemporary with the mounds, then dried mollusc flesh could have been a trade item, and/or large quantities of molluscs, along with the "communion food", cycads (Beaton 1977:165; Meehan 1982a:31-2), may have been processed to feed ceremonial participants. Abandonment of Hope Inlet as a focal occupation area around 550 BP (1400AD) under this hypothesis is consistent with evidence for gradual environmental change resulting in the decline of the mollusc (see below). It is also possible that realignment of coastal groups in response to shifts in the exchange system balance resulting from Macassan contact in Arnhem Land contributed to abandonment of this area (Mitchell 1994a:399).

There are some problems with the hypothesis that exotic stone sites are post-mound building and post-Macassan contact. If the exotic stone artefacts at Hope Inlet were brought in after the mound building period ceased, this begs the question of what was being exchanged and/or used to provide the protein content of the diet of people gathered at Hope Inlet. Without a surplus resource such as molluscs, or foreign Macassan items to trade it seems unlikely that the area would have been a focal part of the expanded exchange network following Macassan contact in Arnhem Land. One could speculate that magpie geese were increasingly exploited, as they became more abundant seasonally with environmental change leading to freshwater conditions and a decline in *Anadara* (see below). The economic importance of geese in the last 1500 years in Arnhem Land is thought to be reflected in frequent rock paintings of magpie geese and their hunters, with bundles of goose spears and goose-wing fans, typifying this freshwater art period (Chaloupka 1985:277-8, 1993:173-188; cf. Lewis 1988:75-7).

Despite Lewis's (1988:76) correct observation that rock art does not provide an all-inclusive record of Aboriginal economy (for example, molluscs are not represented in rock art), spatial as well as temporal differences reflecting the local environment do occur in rock art. Tacon (1993:117) shows in an analysis of western Arnhem Land art that images of fish and macropods are respectively more frequent in northern wetland parts and drier more inland areas. Ethnohistoric observations of geese hunting in northern Australia (Dahl 1927:99; Thomson 1983:101) also suggest that geese figured predominantly in resource procurement strategies. Other unrecognised resources in the Hope Inlet estuarine wetlands may have been available for trade/exchange. Archaeological and ethnohistoric evidence of trade/exchange exists for the South Alligators wetlands regions, of freshwater swamp plant products such as *Phragmites* sp. reeds and bamboo, used as spear shafts, and another seasonally abundant creature Collet's rat (*Rattus colletti*), found in inland sites such as Anbangbang (Brockwell 1996a:99).

But as noted above, the dearth of Macassan or early European material in these sites is not consistent with the hypothesis that the exotic stone resulted from exchange activity in the area post-Macassan contact. Alternatively, exotic stone may have been brought in from elsewhere for mundane use in a non-trade/exchange context in the resource rich Hope Inlet area, though this is difficult to reconcile with the wide range of exotic stone present and the ethnographic data linking exotic stone with exchange. The data thus far seems to support the idea that the Hope Inlet sites represent (in part) the remains of practice of the regional socio-economic systems of alliance and exchange that existed just prior to or at the dawn of their expansion following Macassan contact (Allen 1986:13; Berndt and Berndt 1964; Mitchell 1994a, 1994b). They were possibly small-scale local clan ceremonies held for social intercourse, population redistribution and exchange between neighbouring groups who were in close and frequent contact with each other (Evans and Jones 1997:416; Yengoyan 1976:127).

As Yengoyan (1976:127) points out, this type of social practice can be related to environmental conditions capable of supporting relatively high densities of population groups and their periodic aggregations. Ethnographic, linguistic and genetic studies support White *et al.*'s (1990) idea of a close relationship between the greater habitat diversity and resource richness of the north Australian coast and relatively high population densities, distributed as ecologically bounded, tightly-packed small groups, with high linguistic and genetic diversity (White 1997:73). The predominance of intermarrying of coastal groups is shown by genetic studies (White 1976) to have existed over many generations, indicating that the demographic patterns correlating with environmental factors extend back to pre-European contact (White *et al.* 1990:183).

Thus the archaeological and ethnographic data support the idea of recent cultural change, including small-scale ceremonies being replaced by the large secret male cult *Kunapipi* ceremonies noted at contact, said to be a recent arrival in Arnhem Land from the Roper River region to the south-east (Berndt 1951b:xviii; Evans and Jones 1997:416; Warner 1969:444-5; Yengoyan 1976:127). It could be asked why such large ceremonies and section or subsection systems, described as mechanisms for facilitating alliance and exchange between low density population groups that are spatially or genealogically distant, spread north from the harsh environment of desert areas to the resource rich coast (McConvell 1985; Yengoyan 1976:127)? There is some support for the idea that it could it have been a consequence of population decimation, particularly of women and young children, due to epidemics of foreign disease introduced by Macassan (or other) contact, as Mitchell (1994a, 1995) argues. Foelsche (1882:7-9) suggests that smallpox "the disease most dreaded by the natives" was introduced by Macassan trepang fishers on the north coast hundreds of years ago. He recounts the report of "a native living at Port Darwin", of the last time smallpox appeared around 1860, when so many died "they could not bury them all, but left the corpses lying about....The Port Darwin natives call the disease "Goobimwah", and state it came from the Alligator River tribes, and travelled westward,...no doubt it spread a long distance inland, as pock-marked natives are found among all the inland tribes".

Thus in the Darwin region as elsewhere records indicate the rapid spread of foreign diseases such as smallpox through the Indigenous population well before first direct European contact (Campbell 1983; 1985; Kimber 1988; Mitchell 1994a:415). According to Campbell (1985:349, 357), there may have been fatality rates up to 75% for first smallpox epidemics and 40% for a second outbreak, with the greatest impact on women and young children, leaving surviving Aboriginal populations with a ratio of around 2:1 males to females (and see Kimber 1988). Changes suggested to have occurred in response to large-scale demographic change brought about by epidemics of introduced diseases such as smallpox, include increased male power, particularly in the domain of ceremonial activity, which also intensified, and increased competition for surviving scarce women (Campbell 1985:357-8; Mitchell 1994a:415). In the light of this it is interesting to note Spencer's (1914:193-4, 203) early 20th century comments on considerable modifications to social organisation, totemic systems and marriage customs by northern coastal groups in response to reported decimation of their numbers. Also, it is not unreasonable to assume that earlier contacts also had the potential to bring disease that could have impacted heavily on unexposed populations. There is a high probability that visiting Macassan sailors from the Southeast Asian Archipelago, where smallpox is recorded as endemic from the 16th century, transmitted this disease to north Australian Indigenous groups (Butlin 1985:316).

There is a range of evidence for cultural change in northern and Central Australia centuries prior to European contact, in the last 500 to 1000 years, as Allen (1997:376) argues, from "longstanding but dynamic regional patterns of artefact manufacturing, trade, languages and socio-cultural forms", to widespread social interaction and expansion of large extra-regional ceremonial exchange systems. This includes archaeological evidence for increased populations and ceremonial gatherings, linked to higher environmental productivity through formation of resource rich freshwater wetlands in northern Australia (Jones 1985b). In the central Australian desert carrying capacity is thought to have increased through development of an economy based on grinding of acacia and grass seeds after 1500 BP (Smith 1988b; cf. Veth 1989). Allen (1997:375-6) also outlines ethnohistoric evidence for the recent burgeoning of ceremonies and exchange of exotic goods, and cultural influences on practices such as circumcision, as well as on languages and kinship systems, in response to Macassan and European contacts (eg. Berndt and Berndt 1964:139,424; McConvell 1997:216; Stanner 1933:384,398, 1962-63:244). Evans and Jones (1997:412) liken the spread of large-scale ceremonies such as the Kunapipi in Australia to major types of cultural diffusion elsewhere such as the spread of Islam.

Totemic belief systems

As stated in Chapter Three, according to the anthropologist Elkin (1950, 1970), the recent spread of subsection systems had not been adopted in the north-west corner of the Northern Territory, and the coastal Larrakia possessed no sections. A high degree of linguistic diversity amongst these northern non-Pama-Nyungan language speakers in the coastal and riverine areas of the Top End (McConvell 1997) is interpreted by linguists as indicating a considerable time depth for these languages (Harvey 1997). Development of subsection systems and patrimoieties may be linked to the expansion of the Pama-Nyungun language, whose speakers are known to be predominantly patrilineal (McConvell 1990) and the anteriority of matrilineal dual social organisation in Australia has been demonstrated as logically probable by Testart (1981). McConvell (1985:31) identifies the "Top End" section occupied by the Larrakia and neighbouring groups as an area of matriclans and matrilineal social totemic groups, totemic beliefs being an ideology shared in common with other Australian Aboriginal groups (Spencer 1912:33-4; 1914:204).

As mentioned earlier, totemic belief systems amongst north Australian groups from the northeast coast (Hale and Tindale 1934; Sharp 1939) to the northwest was recorded by Elkin (1978) and Frazer (1937) in the early 20th century as having an economic significance that manifested in totemic sites, where ceremonies or rituals were performed for the maintenance and increase of a species. Sites recorded as "increase centres", which occur for every species of significance to the group, include large stones, stone arrangements or cairns, holes in the ground and mounds of "cockle" shell (Elkin 1978:27-9, Plate 1 opposite p.42,76-7; Frazer 1937:81). Ceremonies for the increase of the totemic species are generally performed annually by the local group, at places where the species is normally abundant, and just before the season in which it becomes plentiful (Frazer 1937:78).

The importance of totemism is that it articulates the Aboriginal concepts of themselves and nature as part of "one moral and psychological system", (Elkin 1969:90), as "human social systems...embedded in the natural world" (McGlade 1994:130). Totemism enables humans and other species and natural happenings to be grouped together as a clan, for example a kangaroo clan includes a line of descent of human beings, and some natural species and phenomena (Elkin 1969:90). Totemism, as "institutionalised religion", as its widespread occurrence amongst Indigenous groups across Australia suggests, could have been a mechanism for cultural evolution of early Pleistocene groups, to societies of increasing cultural complexity (Wenke 1981:113). Wenke (1981:113) argued that such common belief systems facilitated transmission of information – both mundane and ritual, exchange of ideas and people, sanctioning of functional interrelationships and motivation of proscribed behaviour patterns in large numbers of people. If this proposition were believed, then some totemism in all its forms may have had a long history in Aboriginal Australia.

Peterson (1972) proposes that it is a fundamental relationship between Australian Aboriginal totemic beliefs linking humans and other natural beings and phenomena with locality in the landscape, as expressed in totemic practices, which helped maintain foragers in their environment, through its adaptive function as a territorial spacing mechanism. He argues that this mechanism involves people identifying and bonding with a place in the landscape, expressed as cultural symbols such as clan designs and song cycles. In a subsistence economy structured around a population of dispersed groups, and without substantial means of food storage, "The only wealth that can be accumulated is ritual knowledge and wives, the means of production" (Peterson 1972:28). It follows that totemic belief systems and ritual practice, in the creation and upholding of bonds between Indigenous foragers and their country as recorded ethnographically, may have an economic significance that finds expression in terms of occupation patterns and resource procurement strategies. In order to investigate archaeologically the special interaction of past human foragers and their dynamic environment, chronological changes in relative frequencies of different mollusc taxa and in *Anadara* shell size in the three excavated shell mounds is examined in the next chapter.

10 Chronological change in late Holocene resource exploitation

Inter and intra-midden variability and interpretations of chronological change in occupation and subsistence strategies

Introduction

Much has been inferred in north Australian archaeology recently, both on organisation of settlement patterns, subsistence strategies and economic change of forager coastal groups, and on environmental change, based on chronological changes in molluscan fauna in archaeological sites. This applies particularly to changes seen in sites over time, in relative frequency of the mangrove gastropods such as *Terebralia* and *Telescopium*, to the open mudflat bivalves *Anadara* and *Marcia*. Preferred explanations include environmental change, with mollusc procurement strategies said to reflect change in local ecological habitats (eg. Hiscock 1999; O'Connor 1999a), or widespread changes in cultural behaviour, reflecting increased focus on small organisms such as molluscs (Veitch 1999).

Broad scale chronological changes are seen in the archaeological record. Mangrove gastropods such as *Terebralia* and *Telescopium* predominate in surface scatters and the lower levels of stratified middens in western and north-western Australia between 8000 BP and 4000 BP, prior to the appearance of *Anadara* dominated mounds 4000-3000 years BP (Bowdler 1990; Bradshaw 1995; O'Connor 1999a; Veth and O'Brien 1986; Veitch 1999). In Arnhem Land the mangrove gastropods predominate in rockshelter midden deposits near the East Alligator River between 6000 BP and 3000 BP, whilst within this period around 4500 years BP. *Anadara* dominated mounds appear on the coastal plains of the West Alligator River (Mowat 1995:101-6). At Hope Inlet shell and stone scatters of predominantly *Terebralia* on the Hope Inlet saltflats, are on geomorphological grounds possibly less than 1000 years old, while *Anadara* dominated mounds in the area span the period 2000 BP to 550 BP.

O'Connor's (1999a; O'Connor and Sullivan 1994b) interpretation of these chronological changes see the changes in focus on mollusc resources as responses to environmental change on the coast, characterised by an increasing aridity and northward movement of the northwest Australian monsoon. Veitch's (1999) preferred social explanation is linked to the appearance of points in northern Australia, and seen as part of the proposed widespread mid-late Holocene changes in forager economic strategies and population size (see Lourandos 1983, 1985).

Following Hayden (1981a), Veitch (1999) builds a model of an increase in population and sedentism for human groups in the late Holocene, based on a perceived emphasis on mudflat bivalves *Anadara* and *Marcia* that are said to be relatively *r*-selected, as opposed to *k*-selected molluscs such as the mangrove gastropods *Terebralia* and *Telescopium*. This emphasis on *r*-selected mudflat bivalves is linked to an apparent emphasis on *r*-selected grass seeds in the arid zone around the same period. Molluscs and other species that are *r*-selected are relatively small rapid growing individuals that have a relatively short life span, reproduce more rapidly and are invariably a biomass dominant through their ability to rapidly fill suitable habitat niches (Broom 1985:10). They are thus available in larger quantities than slower growing mangrove gastropods such as *Terebralia* and *Telescopium*, which are *k*-selected and produce fewer offspring, invest more energy into their offspring and have a longer life span (Botkin 1980:124-5; Pianka 1970).

Due to the reproductive strategies of *r*-selected mudflat bivalves, they are more resilient than *k*-selected mangrove gastropods to the pressures of human gathering. Population sizes of *r*-strategists are greatly affected by seasonal changes in their environment, and are therefore adapted to recover quickly from stochastic environmental fluctuations such as destructive storms that drastically reduce their numbers (Botkin 1980; Catterall and Poiner 1987; Meehan 1982a). Veitch (1999) argues that people with an economy focussed on *r*-selected molluscs would have a greater potential for relative decreases in mobility than those focussed on *k*-selected species, due to the relatively higher resilience of these molluscs to harvesting.

I would argue that a more significant difference between mudflat bivalves and mangrove gastropods of relevance to long-term environmental change is that the *Potamididae* and *Ellobididiae* mangrove gastropods *Telescopium*, and *Terebralia*, and *Ellobium* and *Cassidula*, are euryhaline. That is, they are much more tolerant of long temporal extremes and wide fluctuations in salinity and temperature than the stenohaline *Arcidae* and *Veneridae* mudflat bivalves such as *Anadara* and *Marcia* (Healy and Wells 1998; Peterson and Wells 1998; Richard Willan pers. comm). The environmental implications of this biological difference in these molluscs are explored below.

Molluscan variability within and between archaeological sites

In this study the relative proportion of mudflat bivalves such as the *r*-selected *Anadara* and *Marcia* to the *k*-selected mangrove gastropods, *Telescopium*, *Terebralia*, *Nerita*, *Cassidula*, *Ellobium* and *Chicoreus*, common to most middens in the area, are calculated from data for excavated sites. The small gastropod *Cerithidea* that dominates the midden deposit at Malangangerr in Arnhem Land (Schrire 1982:89) is not included, as this mollusc occurs only in insignificant quantities in the older mounds HI81 and HI83. *Cerithidea* is found in the most recent shell mound excavated on the saltflats, HI80, dated to the last 700-500 years, and around the outside edges of many other saltflat mounds, and I interpret this as a present day addition brought to the surface of these sites by tidal inundation. This probably relates to local availability of the mollusc as noted by Mowat (1995:150-132).

The results of calculations show that relative proportions of mudflat bivalves to mangrove gastropods are very different for Hope Inlet and Darwin Harbour middens (Table 10.1). For the Hope Inlet mounds the relative proportion of mudflat bivalves to mangrove gastropods is high for middens accumulated around 1800-1400 years BP. In Site HI81 relative proportion of mudflat bivalves to mangrove gastropods increases from 92%:1% in the lower ridge accumulated around 1800 BP, to 97%:0.2% in the upper shell mound, accumulated around 1400 BP. This relative proportion decreases markedly over time to 74%:14% in mound HI80 accumulated around 700 years BP.

For the Darwin Harbour middens spanning 2000 to 700 BP the relative proportion is significantly lower (around 60%:25-30%) than at Hope Inlet. The Hope Inlet saltflats mound HI80, accumulated more recently around 750 to 550 years BP, and Darwin Harbour mounds MA10, which commenced around 400 BP, and the Bayview Haven mound, accumulated around 1400 to 850 BP, exhibit similar relative proportions (around 75-80%:12%). Thus there appears to be no simple relationship between molluscan variability and chronology of these sites.

Two possible explanations can be put forward to explain the molluscan variability in these sites. Firstly, the relative proportion of mollusc species in the middens may be related to different functions of the sites and a reflection of cultural selection and discard practices rather than representative of the environment from which they were procured. Alternatively, the middens may reflect differences in macro-mollusc availability in two different local microenvironments of Middle Arm Peninsula and at Hope Inlet.

There is evidence that the mounds with a much higher relative proportion of mudflat bivalves to mangrove gastropods, HI81 and HI83, may be different site types to the other excavated sites in the Darwin region. Each of these two sites exhibits a much lower rate of shell fragmentation than the other sites and radiocarbon dating indicates they both accumulated relatively rapidly. Statistically at the 95% level the radiocarbon date from the surface of HI81 is the same as the dates from the shell mound HI83, a few hundred metres inland of HI81. This evidence for rapid processing of one mollusc resource and contemporaneity of sites is consistent with a hypothesis of ceremonial status of the large mound complex around 1450 years BP. Similar large mound sites in South Africa, termed megamiddens, have been posited as processing sites for large quantities of molluscs for trade or feeding large aggregations of people during ceremonies, while small mounds and middens are argued to be base camps seasonally occupied by small groups (Jerardino and Yates 1997; Jerardino 1998).

Thus mounds with relatively high relative proportions of mudflat bivalves to mangrove gastropods at Hope Inlet may be large processing sites related to ceremonial activity. It is also possible, if the measure of relative proportions simply indicates different site types, that large mounds around Darwin Harbour similar to those at Hope Inlet may have existed and been destroyed. If the proportions were simply related to size, we would expect a correlation between mound volume and relative proportions, but there does not appear to be any clear relationship between these variables (Table 10.1).

There may be a combination of factors at work here. Ethnographic research indicates that cultural preference plays a significant role in which molluscs are procured. Firstly, of course, macro-molluscs are chosen because they supply a reasonable amount of flesh for consumption in return for the energy expended in their collection. In addition, bivalves are preferred over gastropods due to their availability in large quantities and considered superior taste, and are gathered to feed participants of ceremonial activities (Meehan 1982a:62,70,76,78,142). Bivalves also sometimes provide a higher flesh/shell ratio and protein content than much heavier gastropods (Meehan 1982a:141-2). During everyday mundane activities, ethnographic accounts in Arnhem Land consistently describe gathering of large quantities of *Marcia* and *Anadara*, with variety added by the addition of a few other gastropods from the mangroves (Meehan 1982a:69).

Thus MA1, a small mound ($14m^3$) which radiocarbon dates indicate spanned some 700 years ($0.02m^3/yr$), possibly represents the long-term average composition of local macro-mollusc communities gathered from the Middle Arm mudflats and mangroves between 2050 and 1300 BP (Claassen 1998:134). Mounds HI81 and HI83 may represent the remains of relatively short-term availability of a resource biomass on the Hope Inlet mudflats, intensively harvested for ceremonial feasting and/or trade around 1450 BP. On the other hand HI80 may represent medium-term average composition of local macro-mollusc communities at Hope Inlet around 750 to 550 BP, perhaps as the optimal habitat conditions for the *Anadara* resource were declining (see Chapter Ten).

It is clear that molluscan variability, as represented by relative proportions of mudflat bivalves to mangrove gastropods, varies considerably between some Hope Inlet and Darwin Harbour middens (Table 10.1). At Hope Inlet the variability appears to be related to chronology of the sites, with a marked decrease in relative proportions of mudflat bivalves to mangrove gastropods over time. Similar chronological change in relative proportions of mudflat bivalves to mangrove gastropods in middens can be seen in different areas of the north Australian coast. A study by Hiscock and Mowat (1993; Mowat 1995) on a cluster of three shell mounds at West Alligator Head, shows that the mounds are not contemporary with each other but were built during separate chronological periods. This was correctly predicted by Hiscock and Mowat (1993), who suggested that variability in molluscan composition of the mounds might reflect localised ecological change over time.

Collation of data on the relative proportions of mudflat bivalves to mangrove gastropods in middens on the West Alligator (Hiscock and Mowat 1993:21; Mowat 1995:66), and Bayview Haven (Richardson 1996; Hiscock 1997) are provided in Table 10.1 with data from Hope Inlet and Darwin Harbour. Keeping in mind the relationship between sample size and assemblage richness (Grayson 1984; Rhode 1988), the largest of Mowat's (1995) samples (n=150-250), which are from surface collections, are used. Samples from the Darwin Harbour and Hope Inlet regions are from excavations and are much larger (n=1,700-70,000). As Mowat's (1995) study shows, there is a chronological change between the mound FI-2, dated to the mid-Holocene, which is dominated by mudflat bivalves, and the late Holocene mounds FI-4 and FI-3, which contain a decreasing proportion of bivalves, and corresponding increase in proportions of mangrove gastropods (Table 10.1).

Whilst the data for the Alligator Rivers and Hope Inlet areas each show a directional trend over time, the timing of this differs in each area. At Hope Inlet the marked decrease in relative proportion of mudflat bivalves to mangrove gastropods over time occurs between middens accumulated around 1800-1400 years BP and those accumulated around 700 years BP, rather than between middens accumulated around 4500 years BP and 1200 years BP as at the West Alligator.

The data from the Darwin Harbour area shows differences between separate sections of the shoreline. On one kilometre long section of the Middle Arm shoreline, the relative proportion of mudflat bivalves to mangrove gastropods is very similar for two mounds MA1 and MA7, spanning 2050 to 1300 BP and 1400 to 700 BP respectively (see Chapter Seven). Two other middens, MA10 and Bayview Haven, which also accumulated in different time periods, exhibit similar relative proportions of mudflat bivalves to mangrove gastropods. Site MA10, which accumulated around 400 BP, sits on the northern shoreline of Middle Arm Peninsula, facing the elongated East Arm embayment and, across the harbour, the Bayview Haven midden overlooking Frances Bay (Figure 5.2), which accumulated around 1400-850 BP. It is clear that across the wider region, middens formed around the same time period in different areas do not exhibit the same relative proportions of mudflat bivalves to mangrove gastropods. Yet Darwin Harbour middens of different chronological ages do if they are relatively proximate within the same local environs.

Table 10.1 Chronological change in relative proportion of mudflat bivalves to mangrove gastropods in middens in different areas of the north Australian coast. Data on West Alligator from Hiscock and Mowat (1993:21; Mowat 1995:66,104-9), and Bayview Haven from Richardson (1996) and Hiscock (1997)

Sites	Estimated vol. (m3)	Mudflat bivalves*	Mangrove gastropods*	Calibrated Dates
West Alligator				
Field Island 2 (P 0711): Sample 4	4260	85%	14%	4395 calBP
Field Island 4 (P 0713): Sample 1	118	74%	18%	1243 calBP
Field Island 3 (P 0712): Sample 1	1031	2%	90%	834 calBP
Hope Inlet:				
Hope Inlet 81: Ridge		92%	1%	1814 calBP
Hope Inlet 81: Shell mound	83	97%	0.2%	1427 calBP
Hope Inlet 83	57	92%	2%	1561 calBP - 1448 calBP
Hope Inlet 80	73	74%	14%	733 calBP - 549 calBP
Darwin Harbour:				
Middle Arm 1	14	61%	28%	2056 calBP - 1305 calBP
Middle Arm 7	12	57%	23%	1395 calBP - 742 calBP
Middle Arm 10	7	75%	12%	407 calBP
Bayview Haven: Square 0-113		83%	11%	1395 calBP - 873 calBP

* The mudflat bivalves comprise *Anadara/Marcia/Circe* at the West Alligator sites and *Anadara/Marcia* in the Darwin region

* Taxa included as mangrove gastropods for both areas are *Telescopium, Terebralia, Nerita, Cassidula, Ellobium* and *Chicoreus*. Not all of these taxa are present at each site, in which case only those on the list that are present are included

Table 10.2 Relative percentage of mudflat bivalves to mangrove gastropods

Site	*Anadara*	*Terebralia*	*Telescopium*	*Chicoreus*	*Cassidula*	*Nerita*	% Mudflat bivalves	% Mangrove gastropods
HI49	0	90	1	1	0	2	0	100
HI137	3	20	10	0	0	0	9	91
HI129	0	60	2	0	0	0	0	100
HI138	15	40	10	0	0	0	23	77
HI139	5	22	12	0	0	0	13	87
HI128	0	3	0	0	0	0	0	100
HI112	0	1	1	0	0	0	0	100

Dominance of *k*-selected mangrove gastropods is also seen in surface scatter sites at Hope Inlet, considered on geomorphological grounds to have been deposited within the last 1000 years. Although the sample size is small, taken from one square metre counts, they are representative of the generally low numbers of shells in these site types (Table 10.2). As described in Chapter Five, these sites are situated on the surface of the supra-tidal saltflats in areas of slightly raised topography amongst a network of mangroves, swamps and tributaries that formed during the sedimentary process of build out of the mudflats. This focus on *k*-selected mangrove gastropods occurs within the last 1000 years, following an emphasis on *r*-selected mudflat bivalves, as represented by the *Anadara* mounds, which seemingly proliferated around 1450 BP. If Veitch's (1999:184) logic was followed, the apparent change in focus in mollusc procurement here would represent a decrease in productivity and return to a more generalised foraging pattern, after a period of increased population, greater regionalisation or decreased mobility. I prefer the more parsimonious explanation that the sites represent the partial (visible) remains of resource procurement strategies that were acutely responsive to localised environmental change – people with their finger on the biological pulse, so to speak (Braun 1974:594).

Discussion

The evidence for the Darwin region indicates that chronological trends in relative proportions of the *r*-selected mudflat bivalves to the *k*-selected mangrove gastropods are particular to middens in close proximity within the same local environment. This provides support for Hiscock and Mowat's (1993:24) hypothesis that temporal trends in mollusc procurement are a very localised phenomenon rather than widespread within the north Australian wetlands, as previous studies in Kakadu (Schrire 1982; Jones 1985b; Allen and Barton n.d.) have tended to suggest.

Both short and long term morphological and vegetation change of the Alligator river systems in the sinuous/cuspate phases between about 5300 and 2500 years BP, through ongoing infill and erosion (Chappell and Woodroffe 1985; Davie 1985), would have meant that local availability of certain resources was constantly changing. This availability would have varied in different localities, according to the timing of the environmental change in each river system (Mowat 1995:137,162; Woodroffe *et al.* 1988). Thus between about 4500 and 2500 years BP people camped nearby would have been collecting the *k*-selected mangrove gastropods and bivalves from the mangrove forests that existed at the time in the meandering channel segment on the inside of the sinuous bends of the South Alligator rivers (Woodroffe *et al.* 1988). Within this same period, people would have been gathering *r*-selected mudflat bivalves from the open intertidal flats on the coast of the West Alligator River where they were available (Mowat 1995). In much the same way people in modern times were collecting from the different mollusc environments of mangroves and sandy mudflats in the Blyth River area of Eastern Arnhem Land (Meehan 1982a:59,63).

Chronological change in the relative frequencies of molluscan taxa in these middens strongly supports an interpretation that they represent a succession of short-term responses to local changes in the environment (Bird and Frankel 1991a, 199b; and see O'Connor and Sullivan 1994b). This interpretation does not negate the evidence that suggests some mounds may be a specific site type, or a megamidden (Jerardino and Yates 1997; Jerardino 1998), characterised by rapid accumulation and a significantly higher proportion of mudflat bivalves to mangrove gastropods than that seen in most mounds.

Variability in non-molluscan fauna

Interpretations of economic change in coastal forager groups have also been premised on chronological changes in non-molluscan fauna. Here at Hope Inlet chronological differences in the non-molluscan faunal component also occur between the shell mounds. The saltflats mound HI80, formed post-900 BP, contains an increase in species diversity of fish bone, compared with the mounds on the hinterland and margin, HI83 and HI81, formed between 1800-1400 BP. There is also an increase over time in the average size of both catfish and jewfish, based on otolith

Table 10.3 Changes in size of catfish and jewfish otoliths in Hope Inlet and Darwin Harbour shell mounds

Site	Catfish			Jewfish		
HOPE INLET	wt(g)	n	av.wt(g)	wt(g)	n	av.wt(g)
HI81 (1800-1450BP)	4.8	26	0.2	0.7	2	0.4
HI83 (1550-1450BP)	1.3	6	0.2	0.6	1	0.6
HI80 (750-550BP)	15.6	37	0.4	11.1	20	0.6
DARWIN HARBOUR						
MA1 (2050-1300BP)	1.4	4	0.4			
MA7 (1400-700BP)	0.5	3	0.2			

size. Average otolith size, of catfish increase from 0.2g in both HI81 and HI83 to 0.4g in the more recently formed HI80, and of jewfish from 0.4g in HI81 to 0.6g in HI80, though one jewfish otolith in HI83 is also 0.6g (Table 10.3). Nonetheless overall, unlike the changes observed in some other regions, the fish bone and otolith remains in these mounds increase in size over time.

Some Australian researchers (eg. Bowdler 1976:255; Rowland 1999:141) have associated reduction in the size of fish caught with introduction of fishhooks. In coastal California Broughton (1997) argues that a decrease in large-sized prey–fish and other vertebrates over time in the Emeryville shellmound and other sites is evidence for an increase in human population densities and possible overpredation. In this case the trend is to an increase rather than reduction in size of fish over time, as represented by otoliths in the middens. No fishhooks are present in any remains analysed thus far in this area.

The explanation favoured here is an micro-environmental one, with regard to the site location and availability of fish. At the time HI81 and HI83 were formed, the local environmental context would have differed to the time HI80 formed on the saltflats, some 700 years later, after a further phase of progradation of the Hope Inlet chenier plains (Woodroffe and Grime 1999). Large barra and threadfin salmon remains are found in HI80, accumulated around 750 to 550 years BP. These have been recorded ethnographically as easy to spear when they move on to shallow water on saltpans flooded by king tides (Baker 1981:51-2; Davis 1985:303), but may not have been as accessible to sites HI81 and HI83. These deposits, situated on the hinterland promontory and its margins, accumulated around 1800 to 1400 BP, on what may have been at the time beach ridges overlooking tidal flats (see Chapter Two).

An explanation of local availability is consistent with data on average size of catfish otoliths from the Darwin Harbour mounds MA1 and MA7. Although numbers are small, there is a slight decrease in average size, from 0.3g in the older mound MA1, formed from 2000-1300 BP, to 0.2g in MA7 formed between 1400-700 BP. This is an opposing trend over time to that found at Hope Inlet, suggesting that the differences are specific to the particular environmental context of the sites. Another example in support of this hypothesis can be seen in two sites HI80 and HI66, for which dates around 700 BP were obtained, where the range of fish species in HI80 is much greater than in HI66 (see Chapter Eight), consistent with the more seaward location of that saltflats mound, compared with the more inland sub-surface midden.

Environmental change, human/environment interaction and the appearance of Anadara in the late Holocene

It is important if arguments for the nature of human occupation are linked to environmental changes, to ensure that the reconstructions of past environment and timing of changes in availability of mollusc fauna are as accurate as possible. Interpretations on timing of environmental change and associated subsistence strategies on the Arnhem Land coast are based on relatively few radiocarbon dates. Schrire's (1982:51, 85, 118) ages for middens in rockshelter deposits at Paribari (Badi Badi), Malangangerr and Nawamoyn in the Alligator Rivers area are based on radiocarbon dates obtained on charcoal associated with the midden shell. I have already discussed the problems with using charcoal to date associated material in a deposit. As Beaton (1985:14) says, the best indicator of the use of shellfish is the shell itself, and this is the preferred material for dating (not available at the time of Schrire's work) when the midden material was deposited.

Another well-known site in the Alligator Rivers region, the midden deposit at Malakunanja II, was also dated on charcoal (Kamminga and Allen 1973), and a comparable date on shell obtained, but that date (SUA-264/S1) of 6360±100 BP (Gillespie and Temple 1977:34-5) was on *Pitar* sp. (family *Veneridae*). This shell taxon was also possibly used for another date for Malakunanja II of 4680±110 (SUA-263/S1), also obtained on shell. Other taxa listed by Kamminga and Allen (1973:45-6) in this midden are (mainly) *Cerithidea*, *Telescopium*, *Ellobium*, *Neritina*, *Dosinia*, *Batissa*, *Xanthomelon* and in the top layers freshwater mussel, but no *Anadara*. The date of 6240±100 for Kapalga P on the South Alligator is obtained on several *Telescopium* cemented together (Woodroffe et al 1988:97-9).

I would argue, following Beaton (1985:14), and as Schrire (1982:231) herself says, that the date of 7110±130 BP at Namawoyn rockshelter (Schrire 1982:118) is not a reliable indicator of the first appearance of estuarine conditions on the Alligator Rivers coast. Nor is a date of 6360±100 BP obtained on *Pitar* sp. a reliable indicator of the first appearance of *Anadara granosa* at Malakunanja II (contra Veitch 1999:54). Schrire (1982:231) supports dates of around 6000 BP, obtained for Malakunanja II and Malangangerr, as representing the first appearance of estuarine conditions in the area. Estuarine conditions, in this instance however, would refer to the phase of extensive *Rhizophora*-dominated mangrove forests, the habitat of gastropod taxa such as *Telescopium*, *Terebralia*, *Cerithidea*, *Ellobium*, *Nerita* and the bivalve *Geloina*, which preceded the build-up of mudflats, the habitat of bivalves such as *Anadara*. She (Schrire 1982:120) noted that marine shells such as *Anadara* in midden deposits at Nawamoyn, in the Alligator Rivers area, are present only as rare species. No *Anadara* was recorded at any of the other western Arnhem Land sites, - Malangangerr, Ngarradj, Malakunanja II, and Paribari - where, as at Nawamoyn, mangrove gastropod taxa also dominate the assemblage (Allen and Barton n.d.:88; Schrire (1982:87). As Allen (1986, 1989, 1996) has consistently pointed out, these sites reflect the gathering of molluscs from the landward fringe of the upper tidal limit of mangrove communities.

The earliest date specifically on *Anadara* shell from Arnhem Land is from Field Island 1, an *Anadara* dominated mound located on a beach ridge of the coastal plains of the West Alligator River (Mowat 1995:101). A radiocarbon date of 4570±80 (Beta-58405), calibrated to 4805 calBP. was obtained on this sample (Mowat 1995: 123). A similar date of 4760±90 BP at Princess Charlotte Bay in north Queensland is the earliest date obtained on *Anadara* from the base of an 80cm deep midden deposit in Walaemini rockshelter (Beaton 1985). Although 1500 years after marine transgression ca. 6000 years, identified by Beaton (1985) as a time-lag in economic adaptation to environmental change, this date coincides more closely with the beginning of chenier formation ca. 4000 yrs (Beaton 1985:9). Cultural *Anadara* mound deposition begins around 2000 years BP and continues to 500-600 years BP, around the same period as chenier formation ceased (Beaton 1985:5, see also Chappell and Grindrod 1984:205).

A decade earlier than Beaton's (1985) work, an analogous situation in Eastern New England of the United States, of apparent time-lags in economic adaptation to environmental change, are argued, based on palaeoclimatic, palaeoecological and archaeological data, to be actually closely linked to geomorphic change (Braun 1974). Research in this region showed mollusc exploitation appearing from 5000 BP, while shell midden formation began around 4000 BP in some areas, with increased size and frequency of middens around 2500 BP. Changes in patterns of mollusc exploitation are seen to directly and closely reflect changes in species availability due to local environmental change in suitable habitats. Braun (1974:594) argues that there is an immediate response to local environmental change and exploitation of new resources that indicates "closely ecologically constrained, fairly flexible adaptive strategies".

There is no unequivocal evidence in the form of direct dating of this shell, that *Anadara granosa* was exploited or even an available resource in northern Australia prior to 5000 BP (contra Veitch 1999:54). However, the mollusc *Pitar acuminata* is considered a good indicator species for selecting suitable areas to establish culture beds of *Anadara* in Malaysia (Pathansali 1966:91). Perhaps the *Pitar* sp. collected and deposited at Malakunanja II around 6000 BP heralded the onset of suitable habitat conditions for colonisation by *Anadara* some 1000-1500 years later. I share Cribb's (1986) opinion that if due consideration is given to environmental factors at Princess Charlotte Bay, utilisation of *Anadara* begins around 4000 BP., after eustatic stabilisation of the coast and suitable habitat conditions for this mollusc had established. Intensive harvesting begins around 2000 BP, when optimal habitat conditions had developed for proliferation of *Anadara* as a resource biomass (cf. Beaton 1985).

The time of the mudflat bivalves

On the north coasts of the Top End the dominance of sites by either *r*-selected mudflat bivalves or *k*-selected mangrove gastropods can be related to their localised ecological setting. Thus the East Alligator River rockshelter midden deposits dated between 6000 BP and 3000 BP are dominated by mangrove gastropods from the nearby habitat of mangrove forests that existed at the time. Three open midden sites, Field Island 2, 3 and 4, situated on the West Alligator coastal plains close to the mouth of the river exhibit a change over time, from dominance by *r*-selected mudflat bivalves around 4500 years BP to dominance by *k*-selected mangrove gastropods by around 850 years BP. This change reflects very localised change in ecological conditions around the site locations, from open coast to mangrove forests, with infilling of the river valley and possibly progradation increasing the distance of the sites from the shoreline (Mowat 1995:138).

Similarly at Hope Inlet, open midden sites exhibit a gradual decrease in dominance by *r*-selected mudflat bivalves between 1800 and 750 BP with continued progradation of the coastal plains and expansion of the local mangrove forest. Around Darwin Harbour there appears to be no clear change over time, but similar relative proportions of mudflat bivalves to mangrove gastropods for middens found in the vicinity of the same section of shoreline. Surface scatter sites of stone and shell on saltflats formed within the last 1000 years are more markedly dominated by *k*-selected mangrove gastropods from the mangrove forests, after the passing of this time of the mudflat bivalves.

The evidence from this study, and from direct dating of *Anadara* shell, that this mollusc was not exploited or even available in northern Australia prior to 5000 BP, provides support for the explanation that its earlier absence in archaeological sites is because suitable ecological conditions did not exist at the time. Alternatively, the material remains have not survived, are buried or are under the sea. The data fails to uphold the case for alternative explanations of increased populations and demographic or social pressure, leading to changes in mid-Holocene resource procurement strategies to include an increased focus on small organisms such as mollusc bivalves (Bailey 1983, 1993:15; O'Connor 1999a; cf. Beaton 1985; Veitch 1999).

I argue that, due to the intimate knowledge foraging peoples customarily have of their country and thus of possible net returns from food resources in their local area (Buchanan 1988:101), their mollusc procurement strategies are characterised by a focus on any available resource biomass with the addition of small quantities of less abundant resources. This is observed ethnographically (eg. Meehan 1982a) and is consistently reflected in the archaeological mollusc assemblages. Evidence from changes in fauna in these sites indicate that exploitation of *Anadara* represents a continuation of past, well established localised resource procurement strategies, which are intimately responsive to seasonal cycles and longer term changing environmental conditions over thousands of years (Freslov and Frankel 1999; Rowland 1983). Below I explore chronological changes in *Anadara* shell size in deposits in this area as another indicator of the nature of the relationship between these past foragers and their environment.

Shell size, human exploitation and environmental change

Introduction

A variety of biological factors are known to affect shell size of molluscs (see Claassen 1998:133-137). Spennemann (1987) discusses some of these factors individually with regard to Tongan mounds in the South Pacific. They include seasonal changes, the availability of calcium carbonate in the water, water temperature, salinity, stand location and bottom and the population density of the mollusc itself. These environmental variables are considered below within the constraints of the limited long-term data available for the local Darwin area. The relative role of environmental factors and human exploitation is taken into account by considering coarse-grained data on short and long-term trends in shell size in conjunction with other data such as shell fragmentation and any changes in relative proportions of other exploited molluscs or of other resources.

This study follows previous research elsewhere on cultural shell deposits, which use shell size as an indicator to examine extent of human exploitation of molluscan resources and/or environmental change over time (eg. Jerardino 1997; Spennemann 1987; Swadling 1976). For this analysis, average weight per shell (giving an indication of shell size) over time, as represented by vertical distribution through the three shell deposits, was calculated for each spit of the three excavated mounds (Figure 10.1). This was done by dividing the weight by the minimum number of individuals (Wt/MNI), of unbroken *Anadara* shell (either left or right valves) from each spit. Confirmation of this method was obtained by measuring maximum valve length (to the nearest mm, following Spennemann 1987:85), of whole *Anadara* shell from three spits, at the base, middle and top, from each of the shell mounds (Figure 10.2), which correlated with these calculations, showing the same trends in average shell size (see below).

Shell size and human exploitation

Results of the above method show an alternating variation in shell size over time for each of these mounds (Figure 10.1). Spennemann (1987:86-7) also notes this pattern for Tongan middens, which he suggests may be due to seasonal changes. This interpretation would only apply if the shells were collected at a variety of times during the year (seasons). The mound would have to accumulate sufficiently fast so that shells from one season formed a layer with enough depth to distinguish it from the next layer collected during a different season. Given the dating and rate of fragmentation for these mounds, this is a possibility (see below).

It is also possible that longer-term cyclic environmental change (climatic) is responsible for the fluctuation in shell size noted in the deposits. Molluscs are known to be sensitive environmental indicators, and the oscillating pattern exhibited by average shell size in these mounds over time may reflect changes in the ecosystem of which they are a part. This oscillating pattern resembles that described by Butzer (1982:21-3) as "Steady-State Equilibrium with no net change in equilibrium level"; a type of steady state maintained by negative feedback, characteristic of natural ecosystems.

Significantly, the results clearly show overall trends in *Anadara* shell size during the accumulation of these mounds. In the sample from HI83 in the woodlands, shell size remains more or less constant over time, while in both HI81 on the hinterland margin and HI80 on the saltflats there is a definite trend of decreasing shell size over time (Figure 10.1). At first glance this would appear to suggest over-exploitation in the latter mounds. However, further analysis suggests that this is not necessarily the case.

Data from Pathansali (1966) on modern *A. granosa* populations in Malaysia, subjected to continuous harvesting which removes larger individuals, has been used by Broom (1985:15) to derive average maximum shell lengths of 30-36mm. Measurement of maximum *Anadara* shell length (see Spennemann 1987:85) from spits at the base, middle and top for each of these mounds confirms the trends on

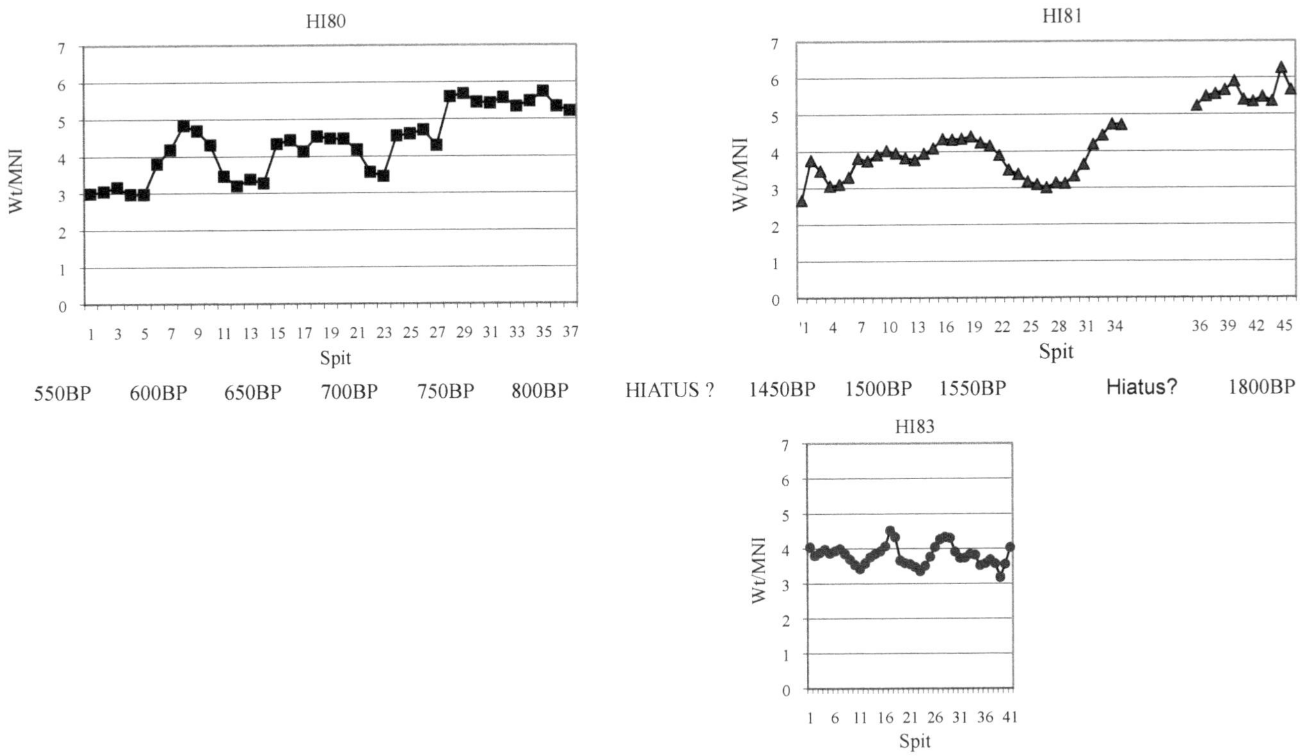

Figure 10.1 Wt/MNI of *Anadara* shell for Sites HI81, HI83 and HI80, showing trends over time

shell size obtained through the Wt/MNI estimates, but also demonstrates that average shell length remains between 27-36mm throughout in each of the mounds. Nor is there any significant decrease in standard deviation, which stays relatively constant throughout (Table 10.4; Figure 10.2).

Whilst Broom's (1985) paper deals with growth models and not mean shell length as in this paper, the closeness of average shell length in the mounds to the maximum possible length of shell of *Anadara* populations, indicates that selective harvesting has been practised. In *A. granosa* molluscs, gonads mature at 18-20mm length (Broom 1985:23). Harvesting of shells close to maximum possible size ensures maturity and breeding success.

When the size measures per spit for each mound are placed in the context of dates obtained for the mounds, it is clear that the changes in shell size are short-term trends specific to each mound (Figure 10.1). For HI81, average shell length measures 35mm at the start of shell accumulation in the laterite ridge ca 1800 BP, which decreases over time to 32mm and then to 29mm in the basal and top layers respectively of the upper shell mound, dated to ca 1400 BP. Thus shell size in the upper layers of HI81 is slightly less than the average length of 30mm throughout accumulation of HI83, dated to around 1450 BP.

Since the dates for the surface of HI81 and for HI83 are statistically indistinguishable at the 95% level, within the limits of radiocarbon dating accumulation of HI83 overlaps with the terminal accumulation period of HI81. Over 500 years later sometime after 900 BP, average shell length has increased again to measure 34mm at the start of accumulation of HI80. This decreases over time to an average shell length of 32mm in the middle of the deposit ca 600-700 BP, and to 28mm in the terminal accumulation period, around 550 BP (Figures 10.1 and 10.2).

Furthermore, calculations show that in HI83 in the woodlands, the proportion of individual *Anadara* shells larger than the average maximum of 36mm is small at the start of accumulation and remains relatively constant over time. For both HI81 on the hinterland margin and HI80 on the saltflats, the proportion of shells larger than the average maximum is relatively high at the start, and drops sharply during mound accumulation (Table 10.5).

The data for HI83 is consistent with a source population subjected to continuous harvesting, which has removed the larger individuals, prior to and during the formation of the mound. For mounds HI81 and HI80, formed before and after HI83 respectively, the data are consistent with a hiatus in harvesting prior to their commencement, which allowed the growth of larger shells. This may indicate reoccupation of the area after an absence of more than 15 months, given a growth rate of around 15 months for *Anadara* in natural beds to reach a size of 18-32mm (Broom 1985:14). In the case of HI81, the large average shell size in the lower ridge may indicate the start of harvesting of the Hope Inlet *Anadara* beds in this area. Further dating of other mounds in the area would be required to test this hypothesis.

It becomes apparent too, when placing the patterns of oscillating shell size, within time frames as in Figure 10.1, that more dates on each mound would give a clearer picture of mound formation. Yet it must be remembered in interpretations of dates, that there is an inherent margin of error in radiocarbon dating of shell, related to exchange of

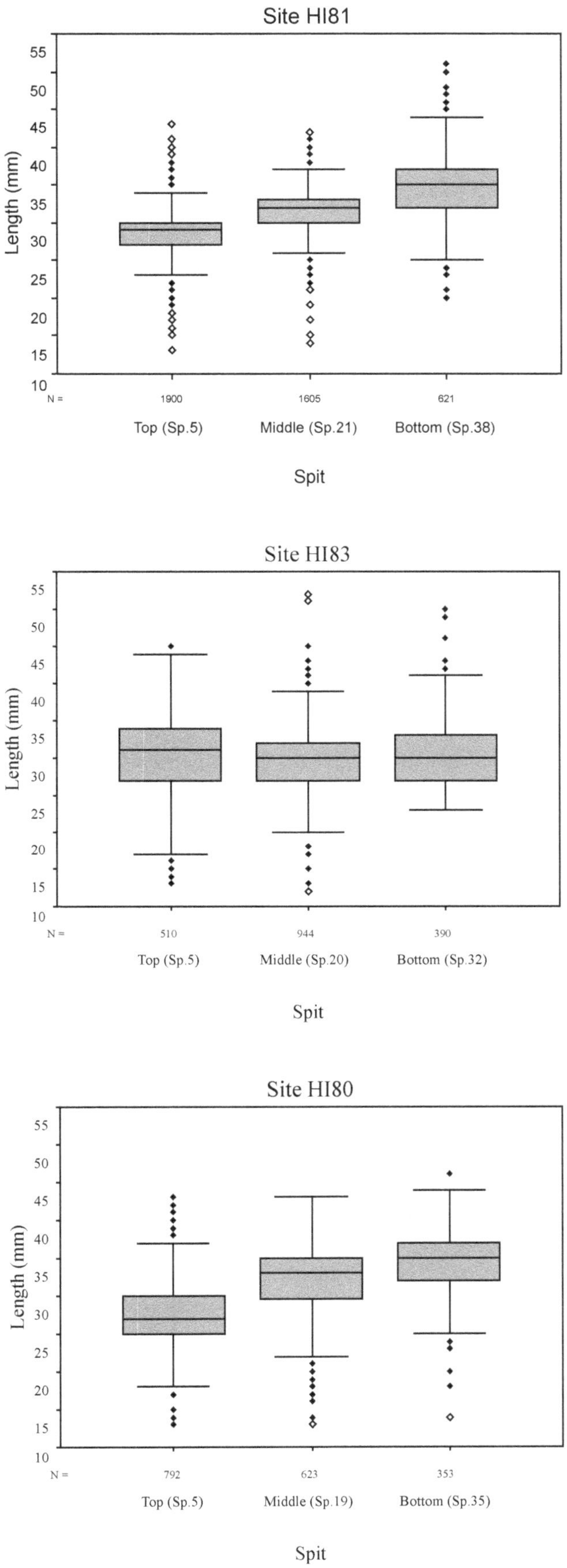

Figure 10.2 Size distribution of *Anadara* shells in Sites HI81, HI83 and HI80, showing the median, interquartile range, outliers and extremes for spits at the top, middle and base of each mound

carbon with the environment (Head *et al.* 1983; Higham and Hogg 1995; and see Ward 1994). I previously interpreted radiocarbon estimates of 1448 calBP and 1369 calBP respectively on the stratigraphically secure shell/charcoal pair from the base of mound HI83 as indicating that the mound accumulated rapidly around 1450 years BP, despite an estimate of 1561 calBP for a surface sample of shell. This was because of the close correspondence of the basal dates and the more negative stable isotopic composition of the also stratigraphically secure surface shell, suggesting that the older date may be due to increased freshwater influence (Head 1991). The low rate of shell fragmentation through the mound also supports an interpretation of rapid deposition.

Thus the placement of the graph for HI83 between 1450 and 1550 BP is probably an overestimate of the time in which this mound accumulated. Similarly, the graphs for HI81 and HI80 probably stretch out the time of accumulation for what were in all likelihood repeated short-term depositional episodes interrupted by hiatus periods. Nonetheless, the graphs are useful in providing an illustration of short-term oscillating shell size and the way in which an apparent overall trend of decreasing shell size over time in one mound is temporary and cyclical, repeated again in a later mound some 700 years later.

Future research to determine if the alternating variation in shell size over time for each of these mounds is due to seasonal changes, could include oxygen isotope analysis. Unfortunately, although this procedure has been shown by various studies to be useful in southern Australia (Godfrey 1988, 1989) and other parts of the world for determining seasonality (eg. Claassen 1986, 1998:152; Cohen *et al.* 1992; Higham 1996; Kennett and Voorhies 1996; Quitmyer and Jones 1997), the cost and uncertainty of its suitability for tropical regions such as Darwin prohibited its use in this one-woman project. Spennemann (1987:87) points out that seasonality studies for tropical fish based on annual growth rings on otoliths have been complicated and less accurate than for temperate fish, because of lower variation in temperature (cf. Fowler 1992). Similarly, tropical coast molluscs subject to estuarine influence may be unsuitable because of the variation in salinity and effects of rainfall and evaporation and lower variation in temperature (Higham 1996:248; Michie *et al.* 1991:4-5). Veitch (1999:69 citing Godfrey pers. comm.) also reports on problems with oxygen isotope analysis for another *Anadara* taxon, *A. trapezia*, related to a seeming lack of differential absorption of O^{16} and O^{18} ratios with temperature variation. Availability and use of this procedure for projects such as this in Australia awaits further research and funding.

My interpretation of the data on shell size, dating and shell fragmentation rates, is that following the commencement of shell deposition on laterite ridge of mound HI81, there was rapid, continuous accumulation of the one metre deep upper shell layer of HI81, and of mound HI83 around the same period. The most recent mound, HI80, appears to have accumulated over a longer period than those two

Table 10.4 Mean length (mm) and standard deviation of *Anadara* shell from spits at the top, middle and base of each mound

Site	Top		Middle		Base	
	Mean	SD	Mean	SD	Mean	SD
HI80	28	4	32	5	34	4
HI83	30	5	30	4	30	4
HI81	29	3	32	3	35	5

Table 10.5 Proportion of *Anadara* shell >36mm from spits at the top, middle and base of each mound

Site	Top	Middle	Base
HI80			
no. of shells	792	623	353
no >36mm	20	86	104
%>36mm	3%	14%	29%
HI83			
no. of shells	425	944	390
no >36mm	38	52	34
%>36mm	9%	6%	9%
HI81			
no. of shells	1900	1605	621
no >36mm	45	76	197
%>36mm	2%	5%	32%

mounds, though still relatively rapidly, when compared with the mounds on Middle Arm Peninsula, Darwin Harbour. Accumulation of HI80 began after a hiatus period, which, I would argue represents a relatively short interval between occupational events in the area, of years rather than centuries. The apparent hiatus in occupation between 1400 BP when HI81 and HI83 had ceased, and after 900 BP when HI80 was formed, would in all likelihood be filled in if more dates were obtained on other mounds in the area.

These data suggest sustainable continuous exploitation of this molluscan resource for around 1500 years. Notwithstanding this, it is possible that the smaller mean shell size during terminal accumulation of the more recent mound HI80 on the saltflats, indicates more pressure on the *Anadara* resource. The overall decrease in average *Anadara* shell size over the time period between 1450 BP and 550 BP, with mean length decreasing from 30mm in HI83 to 28mm in the upper layers of HI80, represents a reduction of 7%, which presumably corresponds with a reduction in return of flesh weight (see Meehan 1982a:142; Siegfried *et al.* 1985:306). Cessation of the mound building phenomenon soon after this mound was accumulated lends weight to this impression. If it is so, this may be due to human predation or environmental change, or a combination of both.

Mollusc habitats and environmental change

Populations of *Anadara granosa* beds sufficient for the construction of the Hope Inlet mounds no longer occur in the Darwin region. The predominance of *Anadara* deposited in large numbers of mounds in this small area suggests that extensive *Anadara* shell beds once existed which could be easily and heavily exploited. Hiscock (1997) has argued that the cessation of the mounds indicates the environment changed from the optimal habitat conditions for *Anadara*, of open sandy mudflats with scattered stands of mangrove, to the extensive mangroves and mud-flats which form a barrier hundreds of metres wide along most of the coastline today.

Some of the other mollusc species recorded in the mounds in this study (Table 9.2) are also now either extinct or rare in the local Darwin area, possibly also as a result of environmental change and/or degradation and over-predation by humans. For example, *Cymbiola flavicans*, a colourful, spotted gastropod collected in small quantities by the Anbarra in the 1970s, from open-sea beds in Arnhem Land, was reported in Shoal Bay four decades ago (McMichael 1960), but not seen since cyclone Tracey in 1974, and is considered locally extinct in the Darwin area (Richard Willan pers. comm.). This mollusc is present in small quantities in mounds HI81 and HI83, dated to around 1800-1400 BP, but is absent from HI80 accumulated some 700 years later. Of the over 100 mollusc taxa recorded in the present day local environment around Darwin (Hanley 1988:145), the most common macro-molluscs suitable for human consumption are the mangrove gastropods *Telescopium*, *Terebralia* and *Nerita* (Caldwell 1983:76; Smith *et al.* 1997).

Detailed knowledge on the ecological parameters of molluscs facilitates archaeological interpretation of middens that are based on perceived responses of the molluscan resource to environmental change. Recently, the ecological context surrounding the formation of the

Anadara mounds of northern Australia, and the implications for past coastal economies has been the subject of some speculation and debate (O'Connor 1999a; Veitch 1999). In order to contribute to this debate, below I provide relevant data on the biology of *Anadara* and other molluscs found in the middens. Unfortunately there is a lack of comparative living populations of *Anadara granosa* and corresponding absence of ecological studies of this mollusc in the Darwin region, so the data are drawn from work in Southeast Asia (eg. Broom 1985; Pathansali 1966), where *Anadara* is grown commercially.

Whilst *Anadara* is known to tolerate a range of conditions, proliferation to abundance would require an optimal habitat setting. Shore elevation, slope of bed and substrate type are some known factors that affect successful *Anadara* colonisation, population density and shell size (Broom 1985; Pathansali 1966:91). Broom (1982:137) notes that at *A. granosa* beds on Malaysian mudflats there was a black, sulphide-rich layer 3-4cm below the surface, and comments (Broom 1985:13) that sandier substrates represent a sub-optimal habitat for this species. *Anadara* bivalves are slow burrowers, preferring the mid-littoral zone in relatively stable, fine sandy/mud silt substrates with little or no current or turbulence (Dodd and Stanton 1981:89; Richard Willan pers. comm.).

Along any section of the coast a number of microenvironments can be found on the mudflats and within the mangrove zone, each supporting different mollusc taxa. For example, the marine bivalve *Marcia hiantina*, often found in association with *Anadara granosa* in middens, occupies a slightly different habitat (Meehan 1982a:62). *Marcia* inhabits coarser, more sandy substrates in high currents in the sub-littoral zone of open sandy beaches (Richard Willan pers. comm.). Another species of *Anadara, A. inequivalvis*, found only in the *Marcia* layer in site HI80, dated to around 700 BP, is also found intertidally in sandy areas (Lim 1966, cited in Broom 1985:5). The oyster *Saccostrea* lives in rocky areas, or on the roots of mangroves like the mangrove tree oyster *Melina*.

Zonation of mangroves and the different molluscs that inhabit them (Reid 1985; Wells and Slack-Smith 1981), and recognition of this habitat zonation by Aboriginal people are well known (eg. Davis 1985:304-5). For example the common gastropod taxa found in the mounds (*Ellobium, Cassidula, Chicoreus, Nerita, Telescopium Terebralia* and *Cerithidea*) each inhabit particular zones of the shady/muddy mangrove forests. *Terebralia* live on firm muds in the upper intertidal seaward fringes amongst *Bruguiera* and *Ceriops* and sometimes *Avicennia* mangroves, associated with drier substrates. *Telescopium* occupy the upper to lower intertidal, on soft muds near pools or creeks, and in *Rhizophora* forests associated with wetter substrates. *Nerita* are found on mangrove roots just above high tide level, and in rocky intertidal and sub-tidal areas and *Chicoreus* live in the intertidal level on mangrove roots and branches. *Cassidula* and *Ellobium* live around and on mangrove roots at the lower levels of the landward fringe and extend down almost to the seaward edges of the *Bruguiera* and *Rhizophora* forests. *Cerithidea* inhabit the high intertidal landward fringe of mainly *Avicennia* and *Bruguiera* mangroves (Crowe 1997; Macnae 1968:218; Wells 1980; Wells and Bryce 1988).

Like the other molluscs represented in the middens, *A. granosa* thrives in tropical areas with an estuarine influence. Unlike some of the mangrove gastropods however, *Anadara* does not tolerate long-term extremes in salinity. Areas colonised by *Anadara* usually have a salinity range of around 26 to 31 parts per thousand (ppt), and are tolerant of large fluctuations following monsoon rains, for a short-term only (Broom 1985:6). Pathansali (1963, cited in Broom 1985:6) showed through experiments that *Anadara* species tolerate salinity levels down to 14 ppt in the short term, but persistent values below this level lead to significant mortalities, as may very high values of 40ppt (Broom 1985:7). The highest population densities of this mollusc are found in the fine soft brackish muds of the open intertidal mudflats bordering, but not within, mangrove swamp areas, and near but not in the mouths of large rivers (Broom 1982; 1985; Pathansali 1966). This microenvironment provides the optimal conditions required for prolific *Anadara* beds; of soft, brackish (salinity between 26 to 31 ppt), fine silt mud substrate, intertidal or marginally subtidal elevation level with a particular slope of bed (Broom 1982:136-7), and temperature of 25 to 32°C.

Although seasonal salinity depression is linked to spawning of *A. granosa*, there are limits to the extent of salinity drop tolerated, observed to be around 18 ppt, and down to 12 ppt for short intervals only for early spat (Broom 1982; 1985:6, 19; Pathansali 1966:90-91). Distribution of larvae during spatfall, which settle between mean high water of neap tides (MHWN) and mean low water of neap tides (MLWN), with peak densities around midtide level (Broom 1985), was observed to be more restricted in areas where there are extensive mangrove forests and numerous tidal creeks that discharge brackish water (Pathansali 1966:6). Pathansali (1966:89-90) noted that in areas where the mangrove belt is narrow salinity remains at higher levels for most of the year except following the wet season rains, when large fluctuations occur over short periods. Thus the effect of the annual monsoons on salinity levels and thus on *Anadara*, may be influenced by the extent of the local mangrove forest.

Formation of habitat conditions suitable for Anadara in the Darwin Region

The type of tide-dominated shoreline characterised by mangroves and adjacent muddy substrates inhabited by *Anadara* beds is influenced by complex local factors, including both seaward deposition and terrestrial sedimentation, and the geometry of the shoreline (Woodroffe 1995). A broad picture of the processes that led to suitable conditions for *Anadara* colonisation is known through the work of Chappell and Grindrod (1984; Grindrod *et al.* 1999:468) and Woodroffe *et al.* (1985, 1988, 1993). Following the last transgression, extensive mangroves developed relatively rapidly on the low gradient north Australian shores. As sea-levels stabilised continued fluvial and marine sedimentary accretion within estuaries resulted in replacement of widespread mangroves with fringing mangroves along estuarine coasts and rivers, with infilling of bays becoming progressively shallower and build out of intertidal mudflats suitable for the establishment of molluscan shell beds, including *Anadara granosa.*

The timing of conditions favourable for *Anadara* colonisation at Hope Inlet can be postulated from archaeological and geomorphic data. The earliest radiocarbon dating of shell from the base of mound HI81, suggests an age range of 1614 to 1986 calBP, some hundreds of years after the phase of progradation of the chenier plains at nearby Camerons Beach around 2300 years BP (Hickey 1981; Woodroffe and Grime 1999). However, Hickey (1981) dated the Camerons Beach chenier ridges using shell identified as *Arcanthocardia tuberculata*, a misspelling and misidentification, as the cockle *Acanthocardia tuberculata* is found from England to the Mediterranean and not in the Indo-Pacific region (Abbott and Dance 1982:327; Richard Willan pers. comm.). Specimens no longer exist for verification, but the shells were possibly small *Anadara granosa*, which are morphologically very similar and are found in the Shoal Bay chenier ridges (Woodroffe and Grime 1999:309). If this were so, the corrected radiocarbon dating results of 2350±120 and 2340±120 years BP obtained on this shell would place the establishment of *Anadara* beds at Shoal Bay soon after progradation of the chenier plain began since 2500 years ago (Woodroffe and Grime 1999).

Environmental factors in the decline of* Anadara *at Hope Inlet and Darwin Harbour

Geomorphic evidence indicates that sedimentation processes that built the habitat conditions suitable for *Anadara* colonisation at Hope Inlet began 2500 years ago. Over time, these processes, which are currently continuing in Shoal Bay (Woodroffe and Grime 1999) would have gradually changed the gradient of the coastal plain, and slope of the shore, eventually reducing tidal inundation (Woodroffe *et al.* 1986), and changing the conditions under which *Anadara* thrived to those seen today. Similar explanations of progressive infilling of the bay, for the demise of large *Anadara* shell beds, have been put forward for Princess Charlotte Bay and at Weipa (Bailey 1999; Beaton 1985; 1986; Chappell and Grindrod 1984). Local Aboriginal people in the Darwin area reported that when they were children (some 30 years ago) they knew of beds with enough *Anadara* to collect a couple of buckets, but that even this quantity had declined in recent years.

Continued accretion of the mudflats and subsequent colonisation by *Avicennia* mangroves still occurs today in the Darwin region (Woodroffe and Grime 1999:319), and would explain the gradual decline of *Anadara* with elevation of the substratum and gradual isolation of the shell beds from the tides. Seaward mangrove growth on an accreting shoreline has been observed to rapidly enroach on *Anadara* culture sites on a Malaysian foreshore (Macintosh 1982:13). As part of this process an important factor that could have contributed to the rarity of *Anadara* shell beds in the present day Darwin region is changes in salinity levels. A study on seasonal variation in salinity and temperature around Darwin Harbour and in Shoal Bay from 1972 to 1978 recorded salinity ranges consistently higher than 31 ppt for seven months of each year. For example, a range of between 34 and 37 ppt was recorded at Middle Arm, and between 32 and 38 ppt at Shoal Bay for the months of June to December in 1973 (Michie *et al.* 1991). Whilst remembering that salinity levels vary along any section of coast (Michie *et al.* 1991:4-6), these levels are consistently higher than those of between 26 and 31 ppt recorded as favourable for *A. granosa* in Malaysia.

Thus a combination of factors today in the Darwin region may produce a difficult environment for molluscs as well as for polychaete worms as suggested by Hanley (1985:248). These factors include low freshwater input in the dry, and a shore topography and substrate type that prevent rapid drainage of seawater after high tides, creating hypersaline conditions after evaporation, and flooding with freshwater of these areas in the wet (Michie *et al.* 1991:5). Hanley (1985) found that in Northern Territory mangroves there is a greater proportion of generalist polychaete species, which have a high tolerance to a wide range of salinity and temperature conditions. Acidity of the substrate may also be an related factor in the dearth of *A. granosa* (Linnaeus, 1758), as Wells and Slack-Smith (1981:273) suggest for Admiralty Gulf, West Australia, where similarly the presence of large middens indicate a previously large population of *Anadara.*

Transformation of the Hope Inlet Estuary?

Evidence of the diverse range and abundant mollusc taxa in the Hope Inlet middens between 2000 and 500 years BP suggests that during this period the estuary was open to the sea and tidal to some distance inland (Michie *et al.* 1991:4; Peterson and Wells 1998:40-41). Recorded in the mounds may be the transformation of the Hope Inlet estuary from a tidal to river-flow dominated ecosystem, through sedimentation processes and progressive isolation from the sea, with increasing hypersalinity, which was previously limited by tidal flushing (Hodgkin and Hesp 1998). As stated above, the *Potamididae* and *Ellobididiae* mangrove gastropods *Telescopium* and *Terebralia* and *Ellobium* and *Cassidula* are euryhaline, and much more tolerant of long temporal extremes and wide fluctuations in salinity and temperature than the stenohaline *Arcidae* mudflat bivalves such as *Anadara* (Healy and Wells 1998; Peterson and Wells 1998; Richard Willan pers. comm.).

Darwin Harbour differs in morphology from the Hope Inlet estuary, being a much larger drowned river valley (Semeniuk 1985), and so would follow a different infilling evolution (Kench 1999:374). The smaller difference in relative proportions of mudflat bivalves to mangrove gastropods exhibited for Darwin Harbour middens over a long period of time, may reflect more gradual development of this estuary. Environmental conditions seen today may be similar on a gross level to those of the late Holocene, though clearly not in terms of habitats favourable for *Anadara*. Today the low freshwater input from the main rivers such as the Elizabeth River into Darwin Harbour except for a few months in the wet season, results in hypersaline conditions, and a hypersaline barrier occurs in the embayment zone where the three arms of the harbour converge. More extreme salinity ranges have been recorded for Hope Inlet due to the localised impact of freshwater run-off during the rainfall months and hypersaline conditions at the end of the dry (Messel *et al.* 1981; Michie *et al.* 1991:4-5).

Thus the decline in *Anadara* and relative increase in the salt-tolerant mangrove gastropods over time is consistent with gradual establishment of hypersaline conditions for these estuaries as described by Hodgkin and Hesp (1998) for the estuaries of south-west Australia. A much less diverse range and abundance of macro-mollusc fauna present in the estuaries today than represented in the middens in the past, suggests this process may be ongoing. Also consistent with this proposition is the possible decrease in abundance of estuarine fish species (see Sheaves 1998), though a lack of any detailed studies frustrates corroboration.

Why mounding and why did it stop?

Whilst environmental change may explain the cessation of a particular resource such as *Anadara*, it does not necessarily follow that the cultural behaviour of mounding shell stopped as a result of this particular mollusc no longer being available. Mounds dominated by other molluscs are known on the north Australian coast. For example, in the Blyth River area on the Arnhem Land coast, Meehan (1982a, 1983) noted mounds dominated both by *Anadara* and by bivalves of the Family *Veneridae*, *Dosinia juvenilis*. Meehan (1983:15) obtained a corrected radiocarbon date of the base of the Kula Kula mound as it is called, of 1440±100 BP (ANU-2024), making it contemporary, if separated geographically, with the Hope Inlet *Anadara* mounds. *Dosinia* are rapid, deep burrowing molluscs that occur in sand and silt areas of good current flow (Harte 1998:358), and were reported by Meehan (1982a:99) to be collected from the mid-littoral zone.

An environmental explanation for the cessation of mounds in this area could be explored through future investigations. If mound building ceased because the main resource, *Anadara*, was no longer available in the required quantities, then it could be asked why did mound building not continue using another resource? Dating of a series of mounds composed predominantly of the mangrove oyster *Melina*, on the banks of the Elizabeth River may also help answer this question. If these mounds were built after cessation of formation of the *Anadara* mounds, then they would provide evidence that inhabitants continued the practice of mound building with molluscs other than *Anadara* in this area.

Regional correlations in the timing of mound building

Thus far the data from this project supports an ecological explanation for cessation of mound formation here as elsewhere in north Australia (Bailey 1999:110; Beaton 1985; 1986; Chappell and Grindrod 1984). Dating of the surface of these mounds extends Hiscock's (1997) estimate of a terminal period of mound building from 700-1000 years, to around 500 years BP., correlating roughly with the cessation of mound building on Cape York Peninsula (Bailey 1999). The correlation in timing of cessation of mound building around the same time – by 500 years ago - across the northern coastline of the continent is consistent with an hypothesis of regional environmental change as a primary cause.

Moreover, differences in timing of mound building periods in other parts of the world is consistent with a general explanation of local environmental conditions as a determining factor in when they appear. For example, the large mounded sambaquis (shell middens) of Brazil, dated between 5500 to 1000 BP, have been shown to be closely associated with changing positions of past shorelines over thousands of years (Suguio *et al.* 1992). In Elands Bay, on the South African South West Cape Coast, Buchanan (1988) notes three occupation and settlement pattern phases linked to environmental change, which bear some similarity to the sequence seen in northern Australia. Middens occur from 4400 to 3000 BP in caves, megamiddens between 3000 and 1800 BP, and from 1800 to 300 BP, there is a proliferation of ephemeral midden sites, often on hills with a view (Buchanan 1988:196-7).

Interesting parallels can also be made with the proliferation of shell mounds in Japan, some of which are dominated by *Anadara* spp., during the late Jomon period between about 4500 and 2300 BP. The growth and decline of these site types has been shown to be related to environmental change specific to the Pacific coasts of eastern Japan during that period (Akazawa 1986a, 1986b; Koike 1986; Suzuki 1986). In a number of publications, Akazawa (1986a, 1986b) has argued that regional variation in procurement systems of Jomon foragers, as characterised by differences in stone and faunal assemblages from the middens, are closely related to ecological parameters of the region. Midden sites, containing evidence of resource procurement strategies for exploiting game in the alluvial lowlands and fish and molluscs in the shallow waters of a combined forest-estuary ecosystem, decreased markedly in number with local environmental change resulting in retreat of estuarine conditions (Akazawa 1986a). However, for the cessation of the Archaic shell mounds of Kentucky's middle Green River area in the United States, around 3500 BP, Claassen (1996:137) favours social over environmental explanations, partly because molluscs are present in the area today in spite of intensive harvesting and pollution.

Environmental and cultural aspects of mound building

Claassen (1998:224) asks what are the implications for social organisation, of the creation of visible shell deposits and their subsequent disappearance? Finding a reason for why people created mounds of shell is difficult simply on archaeological data. Mounding appears to be a fairly universal human behaviour, given the appearance of humanly built shell mounds around the world (see Bailey 1975a; Rowland 1994a). A common theme that emerges in studies of mounded shell deposits is their position in wetland areas of high resource biomass, where the distribution of these resources is non-linear (Rowland 1994:155), as found in these indented estuaries.

Almost 20 years ago Bailey (1983) discussed the unique combination of factors -ecological availability of a prolific supply of molluscs, patterns of shell discard and post-depositional preservation- that led to the survival of particular sites such as the Weipa shell mounds. Could it be possible that the natural tendency for mounds to develop from discarded material, in a landscape shaped by a narrow and restricted length of occupation area adjoining seasonally inundated areas (see Bailey 1975a:VII:43; 1999:109), was recognised and incorporated into early human belief systems as part of the lore of utilising that resource? Maybe the very mobility of shell beds and fluctuation in availability due to environmental perturbations stimulated development of myth and ritual and a practice of mounding shells to mark out territory and construct totemic monuments to prevent the molluscs from leaving. The traditional custom would then have continued for as long as the specific environmental conditions provided feedback, in the form of large resource returns.

Bivalve mudflat molluscs, while available in good seasons in quantities sufficient to provide a protein surplus for feasting, enabled periodic aggregations of kin-related groups, that were always an essential part of forager survival strategies, to occur. Elsewhere, other seasonally abundant resources have been postulated as communion foods for large ritual gatherings, such as bogong moths in the southern highlands (Bowdler 1981a; Flood 1980) and wild seeds in arid central Australia (Smith 1986). In like manner harvesting of seasonally abundant resources easily accessed by coastal groups in the tropical wetlands, such as molluscs, fish and magpie geese and plant staple such as cycad nuts and lily roots, would have been an important ingredient of forager resource procurement strategies. However, non-molluscan resources, while contributing to the diet, did not have the same properties of leaving a large durable skeletal relic that allowed the inhabitants to place their mark upon the landscape

In addition, consumption of large quantities of molluscs at a time when they are "fat" may be related to the noted practice of "gorging" on high-fat foods when they are available, to offset times of seasonal scarcity in calorie (energy) supplies (Lieberman 1987:235; Speth 1987:16). In the tropics the holding of ceremonies around August and October precedes the high stress time of the late dry when solar radiation and humidity increase, water and resources are scarce, and people lethargically await the onset of the wet in December (Jones and Bowler 1980:18).

A ceremonial or symbolic aspect to the large shell mounds, as proposed by some researchers (eg. Claassen 1991a; Luby and Gruber 1999), is an idea that is hard to ignore, given their awesome presence, complemented by contemporary sites of different types. Clearly at the mound complex HI81, people deliberately climbed up the sides of a growing mound several metres high to deposit shells on top. The shell mound may have been a marker or monument, built in the landscape to ensure (re)production (Claassen 1991a:295; Tacon 1994). Tacon (1994) argues that humans socialised the landscape through marking and mythologising rock art and built structures such as stone arrangements and mounds.

As monumental structures the large mounds may be an expression of the articulation of physical, social, religious and economic landscapes (Pickering 1994:151). Myths and stories have been associated with shell mounds, as for most aspects of Aboriginal life (eg. Cribb 1986:145,149) as well as ritual practices associated with totemic beliefs (see above). Turner (1976:189) states that a function of totemism is to link the local group of which one is a member, to two groups related at parent level and four groups at grandparents level. The mounds may be the physical expression of totemic (ritual) links that together with affinal (marriage) and cognatic (birth) links form the basis of Aboriginal social organisation (Turner 1976:189), of which alliance and exchange are essential components.

These large shell mounds of northern Australia are archaeological features that extend over a time period that Bradley (1991:210) refers to as the *longue duree* or "geographical time" that operates at the scale of environmental change. Within this period is the social time that operates at the scale of human history (Bradley 1991). These two scales of time are linked in the mounds, which provide evidence of both subsistence and ritual, with ritual ceremonies performing the function of transmitting information from one generation to the next. Bradley (1991), building on Bloch (1977), argues that by using prescribed dance, language and songs, ritual merges the past with the present, which distinguishes it from mundane conceptions of time and everyday activities such as gathering food. The removal of ritual from everyday experience is said to give it stability and help preserve its legitimacy and timelessness.

In the same vein as Bradley (1991) interprets other monumental structures, one could speculate that the long-term continuity of the mounds may represent ritual time and stability, with ritual words and actions remaining essentially the same over the mound building period. Meanwhile the different types of mounds may represent social time as the scale of ritual changed, along with the locations and number of people taking part (Bradley 1991:211). Thus the elaboration of the large mounds around 1450 BP may represent a period of social change when former traditions of mound building were drawn on to deal with a period of crisis, whether socio-political or environmental (Bradley 1991). There is certainly ample evidence for environmental flux around this period, with dramatic landscape change as freshwater conditions became established. This type of cyclical system, with periodic increased energy investment in mound building elaboration, associated with accumulation of wealth in the form of ritual knowledge and expansion of alliances, would conceivably act as a cultural buffer in an environment of high frequency climatic oscillations (Davidson 1990; Hamilton 1999; Jorde 1977).

In Australia, unlike other places such as the eastern coast of the United States, there was no advent of horticulture to explain the cessation of mound building in terms of changes in religious practices, the symbolic importance of shell related to fertility and death, or changes in scheduling of time related to agricultural requirements (Claassen 1991a). Burials are common in large shell mounds in some places elsewhere in the world. For example, 700 burials were recorded in the Emeryville mound and 160 in the Ellis Landing mound in the San Francisco Bay region during early 20th century excavations (Moratto 1984:227-230). This has led to a recent theory that large shellmounds may be monumental structures associated with burials and mortuary rituals (Luby and Gruber 1999). However, like the "Erteb□lle" shellmounds of Denmark (Clarke 1975, cited in Luby and Gruber 1999:98) and in other parts of Australia (Beaton 1985:4; McCarthy and Setzler 1960:233), burials appear to be rare in shell mounds in the Darwin region. The presence of only a few small skeletal elements recorded in two mounds at Hope Inlet is consistent with local ethnographic practice of secondary burial. A possible test of the hypothesis that these were built as burial mounds is dating of the skeletal material to determine the chronological relationship to the mounds.

Discussion and summary

The link between cessation of mound building and environmental changes in relative abundance of particular mollusc species (Hiscock 1997; Beaton 1985:9; Mitchell 1994a:389), does not rule out cultural change in shell discard behaviour as a contributing factor, perhaps stimulated by foreign contact (cf. Mitchell 1994a:394). As discussed earlier, a smallpox epidemic similar to one reported in this region in the early 19th century contact period, which killed so many people that they could not bury them all (Foelsche 1882), could certainly have had a major impact on cultural practices such as mound building. This would also apply to intensification of regional exchange systems involving possible shifts in exchange relationships between different groups, in favour of those directly in contact with foreigners, able to acquire sought after foreign goods for trade (see Mitchell 1994a:383-8).

Cessation of mound building in this region is consistent with data indicating a combination of particular historic environmental and social conditions. The data indicates that mean shell size of *Anadara* reaches its lowest value in the upper levels of the saltflats shell mound HI80, dated to around 550 BP, after some 1500 years of sustainable harvesting of this resource (Figures 10.1 and 10.2). Combined with a marked decrease in relative proportions of the mudflat bivalve *Anadara* to the mangrove gastropods, and increase in diversity and size of fish resources compared with the older mounds HI83 and HI81 (Tables 10.1 and 10.3), this suggests an explanation of gradual environmental change and declining quantities of *Anadara*. Development of a trade relationship with foreign visitors such as Macassan trepangers around this time would conceivably provide the social conditions at that point in history to sound the death knell of mound building traditions.

In a scenario of gradual, environmentally related, declining yields of bivalve molluscs, in quantities necessary to support the essential gatherings of allied groups, the arrival of more efficient methods of obtaining surplus protein, and calories (energy), would have been eagerly seized upon by the local inhabitants. These would include harvesting the large high-fat animals such as dugong and sea turtle through use of dugout canoes and metal for harpoon points brought by the Macassans (Bradley 1997:285-306). Of course the effects of environmental change would not have been simply one of biological feedback for humans but would have been mediated through cultural perspectives of events in the particular historical context of the time. As discussed above, events included introduction of highly valued exchange goods that affected the balance of exchange, and population decimation through introduced disease that differentially affected women, for which the socio-cultural response was widespread changes in social organisation, with intensified male competitiveness and power, and ceremonial activities (Campbell 1985:357-8; Mitchell 1994a:415).

It is interesting here to reiterate Bowdler's (1981a, 1981b:531) argument that all forager diets were characterised by a dual-component strategy based on division of labour, in which the women generally collect relatively reliable *r*-selected species such as bivalve molluscs, while the men hunt in a more opportunistic fashion for large *k*-selected species such as kangaroos. In this context one could speculate on the possibility of a diminution in numbers and status of women, traditionally associated with gathering of *r*-selected species, around the time of the decline in molluscs and mound building, and a rise in male prestige accompanied by increased hunting of large marine animals.

11 Late Holocene economies in the Darwin region: Interpretation of cultural deposits and intensification

Introduction

Interpretation of the apparent increase in site numbers and more intensive use or occupation of individual sites within the late Holocene period as seen in this study feed into the well-known intensification debate in Australian archaeology. Increases in site numbers and use of sites, usually expressed as increases in discard rates of cultural material, are two of the key traits first posited by Lourandos (1983:82) and subsequently taken up by others others (Lourandos and Ross 1994:58; Morwood 1987:343; Ross 1985) as one of the archaeological features indicative of intensification. Intensification, defined as increased productivity and production involving socio-economic change, is linked to changes in settlement, subsistence and demographic patterns and associated with increasing complexity of social networks in the late Holocene in Australia (Lourandos 1983, 1985a, 1985b).

The archaeological record in the Darwin coastal region can be examined with reference to Freslæv and Frankel's (1999:248) required criteria, which deals in particular with the archaeology of coastal regions, for expected expressions of intensification:

- Greater diversification in the use of resources, so that the region became more densely occupied and the total resources of the region more effectively utilised (the broad spectrum revolution);
- Greater technological investment in the effective exploitation of single resources;
- More intensive use of existing resources (new sites, more deposit, greater predation).

Greater diversification of resources

Site location and broad spectrum resource procurement strategies

Data from this study indicates the shell mound sites in the Darwin region are located in optimal positions to access the range of environmental units in a non-linear landscape (Rowland 1994a:155), extending from the shallow seas to the inland savannah (Figures 7.1, 7.4, 8.2, 8.7, and 8.11). These include the nearshore reefs, tidal flats, estuarine creeks and rivers, swamps and monsoon jungle and mixed forest patches, and more open eucalypt woodland. Bailey (1977:140) noted long ago for the Weipa mounds, that the position of individual sites within the cluster probably relates to resource availability according to season and microenvironment. This interpretation is supported by the data from this study, which shows variation in relative frequency of marine/estuarine resources and terrestrial game according to the micro-environmental location of each site. This includes chronological differences in diversity and size of fish related to micro-environmental context of shell mound sites. (see Chapter Ten).

As demonstrated by this research, a similar range of terrestrial, marine and estuarine foods in excavated sites in the Darwin region occurs throughout the period 2000 BP to 500 BP (Tables 9.2 and 9.3). Evidence of exploitation of a range of resources from these environments consists of the most durable remains that have survived to the present, including the abundant shell, and less robust bones, and teeth of fish and other animals. The remains indicate that in the last 2000 years people were exploiting both *r*-selected mudflat bivalve molluscs and a range of *k*-selected mangrove gastropod molluscs, as well as fish, crabs, possums, reptiles, birds and *k*-selected macropods, from a rich, if patchy, environment that allowed seasonal semi-sedentary residence (cf. Cohen 1985:103).

This broad spectrum of faunal remains as found in the Darwin region mounds is not restricted to the late Holocene period. A similarly diverse range of faunal remains is also exhibited in midden deposits in rockshelters in the Alligator Rivers region spanning the period 3000 to 6000 BP (Schrire 1982), with differences related to chronological/ecological context of each area. Variations in the type and relative proportions of faunal resources utilised within this period are shown to be related to locational micro-environmental factors and/or taphonomy. Thus the mudflat bivalves *Anadara* dominate in sites close to the coast while mangrove taxa like *Telescopium*, *Geloina* and *Cerithidea* dominate in more inland sites near mangrove forests, such as on the lower reaches of East and South Alligator rivers (see Figure 1.2).

The data from this study shows that chronological changes in relative proportion of mudflat bivalves to mangrove gastropods in middens reflects very localised change in ecological conditions around the site locations, as expanding mangrove forests enroach on open coasts (Tables 10.1 and 10.2). This is linked to the different formation processes that occur with different site types and specific local environmental contexts and timing of estuarine development in each area. *Anadara* appears in sites around 4000-4500 BP, soon after establishment of suitable habitat conditions for this mollusc, with a temporary increased focus on these small mudflat bivalves in particular locations when optimal habitat conditions had developed for proliferation. (see Chapters Seven, Eight, and Ten).

Harvesting of molluscs in this context does not appear to be an "inefficient strategy" of reliance on less preferred small-bodied *r*-selected species, imposed onto these foragers because of population increase or declining resources (cf. Cohen 1985:102; Hayden 1981; O'Connell and Hawkes 1981; Veitch 1999). The site settlement patterns of this late Holocene economy are not consistent with the picture, of densely-packed foragers being forced into sub-optimal areas and focussing on less-preferred emergency foods, painted by advocates of the evolving social complexity and intensification theory (Cohen 1985; Lourandos 1985a, 1985b).

Hayden (1981a) proposed the broadening of the resource base to include low-ranked *r*-selected species, driven by periodic resource stress and associated with technological innovation such as the introduction of grinding stones, as a development that occurred at the end of the Pleistocene. Responses to this hypothesis include Dunnell's (1981:533) observation that though species with *r*-selected reproductive strategies are abundant when they occur, they are also "likely to have highly irregular temporal/spatial distributions". Others (eg. Bowdler 1981b; Kamminga 1981) point to evidence of exploitation of molluscs during the Pleistocene in the form of open midden deposits, for example at Willandra Lakes (Allen 1972; Johnston and Clark 1998) and in cave sites including Klasies River Mouth in South Africa (Shackleton 1982), Haua Fteah in Libya (McBurney 1967) and Devils Tower at Gibraltar (Garrod *et al.* 1928), to mention a few. But debate revolves over the extent to which this formed a significant aspect of resource procurement strategies in the Pleistocene compared with the evidently intensive exploitation of this type of small-bodied resource in the Holocene (Hayden 1981b; Waselkov 1987).

This study finds agreement with Gamble (1986) who argues that broad-based subsistence strategies and information sharing designed to allow exploitation of a variable combination of resources appear with modern *Homo sapiens*. According to this view, variation in archaeological remains such as large Holocene middens represents the application of these resource procurement strategies, rather than increased complexity or intensification. Pleistocene sites that may contain further evidence of broad-spectrum subsistence strategies are likely to have been destroyed or lie under the sea (eg. Bowdler 1977, 1981b; Druss 1981; Hughes 1977; Kamminga 1981). Such a view is bolstered by the restricted location of Pleistocene middens on coasts with steeply sloping continental shelves and the close correlation between changing Holocene shorelines and shell middens and mounds (Suguio *et al.* 1992; Swadling 1997; Waselkov 1987:124-5). And increasingly, as research continues both here and abroad, more cultural shell deposits that are found along the coast that date to the mid and early Holocene or earlier, may reveal evidence to support this view (eg. Cann *et al.* 1991; Erlandson *et al.* 1999; Hughes and Sullivan 1999).

Greater technological investment in the use of single resources

The evidence from this research supports the notion that intensive exploitation of a mono-specific resource - *Anadara* - occurred around the period 1450 BP at Hope Inlet. One piece of technology –the bark canoe- may have been used in exploitation of the *Anadara* resource. Historic evidence suggests that canoes may have been used to transport *Anadara* molluscs from the harvested shell beds up the tidal creeks to campsites. This includes the observation during a survey of the coast in Knockers Bay, north of Darwin, in the early 1800s, of a bark canoe some 18 feet long containing a large quantity of bivalve molluscs (*arca scapha,* probably *Anadara*) (King 1969:89) and canoe-loads of molluscs in Arnhem Land (Tindale 1925:82). At Shoal Bay in the late 1800s Aboriginal people in canoes exchanged turtles and shells for damaged flour with Paul Foelsche and the naturalist Frederick Schultze, who were there collecting natural science specimens (Reid 1990:38).

There is historical evidence that the Indigenous inhabitants used canoes extensively to traverse the water routes of the indented Darwin coast at the time of European contact. Goyder noted that mangrove swamps cut off much of the hinterland around Darwin Harbour from the coast, and boats were necessary for access to this land. Many miles of tortuous travel through swamps, monsoon vine thicket, thick scrub and tall speargrass could be avoided by travelling by boats across only a few miles of water to each of the many promontories which indent the harbour shoreline (Kerr 1971:71). Basedow (1925:162-3) noted the presence of clearings in the mangroves fringing Darwin Harbour, which were maintained through regular removal of mangrove seedlings by the Indigenous inhabitants, to provide access and "housing" for canoes, each in its own recognised place.

Although the timing of use of this perishable equipment is not known, these reports indicate that at contact both the sewn bark canoe and dug-out canoe were in use in the Darwin region (Basedow 1925:164; Holland 1976; Kerr 1971:68). What is known is that the bark canoe was associated with the period preceding the introduction of the dugout canoe by the Macassans (Basedow 1907:53; Bradley 1997:286; Foelsche 1882:12; Mitchell 1994a:143,398). However, the investment in this technology would appear to be small. Bark canoes are relatively easily constructed from sheets of abundant bark sewn together, but carry less, do not last as long and sink more easily than dugout canoes. They were reputedly treated very much as disposable items by the Yanyuwa people of the Gulf of Carpentaria (Bradley 1997:285-294; and see Mitchell 1994a:398).

As discussed in Chapter Eight, harvesting of mollusc beds carried out from a boat with a hand dredge, in a manner similar to that described in southeast Asia and Japan (Broom 1985:25-9), would constitute a greater technological investment in the intensive exploitation of a single resource. But, as noted previously, in the rapidly accumulated large mound HI81, there is a relatively low proportion of incidental bycatch such as small juvenile *Anadara* shells, which would be expected to collect through dredging, compared with the much more slowly formed Middle Arm mounds. It is considered likely that the relatively higher proportion of juvenile *Anadara* in some middens is related to the season of collection as suggested by Richard Willan (pers. comm.), and that the equipment used, besides canoes for transport, was hands and digging sticks as observed ethnographically (see Meehan 1982a:97-8).

More intensive use of existing resources

Alternative interpretations below, which are based on the data from this study, provide more coherent explanations for apparent increased site numbers, increased cultural remains and evidence of over-predation in the late Holocene. These are often taken as archaeological signatures of more intensive use of existing resources compared to earlier periods.

Site survival as a factor in increased site numbers

With regard to the late Holocene proliferation of sites at Hope Inlet, the coastal landscape itself is a construct of the last 2500 years and therefore it is not surprising that the archaeological remains date from that period. Earlier sites may be buried in depositional environments and/or of a type not conducive to their preservation in the face of the innumerable climatic and other physical and chemical post-depositional factors that affect site survival (Godfrey 1989; Head 1986; Rowland 1989; Woodroffe *et al.* 1988). Although shell mounds are afforded more protection than other site types, provided by their solid alkaline bulk, this is no guarantee of their survival, or archaeological visibility (see Bird 1992; Head 1983; O'Connor and Sullivan 1994b:27). It is possible that sea level oscillations, for example at around 3000-3500 BP (Flood and Frankel 1989; Rowland 1999:27), could have eroded previous mudflat buildouts in the area, removing any earlier deposits in the process (see Chapter Two). Such an hypothesis is consistent with the gap in stratigraphy of the Shoal Bay chenier plain between 6000 and 2500 years ago (Woodroffe and Grime 1999:313) and gap in coastal stratigraphy around 3400 years ago on van Dieman Gulf shorelines (Colin Woodroffe pers. comm.).

Some researchers abroad reach similar conclusions, based on evidence of the effect of geological processes on preservation of the archaeological record, that it is geomorphic processes, rather than cultural activities, that are more likely to be responsible for the pattern of increased site density in more recent times (Waters and Kuehn 1996:488). In addition to the natural effect of preferential survival of more recent sites, the observed spatio-temporal pattern of mid-late Holocene increased site numbers and use of so-called marginal environments could also partly represent the expansion of archaeological research and dating of previously unexplored areas, as has occurred here (cf. Lourandos 1994).

Increased cultural remains, mounding and ceremonies

Increased intensity in exploitation of the *Anadara* mollusc resource does appear to occur around 1450 BP at Hope Inlet, consistent with ceremonial gatherings. The type of large shell mounds that result from discard of this shell, have been called megamiddens in South Africa (Jerardino 1998; Jerardino and Yates 1997). One theory posed to explain the megamiddens elsewhere in the world, South Africa and West Mexico for example, has been that this site type is the result of bulk processing of an abundant mollusc with perhaps drying of the flesh to extend its storage time. It is suggested that the flesh obtained, after being sun dried, smoked or salted was then exported to other areas for trade with inland or offshore visitors (Henshilwood *et al.* 1994; Shenkel 1971:137-8). The process of drying large quantities of molluscs for storage/trading with inland groups has been recorded ethnographically for many coastal societies in North and South America, Oceania, Asia and Africa (Waselkov 1987:106-7). As discussed in Chapter Eight, while there are no ethnographic reports of sun-drying molluscs in Australia, there are early 20th century reports of drying, mostly plant foods, and isolated references to drying kangaroo meat strips, indicating this practice in some parts of northern Australia (Mulvaney 1975:73, 241; Thomson 1949:23-4; Withnell 1965:20).

Moreover, the archaeological data of low diversity of molluscs, dispersed charcoal and ash in the upper shell mound of HI81 in particular, is consistent with reflecting large-scale cooking, which Henshilwood *et al.* (1994:108) suggest is a feature of this type of large bulk processing site (and see Shenkel 1971:137). The evidence for cooking makes it unlikely that the flesh was used as bait for fishing (cf. Claassen 1991a). While it seems clear that bulk processing was responsible for the large deposits, it is less clear whether the resultant flesh was fed to ceremonial participants in the area, and/or traded as dried meat with inland or offshore visitors. An alternative explanation is one in which the local Indigenous group granted mollusc harvesting rights to offshore visitors in exchange for foreign goods not available locally, in much the same way as trepang trade operated with the Macassans (Mitchell 1994:93-114). Shenkel (1971:153) forwards these alternative hypotheses as possible explanations for large shell mounds in West Mexico, two of which are composed predominantly of a species of *Anadara*, the largest estimated at 182m by 121m at the base and about 30m in height (Shenkel 1971:106). Shenkel (1971:137) states that these mounds "appear to be the result of a concentrated collecting activity, there being no cultural refuse of any kind found with them. Though they are loosely packed and well weathered, there are innumerable small flecks of carbonized twigs mixed all through the shell matrix".

Contemporary accumulation of part of the shell mound HI81 and of HI83, and possibly the earth mound HI97 is consistent with ceremonial gatherings and exchange. However, exotic materials in the stone artefact scatters that are indicators of exchange in the Hope Inlet area are on geomorphological grounds possibly dated to within the last 1000 years, at least 400 years after formation of the large mounds, though they may be contemporary with some more recent mounds.

In support of the theory of intensification, mounding of deposits, in particular earth mounds in southeastern Australia, have been interpreted by some (eg. Lourandos 1983; Williams 1988) as indicating a change in behaviour and social organisation in the late Holocene, associated with decreased residential mobility and increased population. Such a view is not supported by evidence suggesting sequential rather than concurrent occupation of earth mound sites, clustered in wetland areas of southeastern Australia, since, as Bird and Frankel (1991a:8) point out, these are separated in time by up to 1000 years (Williams 1988:94, 139). In northern Australia one of the large Adelaide River earth mounds 25km inland of Darwin, excavated by Schrire (1968, cited in Brockwell 1996c:161), reveals a sequence of use which broadly reflects long-term environmental changes for the region, of the estuarine to freshwater transition, with mangrove mollusc shells *Geloina* sp. in the basal levels, and freshwater mussel shell *Velesunio* sp. in the surface layers of the deposit.

While most Hope Inlet earth mounds, including the excavated mound HI97, differ in form and composition from the Adelaide River earth mounds (Brockwell 1996b, 1996c), in both types of mound the data indicates a pattern of successive periods of use (see Chapters Five and Eight). Use of the Adelaide River mound spans centuries, and the smaller Hope Inlet mound HI97, which appears similar to the small earth "oven mounds" described at Aurukun, Cape York (Cribb *et al.* 1988:68), may represent a shorter time-span, but neither appears to uphold the argument for more intensive production. In recent botanical and experimental studies on earth-ovens in south-central North America, the returns of cooking plant foods that are poisonous and inedible until processed (thus classed as low-ranked resources) in this type of earth oven, has been found to be high-cost and low return in calorie or energy terms (Dering 1999). This has led Dering (1999) to argue that intensive use of earth ovens indicates a broad-spectrum, low return economy and small groups and high residential mobility. Dering (1999) also argues that this means they are used in times of periodic resource stress when high-ranked resources are not available or scarce.

From the perspectives of Indigenous Australians however, cycads, which are also poisonous and inedible resource until processed, are not considered a low-ranked resource, since ethnographic accounts indicate the use of cycads as a seasonal staple, and significantly, as a "communion food" constantly used to feed participants of ceremonies (Beaton 1977:165). The continued ceremonial use of cycad nuts by Indigenous people in Arnhem Land in contemporary times, when many other plant resources have been replaced by European carbohydrates (Meehan 1982b:106-7), supports Beaton's (1977:165) proposition that this resource has a long history as primarily a communion food (see Beck 1992; Jones and Meehan 1989). Use of earth ovens to cook large quantities of food for ceremonial use does not however, necessarily mean decreased mobility and/or increased population. This is so particularly considering the large amount of energy expended in relation to the caloric return (eg. a week's work for one woman equals a 40-50kg cycad loaf) (Jones and Meehan 1989:126). Cycad nuts are laborious to prepare and associated with periods of stress, in that they are a resource that becomes available at the end of the dry season, when other resources are low. Their "special value" to an Indigenous economy is the social use of these nuts to feed the participants of large ceremonial gatherings:

> ... for it enables women to maintain an adequate food supply on ceremonial occasions when hundreds of people are gathered in one camp for weeks or months at a time, who could otherwise not be supported for such periods on local resources (Thomson 1949:23).

A scenario that fits in with the available data is the periodic intensification of resource utilisation, of a local surplus, during large ceremonial gatherings of relatively small neighbouring groups. In this scenario the return from the energy invested is not simply an instant one of calories, but a more intangible, yet vital delayed return of alliances made and genes exchanged (Beaton 1977:196), allowing the pooling of risk through rights of access to resources of neighbouring land holding groups (Davidson 1990:51). As Davidson (1990:50-1) points out, ceremonial institutions and exogamous marriage were the social mechanisms by which small populations of Indigenous Australians in the past coped with environmental variation and the risks it imposed.

I share Kefous' (1988) and Davidson's (1990) opinion that survival was not an easy matter for past Indigenous groups, but argue that the environmental evidence suggests that life for north Australian foragers was possibly easier in the Pleistocene and early Holocene that the late Holocene. Environmental evidence indicates continued climatic amelioration from the late Pleistocene to the mid-Holocene, but increased climatic variability and overall aridity in the late Holocene, with the onset of ENSO around 4000 BP meaning reduced reliability of monsoonal rain and increased likelihood of droughts and floods (see Chapter Two). Arguably this would have meant a more difficult environment for all life forms, including humans, with recovery in rainfall levels only in the last millennium.

Shell size and extent of predation

As discussed in Chapter Ten, decreasing shell size in middens has sometimes been interpreted as evidence of overpredation of the mollusc resource. Within the time period of 2000 to 500 BP in the Hope Inlet area, data on shell size shows no unambiguous trend, although there are a series of short-term trends. This study illustrates the way in which an apparent trend of decreasing shell size over time in one mound is temporary and cyclical, repeated again in a later mound some 700 years later. Although it may be true that all long-term trends are composed of short-term events (Lourandos 1996:17; Ross 1985), this does not necessarily mean that all short-term events lead to a unidirectional long-term trend.

The short-term trends of decreasing shell size within each of these deposits are consistent with common forager practices of harvesting larger, more desirable, more easily procured resources first on (re)occupation of a site location. In keeping with this practice, resource stress commonly occurs with increasing length of occupation period, to a level at which people move on to the next location in their seasonal round (Tomka and Stevenson 1993). However, it may not be diminishing returns of the mollusc resource that determines when a group moves to a new location, but the increasing distance required to procure plant foods. As Kelly (1992:47) points out, plants provide lower returns than large game, (and probably less than the seasonally abundant small game such as fish and molluscs).

The recurrent decrease in shell size as each mound accumulated, is only to an average mean minimum of 28-30mm, very similar to sizes considered to be sustainably marketable in Malaysia in contemporary times (Pathansali 1966:92). The temporary, cyclical trends in shell size in the three Hope Inlet shell mounds spanning some 1500 years, does not suggest that the *Anadara* is an inexhaustible resource targeted to support an increasing population when alternative resources are scarce or in short supply (cf. Veitch 1999:7; Smith 1986). Rather it suggests a managed utilisation of the seasonally abundant *Anadara* resource over millennia - perhaps an example of ecological information transmitted through ritual traditions over many generations. The smaller mean shell size during final accumulation of mound HI80 around 550 BP after this length of time, coinciding with a decrease in relative proportions of the mudflat bivalve *Anadara* to the mangrove gastropods, and increase in diversity and size of fish resources, is consistent with gradual environmental change (see Chapter Ten).

Shell mounds as representing increased productivity, population increase or reorganisation

There is no need to invoke population increase to explain the volume of shell at Hope Inlet, as Beaton (1985) did for Princess Charlotte Bay. A rough, conservative estimate of 12,000m^3 for the total quantity of shell deposited at Hope Inlet over 1300 years, from 1800 to 500 BP is about 9m^3 per year. This figure is not that dissimilar to Meehan's (1982a:166) estimate of 8m^3 collected over one year by the contemporary 40-strong Anbarra group. At 8m^3 per year the combined volume of 140m^3 for the upper shell mound HI81 and mound HI83 would have taken group of 40 around 18 years to accumulate. Any number of permutations on this formula could be proposed, including a hypothesis that 360 people could have accumulated this amount in 2 years, which is consistent with the interpretation of these two deposits as the remains of group aggregations for ceremonies. Importantly, each of the permutations would be consistent with the statistically indistinguishable radiocarbon estimates and low breakage rates of shell in the deposits and my interpretation of rapid accumulation of these deposits.

Other mounds, such as the excavated sites around Darwin Harbour, represent a palimpsest of the lives of generations of relatively small groups over centuries - the remains of a number of short-term discrete occupational events separated by hiatus periods in the dominant shell deposition (Waselkov 1987). Less visible archaeological remains subsumed in the deposits, which become evident with slow down or hiatus of deposition of the major shell being harvested, indicate people continued procuring and cooking resources other than the dominant mollusc during shell mound accumulation and after accumulation of the mound has ceased. Radiocarbon dating and high fragmentation rates for these mounds provide support for this interpretation of these deposits. The archaeological data in this area, of these small mounds spanning long periods both pre and post the period of larger mounds with low fragmentation and rapid accumulation, is consistent with a temporary change in depositional behaviour rather than any unidirectional increase in productivity and/or population (cf. Lourandos 1983; Ross 1985).

If shell mounds do not necessarily represent population increase, does the reverse apply? Does the cessation of shell mounds and decrease in quantities of shell discard present decreased population? Population decrease has been inferred from decreases in rates of stone artefact discard in the last 1000 years in southern Australia (eg. Attenbrow 1987:200-4, 385-6). Some have argued that a decreased rate of artefact discard in some areas reflects reorganisation of groups in space in the last 1000 to 2000 years. In the southeastern Australian highlands, Flood *et al.* (1987) argued for reorganisation of people in the last 2000 years as groups moved into marginal areas, and this was accompanied by decreased population in adjoining heartland areas. In the sub-coastal freshwater wetlands of the south Alligator floodplains, large sites containing high artefact densities and diverse artefact types, and increased site intensity, dated to the last 1000 to 2000 years, have been interpreted as indicative of increased population (Jones 1985b:291-3) and low residential mobility (Brockwell 1989:210; Meehan *et al.* 1985).

Hiscock (1999) has suggested that in northern Australia postulated increased population in some areas represents a reorganisation and redistribution of population from other areas where population seemingly decreased. He suggests that cessation of mound building around Darwin Harbour some 700 years ago may represent movement of people away from the coast to a focus on utilisation of sub-coastal freshwater wetlands. Whilst there is evidence of late Holocene resource exploitation in the sub-coastal freshwater wetlands in the vicinity of Darwin, in the form of earth mounds on the Adelaide River coastal plains (Brockwell 1996b, 1996c), occupation of that area is concurrent with occupation of the coast, spanning the last 3000 years (Sally Brockwell pers. comm.). This study provides evidence of continued occupation around Darwin Harbour after mound building ceased. This includes a shallow non-mounded midden dated to around 400 BP, and surface stone artefact scatters throughout the coastal area, although as Hiscock (1999:99) points out, this site type is poorly defined chronologically. Nonetheless, I submit that further dating of small non-mounded shell deposits and scatters in the Darwin coastal area will reveal more of these very late Holocene sites.

However, on the evidence presented in this study (see Chapter Ten), it does appear that Hope Inlet was an area that was abandoned as a focal occupation area some centuries prior to European settlement of Darwin. Interestingly, recent archaeological research on the northeast coast of Australia has revealed a similar pattern of abandonment in the last 600 years, of the Ngarrabullgan area that had also been previously occupied for millennia. The authors, David and Wilson (1999), interpret this abandonment in terms of changing socio-cultural perceptions of the landscape, and "the emergence of a new cosmology some 600 years ago, akin to the Dreaming as recorded ethnographically" (David and Wilson 1999:184). On the available data, as discussed in Chapter Ten, more than one factor may have contributed to abandonment of Hope Inlet as a focal occupation area around 550 BP. These include environmental change and the decline of *Anadara*, and realignment of coastal groups in response to shifts in the exchange system balance stimulated by Macassan contact in Arnhem Land (Mitchell 1994:399).

One of the questions that arises from this research is the nature of abandonment of Hope Inlet. The wide range of stone artefacts recorded in these surveys must have been left behind when this area was abandoned. This type of cultural remains, left behind in abandoned areas is called *de facto* refuse by Schiffer (1987:89), and variability in the nature of this refuse is determined by a number of factors, in particular whether the abandonment is slow and planned or rapid and unplanned. If planned, a wide range of stone artefacts would not be expected, as the occupants remove or curate items for use elsewhere (Binford 1977, 1979).

In the case of areas occupied on a seasonal basis, as posited for this area, the axes, grinding stones and large numbers of pestles/pounders may have been deposited by the occupants in anticipation of their return to the area (Schiffer 1987:93; and see Gould 1980:71-2 and Kent 1993). Ethnoarchaeological studies, in particular Binford's (1977, 1979) work, has shown that periodic abandonment of regions is part and parcel of forager land-use strategies. However, the apparently relatively intact nature of the assemblage at Hope Inlet, and unusual situation in that cached artefacts were not retrieved for use in neighbouring areas, suggests that abandonment was unanticipated and/ or abrupt, consistent with catastrophic or ritual events (Cameron 1993; Schiffer 1987:89-97).

Catastrophic events could include natural disasters such as cyclones, droughts or fires, which would seem unlikely cause given that they occur periodically, or cultural happenings such as warfare or epidemic disease and death, which could be related to ritual abandonment (Schiffer 1987:92). In light of this it is interesting to note that Shoal Bay adjoins Larrakia country and Wulna country to the north-east, and was probably one of the "neutral zones" referred to by Parkhouse (1895:638), that existed between neighbouring groups. Parkhouse (1895:638-9) comments that groups did not reside in these buffer zones, but were permitted to travel through the area from one territory to another, and take game as required for their journey.

Discussion and summary

Crumley (1994:2) points out the contradictions within Anthropology, in that it is acceptable to explain early human (pre)history in evolutionary and environmental terms, yet socio-cultural explanations are preferred for late (Holocene) human history. This is so in Australian archaeology even when the people are known to be practicing foragers throughout the Pleistocene and Holocene. Just because people operate within the parameters of environmental factors does not mean that they are passive. To recognise the laws of their physical world, of which they consider themselves part, and give meaning and respect to the significance of its power, has long been the way of human foragers. There is no dichotomy in this sense between people living within the laws of their land, which recognise environmental constraints, and being dynamic participants in their natural world (cf. Lourandos 1994).

The apparent increase in site numbers has been shown in this study to be related to the particular set of environmental conditions in a place during a certain period of time. Increased remains and patterns suggesting increases in deposition have been shown to be related to taphonomic formation and post-depositional processes. Evidence from shell size in this study indicates sustainable continuous exploitation of a molluscan resource for around 1500 years, culminating in the eventual decline of the resource consistent with environmental change. Increased intensity in exploitation of the *Anadara* resource around 1450 BP at Hope Inlet is consistent with ceremonial gatherings taking advantage of a localised, temporary seasonal surplus in the resource (see Freslov and Frankel 1999).

12 Summary, discussion and conclusions

When I began this research project my objective was to investigate the extent of, and chronological changes in, human subsistence and settlement in the Darwin coastal region prior to European contact. I hoped to achieve this through investigation of any archaeological sites that were located in the area. Research for my BA Honours thesis had revealed the presence of a number of previously unrecorded shell middens on Middle Arm Peninsula, Darwin Harbour (Burns 1994), suggesting that there may be many more such sites in this region. Thus an aim of this research project was to determine the number, distribution and type of sites in the Darwin coastal region, expected to be predominantly shell middens. Never though, did I imagine the plethora of archaeological remains that lay within easy reach of the city of Darwin in which I had grown up, in blissful ignorance of the existence of another world. This study makes available significant knowledge of an area not previously investigated archaeologically, in which surveying has located numerous shell mound sites comparable to those described elsewhere in Northern Australia.

In this study I have examined the relationship between environmental changes and human prehistoric land use, as reflected by the shell middens. I hope I have contributed to knowledge on the origin, function and chronology of cultural shell deposits and the palaeoenvironmental framework in which they were formed. The period under study of the last 3000 years is one of dynamic environmental change, featuring replacement of mangrove forest with freshwater wetlands, rapid progradation of the coastal plains and increased chenier building (Clark and Guppy 1988). During this period the proliferation of mounds in the Darwin coastal region coincides with an apparent expansion in the number of open sites in the wetlands of the Adelaide River (Brockwell 1996b; 1996c) and western Arnhem Land (Allen 1989:113-4; Meehan *et al.* 1985). After the phase of extensive *Rhizophora* dominated forest, which lasted for several thousand years, the transitional period that followed was one of dynamic transformations in the landscape. Rapid and substantial vegetation changes are thought to have reflected this variable environment, with cutting and infilling of channels, and alternating fresh and saltwater inundation of some areas (Clark and Guppy 1988:682; Russel-Smith 1985). The coastal wetland systems that developed in the high rainfall areas of northern Australia, including around Darwin, are known to be one of the most productive ecosystems of the world (Woodroffe and Bardsley 1988), rich in marine and terrestrial floral and faunal resources (Hanley *et al.* 1997; Woinarski 1992).

Like estuarine systems everywhere, the Hope Inlet and Darwin Harbour estuaries are transient phenomena, forming their present morphology in the last few thousand years, and continuing to develop (Richman 1999:257). The evidence suggests that the estuarine systems of Darwin Harbour and Hope Inlet have evolved through hydrodynamic/sedimentary processes in a relatively arid environment to the morphological and ecological systems seen today. Indigenous Australian groups in the late Holocene, with their highly developed intimate knowledge of and relationship with their environment, a legacy of their Pleistocene ancestors, survived and flourished through this evolution.

I argue as others before me (eg. Cribb 1986b; Hallam 1987) that archaeological data usually taken as evidence for population increase/intensification, is on the contrary and simply the very visible remains of a small part of a broad-based and flexible economy characteristic of the earliest human foragers. Knowledge of the ecology and resources of coastal tropical lowlands would have accompanied the first colonisers moving into Australia from tropical southeast Asia (Hallam 1987:11), and flexible and broad-based subsistence strategies would have facilitated expansion into more arid parts of Australia during the Pleistocene (Edwards and O'Connell 1995; Smith 1987, 1993 cf. Nicholson and Cane 1994; Veth 1989, 1995).

Unfortunately it is difficult to provide data with which to argue against the idea of less complex societies and less intensive production in the early Holocene and Pleistocene compared to the late Holocene. Archaeological evidence that people practised broad-spectrum resource procurement strategies in Pleistocene Australia similar to those seen in the late Holocene sites is hard to come by, due to poor faunal preservation in Australia's acidic soils. However, some evidence for exploitation of a range of resources comes from open sites of the Willandra Lakes region in southwest Australia and rockshelter deposits at North West Cape in northwest Australia, dated to the Pleistocene. Investigations in the Willandra Lakes region have revealed remains indicating procurement of a range of foods, including small mammals, fish, reptiles and molluscs (Johnston 1993; Walshe 1998). Amorphous grindstones indicating knowledge of ground stone technology for processing plants and possibly seeds, occur in these sites and in north Australian rockshelter deposits dated to the late Pleistocene (Allen 1998; Kamminga and Allen 1973; Jones 1985a; cf. Balme 1991). Deposits at North West Cape include the remains of resources such as marine molluscs, macropods and emu eggs (Morse 1993). Evidence that the social mechanisms of ceremonial gatherings and long distance trade and exchange, seen as indicators of social complexity in the Holocene, also existed in the Pleistocene is also scarce, but is present in Pleistocene levels of deposits in northwest Australia. At North West Cape this occurs in the form of small quantities of exotic ochre, and shell ornaments (Morse 1993), and in the Widgingarri shelters in the Kimberleys in the form of flakes from edge-ground axes and baler and pearl shell (O'Connor 1999b:119-121).

And just as coastal does not equal littoral (Hallam 1987), so too littoral does not necessarily equal marine or maritime societies (Yesner 1980). All of the faunal remains in these sites indicate subsistence tied to the land. Molluscs and crabs are all available from the mangrove or intertidal or shallow subtidal areas. Although fish may have comprised an important component of the diet, all of the fish remains in these coastal sites could be obtained from the estuarine creeks and rivers or flooded saltflats, or the shallow intertidal waters or nearshore reefs. The diet of this coastal group is consistent with the findings of stable isotope analysis for human bone collagen from a late Holocene people of the southeast Queensland coast, that land-based resources were a more important component of the diet of coastal peoples than marine foods. (Collier and Hobson 1987; Hobson and Collier 1984). Collier and Hobson (1987:563) note that their findings indicate less of a reliance on marine foods than suggested by the ethnographic data.

The impressive, highly visible mounds at Hope Inlet represent the discarded remains of seasonal semi-sedentary occupation, but the remains are only a small proportion of a diverse and complex economy of a people who left little trace of the bulk of their activities (cf. Bailey 1975; Buchanan 1988). In other words, to paraphrase Butzer (1982:8) "known late Holocene sites are in no way representative of pre-contact settlement patterns, but are largely a function of selective surface preservation of large sites such as shell mounds". Activities that are archaeologically visible within the mounds include cooking, using both shell and rocks as hearth material, of a range of estuarine fish, molluscs and crustacea, wetland birds and reptiles and terrestrial woodland marsupials, as well as the littoral molluscs. Small sized lithic debitage within the mounds are visible evidence of maintenance and manufacture of (mainly) local stone material where it is available, while exotic stone in the surface scatters provide evidence of exchange.

I contend that the reason why macro-mollusc remains in archaeological deposits seem to reflect ecological change is because that is what they do: reflect ecological change, in that those taxa discarded in archaeological remains inhabited the local environment of the time. What they do not represent, however, is the total range of resources present, nor the whole breadth of Aboriginal subsistence activities for that period delineated by dating of shell deposits.

It can be said that if molluscs provided 10 to 50% of calories required for diets, then at least 50% and maybe up to 90% of these calories had to be acquired from elsewhere, that is plants, and this is the invisible aspect of archaeological sites. The much less visible utilisation of plants in the archaeology record, must be inferred from the presence of plants on and around the mounds, and grinding and pounding implements, known through ethnographic studies to be utilised by Aboriginal people. Use of fire, to extend forest margins, clear coastal plains for access to swamps, create green pick areas to attract wallabies and increase yields of some plants such as cycads (Beaton 1982) must be inferred from vegetation studies that show areas of relict stands of the fire-sensitive tree *Callitris intratropica* (see Chapter Three).

When collective groups from the region (re)occupied the promontory at Hope Inlet and deposited large quantities of shell on the large mound HI81, they were continuing a tradition that had already begun, as attested by evidence of a small group's practice of mound building at Darwin Harbour some 2000 years ago, and mounds built 4500 years ago in the West Alligator region. Aggregation sites such as the large Hope Inlet mounds are archaeological indicators of ceremonial gatherings and/or exchange associated with alliance systems or social relationships extended beyond kinship to wider regional systems (Freslov and Frankel 1999:250), which were always an integral part of forager social organisation. The archaeological data from these sites reflects short-term and long-term fluctuations in resource availability. They demonstrate, as (Freslov and Frankel 1999:250) suggest for late Holocene coastal Victoria, that the surplus is short-term, seasonal and localised, which would limit the extent of expansion, or intensification, of the system.

Cribb (1986b) states that the absence of sites doesn't mean absence of people. Similarly, ephemeral sites doesn't necessarily equate with ephemeral occupation, as some studies have shown, with the disjunction between ethno-historic data indicating relatively intense use and occupation, and an absence of archaeological remains (Anderson and Robins 1988; Bowen 1996). As Bowen (1996) notes, this is despite places like the Magela floodplains comprising similar micro-environments and resources to other areas in the vicinity that do contain sites, like the South Alligator floodplains. Here in the Darwin region, as for Princess Charlotte Bay, as Cribb (1986b) points out, the absence of extensive shell beds means that both chenier and mound formation cease, and without extensive shell beds mollusc gathering may continue at a lower level, that is, both prior to (in rockshelter middens) and post mound period (in shallow open middens).

As I have argued in this study, if material such as bone, charcoal and stone artefacts are intrusive into shell deposits, as the evidence suggests, then activities other than shell gathering that occur during periods of hiatus may be subsumed within the deposits. Life and people did not go away when shell gathering stopped, but people continued to fish, hunt, make artefacts and conduct their social relationships in the same area. For over thousands of years in the late Holocene, coastal groups, with perhaps small oscillations in population size, lived, fished, gathered and hunted in tune with the changing seasons, coming together periodically for large ceremonies with neighbouring groups. Whilst shell deposition temporarily ceased at some sites, the cultural custom which led people to pile the shell in mounds continued for around 2000 years. Compliance of this custom by small groups whilst gathering for their material needs evokes a sense of the extent to which spiritual beliefs would have permeated the lives of these foraging peoples.

The people who built the large *Anadara* mounds practised not a specialised marine economy, but a coastal economy as defined by Gaughwin and Fullagar (1995:39), with a very generalised and flexible multi-component subsistence system (Hallam 1987) utilising the resources of the coastal margins, the coastal plains and the coastal hinterland. This suggests a resource procurement system clearly different to that recorded ethnographically for north Australian coastal groups, including in the Darwin area, where sea going dugout canoes and hunting of saltwater turtles and dugong feature predominantly.

The system practised by these coastal peoples was flexible enough to cope with both short and long-term environmental change, including the onset of ENSO, but maybe not extreme perturbations when a number of phenomena coincide. To paraphrase Binford (1983:203), a system changes when people are unable to continue previously successful strategies in the face of major changes in the ecological relationship between humans and their environment. I argue in this study that a combination of environmental and particular historical social conditions resulted in major cultural change, with cessation of the mound building phase of Australian Aboriginal history that had continued for millennia until around 500 years BP, a time of flux, of major perturbations. This could include the combined effects in the last millennium, of reaching a long-term threshold of the ecosystem (Butzer 1982:27) as the Hope Inlet estuary developed towards its final phase, cultural contact (either Macassan or other unknown) leading to introduction of foreign ideas and goods, and introduced disease of epidemic proportions. The interactive human/environment relationship of the last 1000 years then played out according to the particular historical socio-cultural behaviour and to perceptions of the cumulative environmental and social changes that had occurred.

References

Abbott, R.T. and S.P. Dance 1982 *Compendium of Seashells*. E.P. Dutton, Inc., New York.

Abrams, Jr. H.L. 1987 The preference for animal protein and fat: a cross-cultural survey. In M. Harris and E.B. Ross (eds) *Food and Evolution: Toward a Theory of Human Food Habits*, pp.207-23. Temple University Press, Philadelphia.

Akazawa, T. 1986a Hunter-gatherer adaptations and the transition to food production in Japan. In M. Zvelebil (ed.) *Hunters in Transition*, pp.151-65. Cambridge University Press, London.

Akazawa, T. 1986b Regional variation in procurement systems of Jomon Hunter-Gatherers. In T. Akazawa and C.M. Aikens (eds) *Prehistoric Hunter-Gatherers in Japan*, pp.73-89. University of Tokyo Press, Tokyo.

Ahmad, R. 1996 Late Holocene major Australian arid period revealed by direct sedimentological evidence from lakes in the Coorong region of South Australia. *Geology* 24(7):619-22.

Alford, B. 2000 A heritage survey of Mitchell Creek. Unpublished report to Palmerston Town Council, Palmerston.

Allen, F.J. 1969 Archaeology, and the History of Port Essington. Unpublished PhD thesis, Australian National University, Canberra.

Allen, H. 1972 Where the Crow Flies Backwards: Man and Land in the Darling Basin. Unpublished PhD Thesis, The Australian National University, Canberra.

Allen, H. 1987 Holocene mangroves and middens in northern Australia and Southeast Asia. *Bulletin of the Indo-Pacific Prehistory Association* 7:1-16.

Allen, H. 1989 Late Pleistocene and Holocene settlement patterns and environment, Kakadu, Northern Territory, Australia. *Bulletin of the Indo-Pacific Prehistory Association* 9:92-116.

Allen, H. 1996 The time of the mangroves: changes in mid-Holocene estuarine environments and subsistence in Australia and Southeast Asia. In I.C. Glover and P. Bellwood (eds) *Indo-Pacific Prehistory: The Chiang Mai Papers*, *Bulletin of the Indo-Pacific Prehistory Association* 15:193-205.

Allen, H. 1997 The distribution of large blades: evidence for recent changes in Aboriginal ceremonial exchange networks. In P. McConvell and N. Evans (eds) *Archaeology and Linguistics: Aboriginal Australia in Global Perspective*, pp.357-76. Oxford University Press, Melbourne.

Allen, H. 1998 Reinterpreting the 1969-1972 Willandra Lakes archaeological surveys. *Archaeology in Oceania*. 33:207-20.

Allen, H. and G. Barton n.d. *Ngarradj Warde Djobkeng. White Cockatoo Dreaming and the Prehistory of Kakadu*. Oceania Monograph 37. University of Sydney, Sydney.

Ambrose, W.R. 1967 Archaeology and shell middens. *Archaeology and Physical Anthropology in Oceania* 2:169-87.

Amesbury, J.R. 1999 Changes in species composition of archaeological marine shell assemblages in Guam. *Micronesica* 31(2):347-66.

Ammerman, A.J. 1981 Surveys and archaeological research. *Annual Review of Anthropology* 10:63-88.

Amorosi, T., J. Woollett, S. Perdakis and T. McGovern 1996 Regional zooarchaeology and global change: problems and potentials. *World Archaeology* 28(1):126-57.

Anderson, C. and R. Robins 1988 Dismissed due to lack of evidence? Kuku-Yalanji sites and the archaeological record. In B. Meehan and R. Jones (eds) *Archaeology with Ethnography: an Australian Perspective*, pp.182-205. Department of Prehistory, Research School of Pacific Studies, Australian National University, Canberra.

Anderson, J. 1984 *Between Plateau and Plain: Flexible Responses to Varied Environments in Southwestern Australia. Occasional Papers in Prehistory* 4, Department of Prehistory, Research School of Pacific Studies, Australian National University, Canberra.

Aplin, K. 1981 The Kamapuk fauna. Unpublished B.A. Hons thesis, Australian National University, Canberra.

Attenbrow, V.J. 1982 The archaeology of Upper Mangrove Creek catchment: research in progress. In S. Bowdler (ed.) *Coastal Archaeology in Eastern Australia: Proceedings of the 1980 Valla Conference on Australian Prehistory. Occasional Papers in Prehistory* 11, pp.67-78. Department of Prehistory, Research School of Pacific Studies, Australian National University, Canberra.

Attenbrow, V. 1987 The Upper Mangrove Creek Catchment. A study of quantitative changes in the archaeological record. Unpublished PhD thesis, Sydney University.

Attenbrow, V. 1992 Shell bed or shell midden. *Australian Archaeology* 34:3-21.

Austin, T. 1997 *Never Trust a Government Man*, NTU Press, Darwin.

Bailey, G.N. 1975a The Role of Shell Middens in Prehistoric Economies, Unpublished PhD thesis, 2Vol., University of Cambridge, Cambridge.

Bailey, G.N. 1975b The role of molluscs in coastal economies: the results of midden analysis in Australia. *Journal of Archaeological Science* 2:45-62.

Bailey, G.N. 1977 Shell mounds, shell middens, and raised beaches in the Cape York Peninsula. *Mankind* 11:132-43.

Bailey, G.N. 1983 Problems of site formation and the interpretation of spatial and temporal discontinuities in the distribution of coastal middens. In P.M. Masters and N.C. Fleming (eds) *Quaternary Coastlines and Marine Archaeology: Towards the Prehistory of Land Bridges and Continental Shelves*, pp.559-82. Academic Press, London.

Bailey, G.N. 1991 Hen's eggs and cockle shells: Weipa shell mounds reconsidered. *Archaeology in Oceania* 26:21-3.

Bailey, G.N. 1993 Shell Mounds in 1972 and 1992: reflections on recent controversies at Ballina and Weipa. *Australian Archaeology* 37:1-18.

Bailey, G.N. 1994 The Weipa shell mounds: natural or cultural? In M. Sullivan, S. Brockwell and A. Webb (eds), *Archaeology in the North: Proceedings of the 1993 Australian Archaeological Conference*, pp.107-29. North Australian Research Unit, Australian National University, Darwin.

Bailey, G.N. 1999 Shell mounds and coastal archaeology in northern Queensland. In Hall, J. and I. McNiven 1999 (eds) *Australian Coastal Archaeology*, pp.105-12. ANH Publications, Department of Archaeology and Natural History, RSPAS, The Australian National University, Canberra.

Bailey, G. and J. Parkington 1988 The archaeology of prehistoric coastlines: an introduction. In G. Bailey, and J. Parkington (eds) *The Archaeology of Prehistoric Coastlines*, pp. 1-10. Cambridge University Press, Cambridge.

Bailey, G., J. Chappell and R. Cribb 1994 The origin of *Anadara* shell mounds at Weipa, North Queensland, Australia. *Archaeology in Oceania* 29:69-80.

Baker, C.M. 1978 The size effect: an explanation of variability in surface artifact assemblage content. *American Antiquity* 43(2):288-93.

Baker, R. 1981 The Aboriginal and Environmental History of the Chambers Bay Coastal Plains. Unpublished Honours thesis, Department of Prehistory and Anthropology, Australian National University, Canberra.

Baker, V.R. and G. Pickup 1987 Flood geomorphology of the Katherine Gorge, Northern Territory, Australia. *Geological Society of America Bulletin* 98:635-46.

Baker, V.R., G. Pickup and R.H. Webb 1987 Palaeoflood hydrologic analysis at ungauged sites, central and northern Australia. In V.P. Singh (ed.) *Regional Flood Frequency Analysis*, pp.325-38. D. Reidel, Boston.

Balme, J. 1991 The antiquity of grinding stones in semi-arid Western New South Wales. *Australian Archaeology* 32:3-9.

Balme, J. and W. Beck 1996 Earth mounds in southeastern Australia. *Australian Archaeology* 42:39-49.

Barclay, A. 1937 What of Darwin. *Walkabout* 4(2):13-18.

Barker, B. 1991 Nara Inlet 1: coastal resource use and the Holocene marine transgression in the Whitsunday Islands, central Queensland. *Archaeology in Oceania* 26:102-9.

Barker, B. 1999 Coastal occupation in the Holocene: environment, resource use and resource continuity. In Hall, J. and I. McNiven 1999 (eds) *Australian Coastal Archaeology*, pp.119-27. ANH Publications, Department of Archaeology and Natural History, RSPAS, The Australian National University, Canberra.

Barnes, R.S.K. 1974 *Estuarine Biology*. The Institute of Biology's Studies in Biology No.49, Edward Arnold, London.

Barz, R.K. 1977 Some theoretical and practical aspects of midden sampling as applied to a site at St. George's Basin, Jervis Bay, A.C.T., Unpublished BA (Hons) thesis, Australian National University, Canberra.

Basedow, H. 1907 Anthropological notes on the western coastal tribes of the Northern Territory of South Australia. *Transaction and Proceedings and Report of the Royal Society of South Australia*, 31:1-55.

Basedow, H. 1925 *The Australian Aboriginal*, F.W. Preece and Sons, Adelaide.

Bauer, F.H. 1964 *Historical geography of white settlement in part of Northern Australia. Part 2. The Katherine-Darwin Region*. C.S.I.R.O. Canberra.

Bauer, F.H. 1984 What man hath wrought: geography and change in northern Australia. In D. Parkes (ed.) *Northern Australia: The arenas of life and ecosystems on half a continent*, pp.1-20. Academic Press, Sydney.

Bayliss, P. and K.M. Yeomans 1990 Seasonal distribution and abundance of magpie geese, *Anseranas semipalmata* Latham, in the Northern Territory, and their relationship to habitat, 1983-86. *Australian Wildlife Research* 17:15-38.

Beaton, J.M. 1977 Dangerous Harvest: Investigations in the late prehistoric occupation of upland south-east central Queensland. Unpublished PhD thesis, Australian National University, Canberra.

Beaton, J.M. 1982 Fire and water: aspects of Australian Aboriginal management of cycads. *Archaeology in Oceania* 17(1):51-8.

Beaton, J.M. 1983 Does intensification account for changes in the Australian Holocene archaeological record? *Archaeology in Oceania* 18:94-7.

Beaton, J.M. 1985 Evidence for a coastal occupation time-lag at Princess Charlotte Bay (North Queensland) and implications for coastal colonization and population growth theories for Aboriginal Australia. *Archaeology in Oceania* 20:1-20.

Beaton, J.M. 1986 Reply to Cribb. *Archaeology in Oceania* 21:175-6.

Beaton, J.M. 1990 The importance of past population for prehistory. In B. Meehan and N. White (eds) *Hunter-Gatherer Demography, Past and Present, Oceania Monograph* 39, pp.23-40. University of Sydney, Sydney.

Beck, W., A. Clarke and L. Head 1989 Plants in hunter-gatherer archaeology. In W. Beck, A. Clarke and L. Head (eds) *Plants in Australian Archaeology. Tempus* 1, pp.1-10. University of Queensland, St. Lucia, Queensland.

Beck, W., R. Fullagar and N. White 1988 Archaeology from ethnography: the Aboriginal use of cycad as an example. In B. Meehan and R. Jones (eds) *Archaeology with Ethnography: An Australian Perspective*, pp.137-47. Department of Prehistory, Research School of Pacific Studies, Australian National University, Canberra.

Beesley, P.L., G.J.B. Ross and A. Wells (eds) 1998 *Mollusca: The Southern Synthesis. Fauna of Australia* Vol.5. CSIRO Publishing, Melbourne, 2 Vols.

Behrensmeyer, A.K. 1975 The taphonomy and paleoecology of Plio-Pleistocene vertebrate assemblages east of Lake Rudolf, Kenya.

Behrensmeyer, A.K. and A.P. Hill (eds) 1980 *Fossils in the making: vertebrate taphonomy and palaeoecology*. University of Chicago Press: Chicago.

Bellwood, P. 1985 *Prehistory of the Indo-Malaysian Archipelago*. Academic Press, Sydney.

Berndt, R.M. 1951a Ceremonial exchange in western Arnhem Land. *Southwestern Journal of Anthropology* 7(2):156-76.

Berndt, R.M. 1951b *Kunapipi: a Study of an Australian Aboriginal Religious Cult*. F.W. Cheshire, Melbourne.

Berndt, R.M. 1979 *A Profile of Good and Bad in Australian Aboriginal Religion*. Reprint of Colloquium Journal of the A.N.Z.S.T.S., Melbourne.

Berndt, R.M. and C.H. Berndt 1964/1977 *The World of the First Australians*. Revised edition/2nd edition. Ure Smith, Sydney.

Berndt, R.M. and C.H. Berndt 1970 *Man, Land and Myth in North Australia. The Gunwinggu People*. Ure Smith, Sydney.

Binford, L.R. 1977 Forty-seven trips: a case study in the character of archaeological formation processes. In R. Wright (ed.) *Stone Tools as Cultural Markers: change, evolution and complexity*, pp.24-36. Australian Institute of Aboriginal Studies, Canberra.

Binford, L.R. 1979 Organisation and formation processes: looking at curated technologies. *Journal of Anthropological Research* 35:255-73.

Binford, L.R. 1980 Willow smoke and dog's tails: Hunter-gatherer settlement systems and archaeological site formation. *American Antiquity* 45(1):4-20.

Binford, L.R. 1981 *Bones: Ancient Men and Modern Myths*. Academic Press, New York.

Binford, L.R. 1982 The archaeology of place. *Journal of Anthropological Archaeology* 1:5-31.

Binford, L.R. 1983 *In Pursuit of the Past: Decoding the Archaeological Record*. Thames and Hudson, London.

Bird, C. and C. Beeck 1988 Traditional plant foods in the southwest of western Australia: the evidence from salvage ethnography. In B. Meehan and R. Jones (eds) *Archaeology with Ethnography: An Australian Perspective*, pp.113-22. Department of Prehistory, Research School of Pacific Studies, Australian National University, Canberra.

Bird, D.W. and R.L. Bliege-Bird 1997 Contemporary shellfish gathering strategies among the Meriam of the Torres Strait islands, Australia: testing predictions of a central place foraging model. *Journal of Archaeological Science* 24:39-63.

Bird, C.F.M. and D. Frankel 1991a Chronology and explanation in western Victoria and south-east South Australia. *Archaeology in Oceania* 26:1-16.

Bird, C.F.M. and D. Frankel 1991b Problems in constructing a prehistoric regional sequence: Holocene south-east Australia. *World Archaeology* 23(2):179-92.

Bird, E.C.F. 1964 *Coastal Landforms*. The Australian National University, Canberra.

Bird, M.K. 1992 The impact of tropical cyclones on the archaeological record: an Australian example. *Archaeology in Oceania* 27(2):75-85.

Bishop, K.A. and M.A. Forbes 1991 The freshwater fishes of northern Australia. In C.D. Haynes, M.G. Ridpath and M.A.J. Williams (eds) *Monsoonal Australia: Landscape, Ecology and Man in the Northern Lowlands*, pp.79-107. A.A. Balkema, Rotterdam.

Blackburn, H. 1982 *Marine Shells of the Darwin Area*. Museums and Art Galleries of the Northern Territory, Darwin.

Bloch, M. 1977 The past and the present in the present. *Man* 12:278-92.

Boot, P.G. 1993 Analysis of resins and other plant residues on stone artefacts from Graman, New South Wales. In B.L. Fankhauser and J.R. Bird (eds) *Archaeometry: Current Australasian Research, Occasional Papers in Prehistory*, 22, pp.3-12. Department of Prehistory, Research School of Pacific Studies, The Australian National University, Canberra.

Botkin, S. 1980 Effects of human exploitation on shellfish populations at Malibu Creek, California. In T.K. Earle and A.L. Christenson (eds) *Modeling Change in Prehistoric Subsistence Economies*, pp.121-39. Academic Press Inc., New York.

Bowdler, S. 1976 Hook, line, and dilly bag: an interpretation of an Australian coastal shell midden. *Mankind* 10:248-58.

Bowdler, S. 1977 The coastal colonisation of Australia. In J. Allen, J. Golson and R. Jones (eds) *Sunda and Sahul: Prehistoric Studies in Southeast Asia, Melanesia, and Australia*, pp.205-46. Academic Press, New York.

Bowdler, S. 1981a Hunters in the highlands: Aboriginal adaptations in the eastern Australian uplands. *Archaeology in Oceania* 16:99-111.

Bowdler, S. 1981b Comment on B. Hayden 1981 Research and development in the Stone Age: technological transitions among hunter-gatherers. *Current Anthropology* 22(5):531.

Bowdler, S. 1983 Sieving seashells: midden analysis in Australian archaeology. In G. Connah (ed.) *Australian Field Archaeology: A Guide to Techniques*, pp.135-44. Australian Institute of Aboriginal Studies, Canberra.

Bowdler, S. 1990 The Silver Dollar site, Shark Bay: an interim report. *Australian Aboriginal Studies* 2:60-3.

Bowdler, S. and S. O'Connor 1991 The dating of the Australian Small Tool Tradition, with new evidence from the Kimberley, WA. *Australian Aboriginal Studies* 1:53-62.

Bowen, G.M. 1996 Culture shock the prehistory of the Magela floodplain, Kakadu National Park, and its implications for analogy. Unpublished MA Thesis, Northern Territory University.

Bowler, J.M., G.S. Hope, J.N. Jennings, G. Singh and D. Walker 1976 Late Quaternary climates of Australia and New Guinea. *Quaternary Research* 6:359-94.

Bowman, D.M.J.S. 1988 Stability amid turmoil?: towards an ecology of north Australian eucalypt forests. *Proceedings of the Ecological Society of Australia* 15:149-58.

Bowman, D.M.J.S. 1992 Evidence for gradual retreat of dry monsoon forests under a regime of Aboriginal burning, Karslake Peninsula, Melville Island, northern Australia. *Proceedings of the Royal Society of Queensland* 102:25-30.

Bowman, D.M.J.S. and W.J. Panton 1993 Decline of *Callitris intratropica* R.T. Baker & H.G. Smith in the Northern Territory: implications for pre-and post-European colonisation fire regimes. *Journal of Biogeography* 20:373-81.

Bowman, D.M.J.S. and G.M. Wightman 1985 Small scale vegetation pattern associated with a deeply incised gully, Gunn Point, Northern Australia. *Proceedings of the Royal Society of Queensland* 96:63-73.

Bowman, D.M.J.S. and B.A. Wilson 1986 Wetland vegetation pattern on the Adelaide River flood plain, Northern Territory, Australia. *Proceedings of the Royal Society of Queensland* 97:69-77.

Bowman, D.M.J.S., W.J. Panton and J. Head 1999 Abandoned orange-footed Scrubfowl (*Megapodius reinwardt*) nests and coastal rainforest boundary dynamics during the late Holocene in monsoonal Australia. *Quaternary International* 59:27-38.

Bowman, G.M. 1985 Oceanic reservoir correction for marine radiocarbon dates from northwestern Australia. *Australian Archaeology* 20:58-67.

Bowman, G. and N. Harvey 1983 Radiocarbon dating marine shells in South Australia. *Australian Archaeology* 17:113-23.

Boyd, S.E. 1998 Order Arcoida. In P.L. Beesley, G.J.B. Ross and A. Wells (eds) *Mollusca: The Southern Synthesis. Fauna of Australia* Vol.5, Part A, pp.253-61. CSIRO Publishing, Melbourne.

Bradley, J.J. 1997 Li-Anthawirriyarra, people of the sea: Yanyuwa relations with their maritime environment. Unpublished PhD thesis, Northern Territory University, Darwin.

Bradley, R. 1991 Ritual, time and history. *World Archaeology* 23(2):209-219.

Bradshaw, E. 1995 Dates from archaeological excavations on the Pilbara coastline and islands of the Dampier Archipelago, Western Australia. *Australian Archaeology* 41:37-8.

Braithwaite, R.W., G.R. Friend and J.C. Wombey 1991 Reptiles and amphibians. In C.D. Haynes, M.G. Ridpath and M.A.J. Williams (eds) *Monsoonal Australia: Landscape, Ecology and Man in the Northern Lowlands*, pp.109-24. A.A. Balkema, Rotterdam.

Braun, D.P. 1974 Explanatory models for the evolution of coastal adaptation in prehistoric Eastern New England. *American Antiquity* 39(4):582-96.

Brennan, L.A. 1977 The midden is the message. *Archaeology of Eastern North America* 5:122-137.

Brocklehurst, P. and B. Edmeades 1996 *Mangrove Survey of Darwin Harbour, Northern Territory, CCNT/NFI Project 1994-5*. Technical Report No. 96/7. Department of Lands, Planning and Environment, Darwin.

Brockwell, C.J. 1989 Archaeological investigations of the Kakadu Wetlands, Northern Australia. Unpublished MA thesis, Australian National University, Canberra.

Brockwell, S. 1993 Archaeological survey of Lambells Lagoon subdivision and buffer zone. NTUAS Report 20. Unpublished report to Department of Lands, Housing and Local Government, Darwin.

Brockwell, S. 1996a Open Sites of the South Alligator River Wetland, Kakadu. In Veth, P. and Hiscock, P. (eds) *Archaeology of Northern Australia, Tempus* 4, pp.90-105. Anthropology Museum, The University of Queensland, St. Lucia, Queensland.

Brockwell S. 1996b Prehistoric settlement patterns on the Adelaide River, northern Australia. *Australian Aboriginal Studies* 1:45-59.

Brockwell S. 1996c Mound sites on the Adelaide River coastal plains. In I.C. Glover and P. Bellwood (eds) *Indo-Pacific Prehistory: The Chiang Mai Papers, Bulletin of the Indo-Pacific Prehistory Association* 15:159-64.

Broom, M.J. 1982 Structure and seasonality in a Malaysian mudflat community. *Estuarine, Coastal and Shelf Science* 15:135-50.

Broom, M.J. 1985 *The Biology and Culture of Marine Bivalve Mollusca of the Genus Anadara*, International Centre for Living Aquatic Resources Management, Manila, Philippines.

Broughton, J.M. 1997 Widening diet breadth, declining foraging efficiency, and prehistoric harvesting pressure: ichthyofaunal evidence from the Emeryville Shellmound, California. *Antiquity* 71:845-62.

Brown, H.Y.L. 1906 Northern Territory of South Australia. *South Australian Parliamentary Papers* 11(55).

Brown, J.A. 1975 Deep-site excavation strategy as a sampling problem. In J.W. Mueller (ed.) *Sampling in Archaeology*, pp.155-69. University of Arizona Press, Tuscon, Arizona.

Bruce, A.J. 1988 The shrimp fauna of a small tropical reef, the East Point fish reserve, Darwin. In H.K. Larson, M.G. Michie and J.R. Hanley (eds) *Proceedings of the Workshop on Research and Management held in Darwin 2-3 September, 1987*, *Mangrove Monograph* 4, pp.226-45. Australian National University, North Australian Research Unit, Darwin.

Buchanan, W.F. 1988 *Shellfish in Prehistoric Diet: Elands Bay, S. W. Cape Coast, South Africa*, *Cambridge Monographs in African Archaeology* 31, *BAR International Series* 455, Oxford, England.

Burns, T. 1994 Mound Over Matter: Origins of Shell and Earth Mounds of Northern Australia: An Evaluation of Mounds on Channel Island and Middle Arm Mainland, Darwin Harbour. Unpublished Honours thesis, Department of Anthropology, Northern Territory University, Darwin.

Burns, T. 1996a The Northern Territory Archaeological Survey: Stage 1. Darwin Catchment Region. Unpublished report to the Australian Heritage Commission. Heritage Conservation Branch, Department of Lands, Planning and Environment, Darwin.

Burns, T. 1996b Archaeological Survey of the Proposed Darwin-Alice Springs Railway Corridor, Katherine Origin 177.0km to Adelaide River Origin 26.0km. Unpublished *report to the Department of Transport and Works, Transport Infrastructure Branch, Northern Territory. November 1996.*

Burns, T. 1997 A Report on Aboriginal Shell Midden Sites at Winnellie, Darwin Harbour. Unpublished report to the Department of Lands, Planning and Environment, Heritage Conservation Branch. June 1997.

Burns, T. 1999 Stage 1 Archaeological Survey of the Proposed Middle Arm Berry Springs Connector Road and Rural Residential Subdivision. Unpublished *report to the Department of Lands, Planning and Environment,* Land Development Division *July 1999.*

Butlin, N. 1985 Macassans and smallpox. Historical Studies 84(21):315-35.

Butzer, K.W. 1972 *Environment and Archaeology: An Ecological Approach to Prehistory.* 2nd ed. Methuen, London.

Butzer, K.W. 1982 *Archaeology as Human Ecology: Method and Theory for a Contextual Approach.* Cambridge University Press, Melbourne.

Cahen, D. and J. Moeyersons 1977 Subsurface movements of stone artefacts and their implications for the prehistory of Central Africa. *Nature* 266:812-5.

Caldwell Engineers Pty. Ltd. 1983 *Environmental Impact Statement for Channel Island Power Station*, Darwin.

Cameron, C.M. 1993 Abandonment and archaeological interpretation. In C.M. Cameron and S.A. Tomka (eds) *Abandonment of settlements and regions. Ethnoarchaeological and archaeological approaches.* Cambridge University Press, Cambridge, 3-7.

Campbell, J. 1983 Smallpox in Aboriginal Australia, 1829-31. *Historical Studies* 20:536-56.

Campbell, J. 1985 Smallpox in Aboriginal Australia, the early 1830's. *Historical Studies* 21:336-58.

Cann, J.H., P. De Deckker and C.V. Murray-Wallace 1991 Coastal Aboriginal shell middens and their palaeoenvironmental significance, Robe Range, South Australia. *Transactions of the Royal Society of South Australia* 115(4):161-75.

Casteel, R.W. 1976 *Fish Remains in Archaeology and Paleo-environmental Studies.* Academic Press, London.

Castilla, J.C., S.A. Navarrete and J. Lubchenco 1993 Southeastern Pacific coastal environments: main features, large-scale perturbations, and global climate change. In H.A. Mooney, E.R. Fuentes and B.I. Kronberg (eds) *Earth system responses to global change*, pp.167-88. Academic Press, Sydney.

Catterall, C.P. and I.R. Poiner 1984 Distribution, abundance and resilience to gathering of macro-molluscs in the seagrass beds of North Stradbroke Island. In R.J. Coleman, J. Covacevich and P. Davie (eds) *Focus on Stradbroke: New Information on North Stradbroke Island and Surrounding Areas, 1974-1984*, pp.291-9. Stradbroke Island Management Organization, Amity Point, Boolarong Publications, Brisbane.

Catterall, C.P. and I.R. Poiner 1987 The potential impact of human gathering on shellfish populations with reference to some NE Australian intertidal flats. *Oikos* 50:114-122.

Ceci, L. 1984 Shell midden deposits as coastal resources. *World Archaeology* 16:62-74.

Chaloupka, G. 1981 The traditional movement of a band of Aboriginals in Kakadu. In T. Stokes (ed.) *Kakadu National Park Educational Resources*, pp.162-71. Australian National Parks and Wildlife Service, Canberra.

Chaloupka, G. 1985 Chronological sequence of Arnhem Land plateau rock art. In R. Jones (ed.) *Archaeological Research in Kakadu National Park, Special Publication* 13, pp.269-80. Australian National Parks and Wildlife Service, Canberra.

Chaloupka, G. 1993 *Journey in Time*. Reed, Australia.

Chapman, E.C. 1956 Pioneer settlement in Darwin's hinterland 1911-57. *The Australian Geographer* 7(3);113-26.

Chappell, J. 1982 Sea levels and sediments: Some features of the context of coastal archaeological sites in the tropics. *Archaeology in Oceania* 17:69-78.

Chappell, J. and J. Grindrod 1984 Chenier plain formation in northern Australia. In B.G. Thom (ed.) *Coastal Geomorphology in Australia*, pp.197-231. Academic Press, Sydney.

Chappell, J. and C. Woodroffe 1985 Morphodynamics of Northern Territory tidal rivers and floodplains. In K.N. Bardsley, J.D.S. Davie and C.D. Woodroffe (eds) *Coasts and Tidal Wetlands of the Australian Monsoon Region, Mangrove Monograph* 1, 85-96. North Australia Research Unit, Australian National University, Darwin.

Chappell, J., E.G. Rhodes, B.G. Thom and E. Wallensky 1982 Hydro-isostacy and the sea level isobase of 5500BP in north Queensland, Australia. *Marine Geology* 49:81-90.

Chase, A.K. and P. Sutton 1981 Hunter-gatherers in a rich environment: Aboriginal coastal exploitation in Cape York Peninsula. In A. Keast (ed.) *Ecological Biogeography in Australia*, Vol.3, pp.1818-1852. Dr. Junk, The Hague.

Chaseling, W.S. 1957 *Yulengor: Nomads of Arnhem Land.* Epworth Press, London.

Christian, C.S. and G.A. Stewart 1953 *General Report on Survey of Katherine-Darwin Region, 1946, Land Research Series* No.1. *Commonwealth Scientific and Industrial Research Organization*, Melbourne, Australia.

Cities Commission 1974 Lee Point draft environmental impact statement. Unpublished report to the Department of the Northern Territory, Darwin.

Claassen, C. 1986 Shellfishing seasons in the prehistoric southeastern United States. *American Antiquity* 51(1):21-37.

Claassen C. 1991a Gender, shellfishing and the shell mound archaic. In J.M. Gero and M.W. Conkey (eds) *Engendering Archaeology: Women and Prehistory*, pp.276-99. Blackwell, Oxford, UK.

Claassen C. 1991b Normative thinking and shell-bearing sites. In M.B. Schiffer (ed.) *Archaeological Method and Theory* 3, pp.249-98. University of Arizona Press, Tuscon, Arizona.

Claassen, C. 1996 Research problems with shells from Green River shell matrix sites. In C.C. Carstens and P.J. Watson (eds) *Of Caves and Shell Mounds*, pp.132-9. The University of Alabama Press, London.

Claassen, C. 1998 *Shells*. Cambridge University Press, Melbourne.

Clark, P. and M. Barbetti 1982 Fires, hearths and palaeomagnetism. In W. Ambrose and P. Duerden (eds) *Archaeometry: An Australasian Perspective*, pp.144-50. Department of Prehistory, Research School of Pacific Studies, The Australian National University, Canberra.

Clark, R.L. and J.C Guppy 1988 A transition from mangrove forest to freshwater wetland in the monsoon tropics of Australia. *Journal of Biogeography* 15:665-84.

Clark, R.L., J. Guppy, D. Mahon, P. McBride and R.J. Wasson 1992a Late Quaternary stratigraphy of the Magela plain. In R.J. Wasson (ed.) *Modern Sedimentation and Late Quaternary Evolution of the Magela Creek Plain*, pp. 28-80. Supervising scientist for the Alligator Rivers region, Research Report 6, Darwin.

Clark, R.L., J. Guppy, D. Mahon, P. McBride and R.J. Wasson 1992b Late Quaternary evolution of the Magela plain. In R.J. Wasson (ed.) *Modern Sedimentation and Late Quaternary Evolution of the Magela Creek Plain*, pp.81-157. Supervising scientist for the Alligator Rivers region, Research Report 6, Darwin.

Clarke, A. 1985 A preliminary archaeobotanical analysis of the Anbangbang 1 site. In R. Jones, (ed.) *Archaeological Research in Kakadu National Park, Special Publication* 13, pp.77-96. Australian National Parks and Wildlife Service, Canberra.

Clarke, A. 1988 Archaeological and ethnobotanical interpretations of plant remains from Kakadu National Park, Northern Territory. In B. Meehan and R. Jones (eds) *Archaeology with Ethnography: An Australian Perspective*, pp.123-136. Department of Prehistory, Research School of Pacific Studies, Australian National University, Canberra.

Clarke, A. 1989 Macroscopic plant remains. In W. Beck, A. Clarke and L. Head (eds) *Plants in Australian Archaeology. Tempus.*1, pp.54-89. University of Queensland, St. Lucia, Queensland.

Clarke, A.F. 1994 Winds of Change: An archaeology of contact in the Groote Eylandt archipelago, Northern Australia. Unpublished PhD thesis, Australian National University, Canberra.

Clarke, A. 1999 Preservation profiles: a case study from Anbangbang 1, Kakadu National Park, Northern Territory. In M.J. Mountain and D. Bowdery (eds) *Taphonomy: The Analysis of Processes from Phytoliths to Megafauna*, pp.83-91. ANH Publications, Department of Archaeology and Natural History, Research School of Pacific and Asian Studies, The Australian National University, Canberra.

Clarke, M.F., R.J. Wasson and M.A.J. Williams 1979 Point Stuart chenier and Holocene sea levels in northern Australia. *Search* 10(3):90-2.

Clinch, M. 1990 On the edge of the Paperbark swamp: the role of the Buffalo in the NT 1920-1956. *Northern Perspective* 13(1):14-24.

Cogger, H.C. 1992 *Reptiles and Amphibians of Australia.* Reed Books, Sydney.

Cogger, H.G. and D.A. Lindner 1969 Marine turtles in northern Australia. *Australian Zoologist* 15(2):150-9.

Cohen, M.N. 1985 Prehistoric hunter-gatherers: the meaning of social complexity. In T.D. Price and J.A. Brown (eds) *Prehistoric Hunter-Gatherers: The Emergence of Cultural Complexity*, pp.99-115. Academic Press, Orlando, Florida.

Cohen, A., J. Parkington, G. Brundrit and N. van der Merwe 1992 A Holocene marine climate record in mollusc shells from the southwest African coast. *Quaternary Research* 38:379-83.

Colley, S.M. 1990 The analysis and interpretation of archaeological fish remains. In M.B. Schiffer (ed.) *Archaeological Method and Theory.*2, pp.207-53. The University of Arizona Press, Tuscon.

Colley, S.M. 1997 A pre- and post-contact Aboriginal midden at Disaster Bay, New South Wales south coast. *Australian Archaeology* 45:1-19.

Collier, S. and K. Hobson 1987 The importance of marine protein in the diet of coastal Australian Aborigines. *Current Anthropology* 28(4):559-64.

Conservation Commission of the Northern Territory n.d. Gunn Point Area. Draft plan of management (CCNT 9/19 Environment Unit), Museums and Art Galleries of the Northern Territory, Darwin.

Conservation Commission of the Northern Territory 1990a Howard Springs nature park and hunting reserve. Draft plan of management, June 1990, Museums and Art Galleries of the Northern Territory, Darwin.

Conservation Commission of the Northern Territory 1990b Top End wetlands conservation and recreation development strategy. Museums and Art Galleries of the Northern Territory, Darwin.

Cook, P.J. and H.A. Polach 1973 A chenier sequence at Broad Sound, Queensland, and evidence against a Holocene high sea level. *Marine Geology* 14:253-68.

Cotterell, B. and J. Kamminga 1990 *Mechanics of Pre-Industrial Technology*. Cambridge University Press, Cambridge.

Coutts, P.J.F. 1966 Features of prehistoric campsites in Australia. *Mankind* 6(8):338-46.

Coutts, P.J.F. 1970 *The Archaeology of Wilson's Promontory*, Australian Institute of Aboriginal Studies, Canberra.

Coutts, P.J.F., D. Witter, M. McIlwraith and R. Frank 1976 The mound people of Western Victoria. *Records of the Victorian Archaeological Survey 1.*

Cowlishaw, G. 1992 Studying Aborigines: changing canons in anthropology and history. In B. Attwood and J. Arnold (eds) *Power, Knowledge and Aborigines, Journal of Australian Studies, Special Edition* 35, pp.20-31. La Trobe University, Melbourne.

Crassweller, C. 1996 Chronological changes in the archaeological material at Scotch Creek I, an open site near the Adelaide River, N.T. Unpublished BA (Hons) thesis, Department of Anthropology, Northern Territory University, Darwin.

Cribb, R. 1986a A preliminary report on archaeological findings in Aurukun Shire, western Cape York. *Queensland Archaeological Research* 3:133-58.

Cribb, R. 1986b Comment: Sites, people and archaeological information traps: a further transgressive episode from Cape York. *Archaeology in Oceania* 21:166-71.

Cribb, R. 1991 Getting into a flap about shell mounds in northern Australia: a reply to Stone. *Archaeology in Oceania* 26:23-5.

Cribb, R. 1996 Shell mounds, domiculture and ecosystem manipulation on western Cape York Peninsula. In P. Veth, and P. Hiscock (eds) *Archaeology of Northern Australia Tempus* 4, pp.150-73. Archaeology and Material Culture Studies in Anthropology, Anthropology Museum, University of Queensland, St. Lucia, Queensland.

Crome, F.H.J. and H.E. Brown 1979 Notes on social organisation and breeding of the Orange-footed Scrubfowl *Megapodius reinwardt*. *Emu* 79:111-9.

Cribb, R., R. Walmbeng, R. Wolmby, and C. Tasman 1988 Landscape as cultural artefact: Shell mounds and plants in Aurukun, Cape York Peninsula. *Australian Aboriginal Studies* 2:60-73.

Crowe, T.P. 1997 The distribution of *Terebralia palustris* (Linne) with respect to microhabitat in mangrove forests of Darwin Harbour 1: description of patterns. In H.K. Larson, M.G. Michie and J.R. Hanley (eds) *Proceedings of the Workshop on Research and Management held in Darwin 2-3 September, 1987, Mangrove Monograph* 4, pp.425-33. Australian National University, North Australian Research Unit, Darwin.

Crumley, C.L. 1994 Historical ecology: a multidimensional ecological orientation. In C.L. Crumley (ed.) *Historical Ecology: Cultural Knowledge and Changing Landscapes*, pp.1-16. School of American Research Press, Santa Fe, New Mexico.

Curr, E.M. 1886 *The Australian Race.* Vol. 1. John Ferres, Melbourne.

Dahl, K. 1927 *In Savage Australia: An account of a hunting and collecting expedition to Arnhem Land and Dampier Land.* Houghton Mifflin Company, Boston.

David, B. 1990 How was this bone burnt? In S. Solomon, I. Davidson and D. Watson (eds) *Problem Solving in Taphonomy, Tempus*.2, pp.65-79. The University of Queensland, St. Lucia, Queensland.

David, B. 1991 Fern cave, rock art and social formations: rock art regionalisation and demographic models in southeast Cape York Peninsula. *Archaeology in Oceania* 26(2):41-57.

David, B. and D. Chant 1995 Rock art and regionalisation in north Queensland prehistory. *Memoirs of the Queensland Museum* 37(2):357-528.

David, B. and H. Lourandos 1998 Rock art and socio-demography in northeast Australia. *World Archaeology* 30(2):193-219.

David, B. and J. Stanisic 1991 Land snails in Australian archaeology: initial results from Echidna's Rest (north Queensland). *The Artefact* 14:19-24.

David, B. and M. Wilson 1999 Re-reading the landscape: place and identity in NE Australia during the late Holocene. *Cambridge Archaeological Journal* 9(2):163-88.

Davidson, D.S. 1935 Archaeological problems of northern Australia. *Journal of the Royal Anthropological Institute of Great Britain and Ireland* 65:145-85.

Davidson, I. 1988 The naming of parts: ethnography and the interpretation of Australian prehistory. In B. Meehan and R. Jones (eds) *Archaeology with Ethnography: An Australian Perspective*, pp.17-32. Department of Prehistory, Research School of Pacific Studies, Australian National University, Canberra.

Davidson, I. 1990 Prehistoric Australian demography. In B. Meehan and N. White (eds) *Hunter-Gatherer Demography, Past and Present Oceania Monograph* 39, pp.41-58. University of Sydney.

Davie, J.D.S. 1985 The mangrove vegetation of the South Alligator River, Northern Australia. In K.N. Bardsley, J.D.S. Davie and C.D. Woodroffe (eds) *Coasts and Tidal Wetlands of the Australian Monsoon Region, Mangrove Monograph* 1, pp.133-51. North Australia Research Unit, Australian National University, Darwin.

Davies, J.L. 1972 *Geographical Variation in Coastal Development.* Oliver and Boyd, Edinburgh.

Davis, S. 1984 *The hunter for all seasons: an Aboriginal perspective of the natural environ.* Milingimbi School Literature Production Centre, Milingimbi, Northern Territory.

Davis, S. 1985 Aboriginal knowledge and use of the coast and sea in northern Arnhem Land. In K.N. Bardsley, J.D.S. Davie and C.D. Woodroffe (eds) *Coasts and Tidal Wetlands of the Australian Monsoon Region,. Mangrove Monograph* 1, pp.297-312. North Australia Research Unit, Australian National University, Darwin.

Davis, T.L.O. 1985 Seasonal changes in gonad maturity, and abundance of larvae and early juveniles of Barramundi, *Lates calcarifer* (Bloch), in Van Dieman Gulf and the Gulf of Carpentaria. *Australian Journal of Marine and Freshwater Research* 36:177-90.

Deith, M.R. and J.C. Shackleton 1988 The contribution of shells to site interpretation: Approaches to shell material from Franchthi Cave. In J.L. Bintliff, D.A. Davidson and E.G. Grant (eds) *Conceptual Issues in Environmental Archaeology*, pp.49-58. Edinburgh University Press.

Dering, P. 1999 Earth-oven plant processing in Archaic Period economies: an example from a semi-arid savannah in South-Central North America. *American Antiquity* 64(4):659-74.

Dickson, F.P. 1977 Quartz flaking. In R. Wright (ed.) *Stone Tools as Cultural Markers: change, evolution and complexity*, pp.97-103. Australian Institute of Aboriginal Studies, Canberra.

Dickson, F.P. 1981 *Australian stone hatchets: A study in design and dynamics*. Academic Press, Sydney.

Dixon, J.M. and L. Huxley (eds) 1985 *Donald Thomson's Mammals and Fishes of Northern Australia*. Nelson, Melbourne.

Dodd, J.R. and R.J. Stanton 1981 *Paleoecology, concepts and applications*. Wiley, New York.

Douglas, B. 1871 Letter from Government Resident, Northern Territory, July 28th 1871. *South Australian Parliamentary Papers,* No.35.

Drennan, R.D. 1996 *Statistics for Archaeologists: A Commonsense Approach*. Plenum Press, New York.

Druss, M. 1981 Comment on B. Hayden 1981 Research and development in the Stone Age: technological transitions among hunter-gatherers. *Current Anthropology* 22(5):532-3.

Dunnell, R.C. 1981 Comment on B. Hayden 1981 Research and development in the Stone Age: technological transitions among hunter-gatherers. *Current Anthropology* 22(5):533.

Dunnell, R.C. and W.S. Dancey 1983 The siteless survey: A regional scale data collection strategy. *Advances in Archaeological Method and Theory*, 6:267-85.

Dwyer, P.D. 1988 Companion foods, schedules and archaeology. In B. Meehan and R. Jones (eds) *Archaeology with Ethnography: An Australian Perspective*, pp.148-55. Department of Prehistory, Research School of Pacific Studies, Australian National University, Canberra.

Dwyer, P.G., M. Minnegal and J. Thompson 1985 Odds and ends: Bower birds as taphonomic agents. *Australian Archaeology* 21:1-10.

Edwards, R. 1967 Aboriginal quartzite quarry in the Katherine region, Northern Territory. *Journal of the Anthropological Society of South Australia* 5:11-14.

Edwards, D.A. and J.F. O'Connell 1995 Broad spectrum diets in arid Australia. *Antiquity* 69:769-83.

Elkin, A.P. 1950 The complexity of social organisation in Arnhem Land. *Southwestern Journal of Anthropology* 6:1-20.

Elkin, A.P. 1969 Elements of Australian Aboriginal Philosophy. *Oceania* XL(2):85-98.

Elkin, A.P. 1970 The Aborigines of Australia: one in word, thought and deed. In S.A. Wurm and D.C. Laycock (eds) *Pacific Linguistic Studies in Honour of Arthur Capell*. Pacific Linguistics, Canberra.

Elkin, A.P. 1978 [1933] *Studies in Australian Totemism. The Oceania Mongraphs,* No.2. AMS Press, New York.

Engelhardt, R.A. and P. R. Rogers 1997 Maritime adaptive strategies in Post-Pleistocene Southeast Asia: an ethnoarchaeological model for the nature and distribution of archaeological sites. In P. Bellwood and D. Tillotson (eds), *Indo-Pacific Prehistory: The Chiang Mai Papers, Bulletin of the Indo-Pacific Prehistory Association* 15:177-92.

Erlandson, J.M. 1988 The role of shellfish in prehistoric economies: a protein perspective. *American Antiquity* 53(1):102-9.

Erlandson, J.M. 1994 *Early Hunter-Gatherers of the California Coast*. Plenum Press, New York.

Erlandson, J.M. and T.R. Rockwell 1987 Radiocarbon reversals and stratigraphic discontinuities: the effects of natural formation processes on coastal California archaeological sites. In D.T. Nash and M.D. Petraglia (eds) *Natural Formation Processes and the Archaeological Record, BAR International Series* 352, pp.51-73. Osney Mead, Oxford, England.

Erlandson, J.M., T.C. Rick, R.L. Vellanoweth and D.J. Kennett 1999 Maritime subsistence at a 9300 year old shell midden on Santa Rosa Island, California. *Journal of Field Archaeology* 26(3):255-65.

Ernst, C.H. and R.W. Barbour 1989 *Turtles of the World.* Smithsonian Institution Press, London.

Evans, N. and R. Jones 1997 The cradle of the Pama-Nyungans: archaeological and linguistic speculations. In P. McConvell and N. Evans (eds) *Archaeology and Linguistics: Aboriginal Australia in Global Perspective*, pp.385-417. Oxford University Press, Melbourne.

Eylmann, E. 1994 *Die Eingeborenen Der Kolonie Sudaustralien*, [selected chapters translated by R. Hubel and compiled by R. Hodgson] Museum and Art Gallery of the Northern Territory.

Flood, J. 1980 *The Moth Hunters: Aboriginal prehistory of the Australian Alps*. Australian Institute of Aboriginal Studies, Canberra.

Flood, J. 1990 *The Riches of Ancient Australia: A Journey into Prehistory*. University of Queensland Press, Brisbane.

Flood, P.G. and E. Frankel 1989 Late Holocene higher sea level indicators from eastern Australia. *Marine Geology* 90:193-5.

Flood, J.M., B. David, J. Magee and B. English 1987 Birrigai: a Pleistocene site in the southeastern highlands. *Archaeology in Oceania* 22:9-26.

Foelsche, P. 1882 Notes on the Aborigines of North Australia, *Transactions of the Royal Society of South Australia* 5, 1-18.

Foelsche, P. 1886 Port Darwin: The Larrakia Tribe. In E.M. Curr, *The Australian Race: Its Origin, Languages, Customs, Place Of Landing In Australia, And The Routes By Which It Spread Itself Over That Continent*, Vol.I, pp.250-59. John Ferres, Government Printer, Melbourne.

Fogarty, P., D. Howe and C. Dunlop 1979 *The Land Resources of the Darwin Area.* Unpublished Report, Land Conservation Unit, Territory Parks and Wildlife Commission, Darwin.

Fogarty, P.J., B. Lynch and B. Wood 1984 *The Land Resources of the Elizabeth, Darwin and Blackmore Rivers*. Technical Report Number 15, Land Conservation Unit, Conservation Commission of the Northern Territory.

Foley, R. 1981 Off-site archaeology: an alternative approach for the short-sited. In I. Hodder, G. Isaac and N. Hammond (ed.) *Patterns of the Past*, pp.157-83. Cambridge University Press, Cambridge.

Ford, P.J. 1989 Molluscan assemblages from archaeological deposits. *Geoarchaeology: An International Journal* 4(2):157-73.

Ford, P.J. 1992 Interpreting the grain size distributions of archaeological shell. In J.K. Stein, (ed.) *Deciphering a Shell Midden*, pp.283-327. Academic Press, Inc., San Diego, California.

Fowler, A.J. 1992 Attempted validation and ageing of small tropical fish using otoliths. In D.A. Hancock (ed.) *The measurement of age and growth in fish and shellfish: Australian Society for Fish Biology Workshop, Lorne, Victoria, 22-23 August 1990*, pp.34-44. Bureau of Rural Resources, Canberra.

Frankel, D. 1988 Characterising change in prehistoric sequences: a view from Australia. *Archaeology in Oceania* 23:41-8.

Frankel, D. 1993 Pleistocene chronological structures and explanations: a challenge. In M.A. Smith, M. Spriggs and B. Fankhauser (eds) *Sahul In Review*, pp.24-33. Department of Prehistory, Research School of Pacific Studies, The Australian National University, Canberra.

Frazer, J.G. 1910 *Totemism and Exogamy: A Treatise on Certain Early Forms of Superstition and Society*, Vol.1. MacMillan and Co. Ltd., London.

Frazer, J.G. 1937 *Totemica: A supplement to Totemism and Exogamy.* MacMillan and Co. Ltd., London.

Freslov, J. and D. Frankel 1999 Abundant fields? A review of coastal archaeology in Victoria. In Hall, J. and I. McNiven (eds) *Australian Coastal Archaeology*, pp.239-54. ANH Publications, Department of Archaeology and Natural History, RSPAS, The Australian National University, Canberra.

Frith, H.J. 1973 *Wildlife Conservation.* Angus and Robertson, Sydney.

Fullagar, R. 1993 Taphonomy and tool-use: a role for phytoliths in use-wear and residues analysis. In B.L. Fankhauser and J.R. Bird (eds) *Archaeometry: Current Australasian Research, Occasional Papers in Prehistory* 22, pp.21-7. Department of Prehistory, Research School of Pacific Studies, The Australian National University, Canberra.

Galloway, R.W. 1982 Distribution and physiographic patterns of Australian mangroves. In B.F. Clough (ed.) *Mangrove Ecosystems in Australia: Structure, Function and Management*, pp.31-54. Australian Institute of Marine Science in association with Australian National University Press, Canberra.

Gamble, C. 1986 The mesolithic sandwich: ecological approaches and the archaeological record of the early postglacial. In M. Zvelebil (ed.) *Hunters in Transition*, pp.33-42. Cambridge University Press, London.

Gargett R. and B. Hayden 1991 Site structure, kinship, and sharing in Aboriginal Australia: implications for archaeology. In E.M. Kroll and T.D. Price (eds), *The Interpretation of Archaeological Spatial Patterning*, pp.11-32. Plenum Press, New York.

Garrod, D.A.E., L.H.D. Buxton, G. Elliot Smith and D.M.A. Bates 1928 Excavation of a Mousterian rockshelter at Devil's Tower, Gibraltar. *Journal of the Royal Anthropological Institute* 58:33-113.

Gaughwin, D. and R. Fullagar 1995 Victorian offshore islands in a mainland coastal economy. *Australian Archaeology* 40:38-50.

Gifford-Gonzales, D.P., D.B. Damrosch, D.R. Damrosch, J. Pryor and R.L. Thunen 1985 The third dimension in site structure: an experiment in trampling and vertical dispersal. *American Antiquity* 50(4):803-18.

Gill, E.D. 1954 Aboriginal kitchen middens and marine shell beds. *Mankind* 4(6):249-54.

Gillespie, R. and R.B. Temple 1977 Radiocarbon dating of shell middens. *Archaeology and Physical Anthropology in Oceania* XII(1):26-37.

Gillespie, R. and H.A. Polach 1979 The suitability of marine shells for radiocarbon dating of Australian prehistory. In R. Berger and H. Suess (eds) *Radiocarbon dating: proceedings of the 9th international conference on radiocarbon dating*, pp.404-21. University of California Press, Los Angeles.

Gillieson, D., D.I. Smith, M. Greenaway and M. Ellaway 1991 Flood history of the limestone ranges in the Kimberleys, western Australia. *Applied Geography* 11:105-23.

Glassow, M.A. 2000 Weighing vs. counting shellfish remains: a comment on Mason, Peterson, and Tiffany. *American Antiquity* 65(2):407-14.

Godfrey, M.C.S. 1988 Oxygen Isotope Analysis: a means for determining the seasonal gathering of the Pipi (*Donax deltoides*) by Aborigines in prehistoric Australia. *Archaeology in Oceania* 23:17-21.

Godfrey, M.C.S. 1989 Shell midden chronology in southwestern Victoria: reflections of change in prehistoric population and subsistence. *Archaeology in Oceania* 24:65-9.

Godwin, L. 1997 Little Big Men: alliance and schism in north-eastern New South Wales during the late Holocene. In P. McConvell and N. Evans (eds) *Archaeology and Linguistics: Aboriginal Australia in Global Perspective*, pp.297-309. Oxford University Press, Melbourne.

Goodfellow, D. 1993 *Fauna of Kakadu and the Top End.* Denise Goodfellow in association with Wakefield Press, Darwin.

Gordon, E.A. 1993 Screen size and differential faunal recovery: a Hawaiian example. *Journal of Field Archaeology* 20:453-60.

Gould, R.A. 1980 *Living Archaeology*. Cambridge University Press, Cambridge.

Gow, G.F. 1978 *Howard Springs Nature Park, amphibians and reptiles*. Northern Territory Parks and Wildlife Commission, Darwin.

Grayson, D.K. 1984 *Quantitative Zooarchaeology: Topics in the Analysis of Archaeological Faunas*. Academic Press, Orlando, Florida.

Greenwood, R. 1961 Quantitative analysis of shells from a site in Goleta, California, *American Antiquity* 26:416-20.

Gregory, R.L. 1998 Aboriginal Settlement Patterns in the Ord-Victoria River Region. Unpublished PhD thesis, Northern Territory University, Darwin.

Grey, D., R. Lea and C. Calogeras 1988 Recreational and commercial fishing in Darwin Harbour. In H.K. Larson, M.G. Michie and J.R. Hanley (eds) *Proceedings of the Workshop on Research and Management held in Darwin 2-3 September, 1987, Mangrove Monograph* 4, pp.165-79. Australian National University, North Australian Research Unit, Darwin.

Grindrod, J., P. Moss and S. van der Kaars 1999 Late quaternary cycles of mangrove development and decline on the north Australian continental shelf. *Journal of Quaternary Science* 14(5):465-70.

Guse, D. 1992 Predictive Models of Prehistoric Settlement and Subsistence Patterns for the South Alligator and Mary River Wetlands, Northern Territory. Unpublished BA(Hons) thesis, Department of Anthropology, Northern Territory University, Darwin.

Guse, D. 1995 An Archaeological Investigation of the Hopewell Road Horticultural Subdivisions, Berry Springs, Northern Territory, NTUAS Report 35. Unpublished report to the Department of Lands, Housing and Local Government, June 1995.

Guse, D. 1996 A report on the archaeological salvage of a prehistoric quartz scatter/quarry in the Palmerston area, NT. Unpublished report to Department of Lands, Housing and Local Government, Darwin.

Guse, D. 1998 A Report on the Collection of Artefact Scatters from the Hopewell Road Horticultural Subdivisions, Berry Springs, NT. Unpublished report to the Department of Lands, Planning and Environment, Darwin.

Guse, D. n.d. Litchfield Park and Finnis River Wetlands Prehistoric Settlement Patterns. *Unpublished paper presented at The Pre-Contact Archaeology of Darwin and its Hinterland. A workshop held at North Australian Research Unit, Darwin, 1 November 1996.*

Guse, D. and R. Gregory 1994 Archaeological Survey of the Proposed Realignment of the North Australia Railway Bridge Creek to Howley Siding. NTUAS Report No.25. Unpublished report for Australian National Rail, October 1994.

Guse, D. and P. Hiscock 1993 Archaeological Survey of the Proposed Bayview Haven Canal Estate. NTUAS, Report 12. Unpublished report to Henry and Walker Contracting Pty Ltd., Darwin.

Guse, D. and F. Mowat 1993 Preliminary site report for the excavation of the Bayview Haven Estate shell midden. In N. Richardson (ed.) Archaeological Site Survey and Documentation – Darwin Region. Unpublished report to the Australian Heritage Commission, Museum and Art Galleries of the Northern Territory, Darwin.

Hale, H.M. and N.B. Tindale 1934 Aborigines of Princess Charlotte Bay. *South Australian Museum Records* Vol.5 Nos. 1 and 2.

Hall, J. and I. McNiven 1995 Australian coastal archaeology symposium: current research and future directions. *Australian Archaeology* 40:65-6.

Hall, J. and I. McNiven 1999 Australian coastal archaeology: introduction. In Hall, J. and I. McNiven (eds) *Australian Coastal Archaeology*, pp.1-5. ANH Publications, Department of Archaeology and Natural History, RSPAS, The Australian National University, Canberra.

Hall, J., S. Higgins and R. Fullagar 1989 Plant residues on stone tools. In W. Beck, A. Clarke and L. Head (eds) *Plants in Australian Archaeology, Tempus*.1, pp.136-55. University of Queensland, St. Lucia, Queensland.

Hallam, S.J. 1986 Yams, alluvium and "villages" on the west coastal plain. In G.K. Ward (ed.) *Archaeology at ANZAAS, Canberra*. Australian Institutes of Aboriginal Studies, Canberra, 116-32.

Hallam, S.J. 1987 Coastal Does Not Equal Littoral, *Australian Archaeology* 25, 10-29.

Hamilton, F.E. 1999 Southeastern Archaic Mounds: examples of elaboration in a temporally fluctuating environment. *Journal of Anthropological Archaeology* 18:344-55.

Hanley, J.R. 1985 Why are there so few polychaetes (Annelida) in northern Australian mangroves? In K.N. Bardsley, J.D.S. Davie and C.D. Woodroffe (eds) *Coasts and Tidal Wetlands of the Australian Monsoon Region, Mangrove Monograph* 1, pp.239-50. North Australian Research Unit, Darwin.

Hanley, J.R. 1988 Invertebrate fauna of marine habitats in Darwin Harbour. In H.K. Larson, M.G. Michie and J.R. Hanley (eds) *Proceedings of the Workshop on Research and Management Held in Darwin, 2-3 September 1987. Mangrove Monograph* 4, pp.135-52. Australian National University, North Australian Research Unit, Darwin.

Hanley, J.R. 1993 Darwin south stage 1 environmental studies: mangrove benthic invertebrate fauna. Unpublished technical report 93/2, Marine Ecology Unit, Museums and Art Galleries of the Northern Territory, Darwin.

Hanley, J.R., G. Caswell, D. Megirian and H.K. Larson (eds) 1997 *The Marine Flora and Fauna of Darwin Harbour, Northern Territory, Australia: Proceedings of the Sixth International Marine Biological Workshop*. Museum and Art Galleries of the Northern Territory and the Marine Sciences Association, Darwin.

Hare, P.E. 1980 Organic geochemistry of bone and its relation to the survival of bone in the natural environment. In A.K. Behrensmeyer and A.P. Hill (eds) *Fossils in the Making: Vertebrate Taphonomy and Palaeoecology*, pp.208-19. University of Chicago Press, Chicago.

Harney, W.E. 1951 Australian Aboriginal cooking methods. *Mankind* 4(6):242-6.

Hart, C.W.M., A.R. Pilling and J.C. Goodale 1988 *The Tiwi of North Australia.* Holt, Rinehart and Winston, New York.

Harte, M.E. 1998 Superfamily Veneroidea. In P.L Beesley, G.J.B. Ross and A. Wells (eds) *Mollusca: The Southern Synthesis. Fauna of Australia* Vol.5. Part A, pp.355-62. CSIRO Publishing, Melbourne.

Harvey, M. 1997 The temporal interpretation of linguistic diversity in the Top End. In P. McConvell and N. Evans (eds) *Archaeology and Linguistics: Aboriginal Australia in Global Perspective*, pp.179-85. Oxford University Press, Melbourne.

Hayden, B. 1979 *Palaeolithic reflections: Lithic technology and ethnographic excavation among Australian Aborigines.* Australian Institute of Aboriginal Studies, Canberra.

Hayden, B. 1981a Research and development in the Stone Age: technological transitions among hunter-gatherers. *Current Anthropology* 22(5):519-31.

Hayden, B. 1981b Reply to comments. *Current Anthropology* 22(5):541-5.

Haynes, C.D. 1985 The pattern and ecology of munwag: traditional Aboriginal fire regimes in north-central Arnhem Land. *Proceedings of the Ecological Society of Australia* 13:203-14.

Haynes, C.D., M.G. Ridpath and M.A.J. Williams (eds) 1991 Introduction: The vertebrate animals. In *Monsoonal Australia: Landscape, Ecology and Man in the Northern Lowlands*, pp.75-7. A.A. Balkema, Rotterdam.

Hays, J. 1967 Land surfaces and laterites in the north of the Northern Territory. In J.N. Jennings and J.A. Mabbutt (eds) *Landform Studies from Australia and New Guinea*, pp.182-210. Australian National University Press, Canberra.

Head, J. 1991 The radiocarbon dating of freshwater and marine shells. In R. Gillespie (ed.) *Quaternary Dating Workshop 1990*, pp.16-18. Department of Biogeography and Geomorphology, Australian National University, Canberra.

Head, J., R. Jones and J. Allen 1983 Calculation of the "Marine Reservoir Effect" from the dating of shell-charcoal paired samples from an Aboriginal midden on Great Glennie Island, Bass Strait. *Australian Archaeology* 17:99-112.

Head, L. 1983 Environment as artefact: a geographic perspective on the Holocene occupation of southwestern Victoria. *Archaeology in Oceania* 18:73-80.

Head, L. 1986 Palaeoecological contributions to Australian prehistory. *Archaeology in Oceania* 21:121-9.

Healy, B. and K.A. Coates 1997 Enchytraeids (Oligochaeta: Annelida) of the mid and upper intertidal of Darwin Harbour, Northern Territory, Australia. In J.R. Hanley, G. Caswell, D. Megirian and H.K. Larson (eds) *The Marine Flora and Fauna of Darwin Harbour, Northern Territory, Australia, Proceedings of the Sixth International Marine Biological Workshop*, pp.81-97. Museum and Art Galleries of the Northern Territory and the Marine Sciences Association, Darwin.

Healy, J.M. and F.E. Wells 1998 Superfamily Cerithioidea. In P.L. Beesley, G.J.B. Ross and A. Wells (eds) *Mollusca: The Southern Synthesis. Fauna of Australia* Vol.5, Part B, pp.707-33. CSIRO Publishing, Melbourne.

Henshilwood, C., P. Nilssen and J. Parkington 1994 Mussel drying and food storage in the Late Holocene, SW Cape, South Africa. *Journal of Field Archaeology* 21:103-9.

Heritage Surveys 1995 Archaeological Survey of the Proposed Darwin-Alice Springs Railway, Adelaide River Origin 26.0 to 101.3km. Unpublished report for Australian National Rail. November 1995.

Heritage Surveys 1997 Wickham Point Archaeological Survey, Darwin Harbour, Northern Territory, Appendix L. Unpublished report for Phillips Oil Company, Australia.

Heritage Surveys 2000 Sunrise gas project, Gunn Point Peninsula, Northern Territory Archaeological survey. Unpublished report to Woodside Energy Ltd.

Hickey, S.H. 1981 Preliminary Investigation of Stranded Beach Ridges, Shoal Bay, Northern Territory, A small chenier plain? Northern Territory Geological Survey, Technical Report GS81/1, Department of Mines and Energy, Darwin.

Higham, T. 1996 Shellfish and seasonality. In A. Anderson, B. Allingham and I. Smith (eds) *Shag River Mouth: The Archaeology of an Early Southern Maori Village, Research Papers in Archaeology and Natural History* 27, pp.245-53. ANH Publications, RSPAS, The Australian National University, Canberra.

Higham, T.F.G. and A.G. Hogg 1995 Radiocarbon dating of prehistoric shell from New Zealand and calculation of the ΔR value using fish otoliths. *Radiocarbon* 37(2):409-16.

Hiscock, P. 1984 A preliminary report on the stone artefacts from Colless Creek Cave, Northwest Queensland. *Queensland Archaeological Research* 1:120-51.

Hiscock, P. 1985 The need for a taphonomic perspective in stone artefact analysis. *Queensland Archaeological Research* 2:82-97.

Hiscock, P. 1986 Technological change in the Hunter River valley and the interpretation of late Holocene change in Australia. *Archaeology in Oceania* 21(1):40-50.

Hiscock, P. 1990 A study in scarlett: taphonomy of inorganic artefacts. In S. Solomon, I. Davidson and D. Watson (eds) *Problem Solving in Taphonomy, Tempus* 2, pp.34-49. The University of Queensland, St. Lucia, Queensland.

Hiscock, P. (ed.) 1991 Preliminary report on exploring prehistoric cultural ecology in the Kakadu wetlands: methodological considerations. Unpublished report to Australian National Parks and Wildlife Services, NTUAS Report No.3, Northern Territory University, Darwin.

Hiscock, P. 1992 Archaeological evaluation of a locality within the proposed Bayview Haven Canal Estate. NTUAS Report 11. Unpublished report to Dover Investments through Dames and Moore. July 1992.

Hiscock, P. 1993a Interpreting the vertical distribution of stone points within Nauwalabila 1, Arnhem Land. *The Beagle* 10(1):173-8.

Hiscock, P. 1993b Archaeological Survey of the proposed Micket Creek Shooting complex. NTUAS Report 16. Unpublished report to Dames and Moore Pty Ltd., Darwin.

Hiscock, P. 1994a Coming to terms with the Northern Territory Heritage Conservation Act 1991. *Australian Archaeology* 38:55-60.

Hiscock, P. 1994b Technological responses to risk in Holocene Australia. *Journal of World Prehistory* 8(3):267-92.

Hiscock, P. 1995a Archaeological excavations at Bayview Haven 3, Bayview Haven Canal Estate. Unpublished report to Dover Investments Pty. Ltd. Northern Territory University Archaeological Services Report 30, May 1995, Northern Territory University, Darwin.

Hiscock, P. 1995b Archaeological investigations at the Palmerston suburbs of Bakewell and Rosebery. NTUAS Report 31. Unpublished report to the Department of Lands, Housing and Local Government, Darwin.

Hiscock, P. 1996 Mobility and technology in the Kakadu coastal wetlands. In I.C. Glover and P. Bellwood (eds) *Indo-Pacific Prehistory: The Chiang Mai Papers, Bulletin of the Indo-Pacific Prehistory Association* 15:151-7.

Hiscock, P. 1997 Archaeological evidence for environmental change in Darwin Harbour. In J.R. Hanley, G. Caswell, D. Megirian and H.K. Larson (eds) *The Marine Flora and Fauna of Darwin Harbour, Northern Territory, Australia. Proceedings of the Sixth International Marine Biological Workshop*, pp.445-9. Museum and Art Galleries of the Northern Territory and the Marine Sciences Association, Darwin.

Hiscock, P. 1999 Holocene coastal occupation of western Arnhem Land. In Hall, J. and I. McNiven (eds) *Australian Coastal Archaeology*, pp.91-103. ANH Publications, Department of Archaeology and Natural History, RSPAS, The Australian National University, Canberra.

Hiscock, P. and P. Hughes in press Prehistoric and World War Two use of shell mounds in Darwin Harbour. *The Beagle*.

Hiscock, P. and P. Kershaw 1992 Palaeoenvironments and prehistory of Australia's tropical Top End. In J. Dodson (ed.) *The Naive Lands*, pp.43-75. Longman Cheshire, Melbourne.

Hiscock, P. and S. Mitchell 1991 Baseline archaeological survey of the proposed Telecom Darwin- Katherine optical fibre route. NTUAS Report 2. A report to Telecom Australia through Kinhill Engineers Pty Ltd. January 1991.

Hiscock, P. and S. Mitchell 1992 Stage II Archaeological Survey of the Proposed Telecom Darwin- Katherine Optical Fibre Route. Northern Territory University Archaeological Services, Report 2. Unpublished report to Telecom Australia through Kinhill Engineers Pty Ltd., Darwin.

Hiscock, P. and S. Mitchell 1993 *Stone Artefact Quarries and Reduction Sites in Australia: towards a type profile*. Technical Publication Series No.4, Australian Government Publishing Service, Canberra.

Hiscock, P. and F. Mowat 1993 Midden variability in the coastal portion of the Kakadu region. *Australian Archaeology* 37:18-24.

Hiscock, P. and I. Walters 1994 Archaeology at the Northern Territory University: update May 1993. *Australian Archaeology* 38:51-5.

Hiscock, P., F. Mowat, and D. Guse 1992 Settlement patterns in the Kakadu wetlands: initial data on site size and shape. *Australian Aboriginal Studies*. 2:69-74.

Hobson, K.A. and S. Collier 1984 Marine and terrestrial protein in Australian Aboriginal diets. *Current Anthropology* 25:238-40.

Hockey, G. 1981 Assessing the carrying capacity for buffalo on the coastal plains and environs, from the Adelaide River to the East Alligator. In Best, F.G. (ed.) *Proceedings of the Buffalo Industry Symposium* pp. The Department of Primary Production and the Northern Territory Buffalo Industry Council, Darwin.

Hockey, P.A.R. and A.L. Brosman 1986 Man as intertidal predator in Transkei: disturbance, community convergence and management of a natural food resource *Oikos* 46:3-14.

Hodgkin, E.P. and P. Hesp 1998 Estuaries to salt lakes: Holocene transformation of the estuarine ecosystems of south-western Australia. *Marine and Freshwater Research* 49:183-201.

Hodgson, R. 1997 Aboriginal Use of Natural Resources in the Darwin Region – Past and Present. Unpublished report to the Australian Heritage Commission. Museum and Art Galleries of the Northern Territory.

Hofman, J.L. 1986 Vertical movement of artifacts in alluvial and stratified deposits. *Current Anthropology* 27(2):163-71.

Holland, R. 1976 Distribution and methods of construction of Aboriginal bark canoes. *University of Queensland Occasional Papers on Anthropology* 6:69-87.

Holthius L.B. 1991 *FAO Species Catalogue* 13 *Marine Lobsters of the World FAO Fisheries Synopsis* No.125. Food and Agriculture Organisation of the United Nations, Rome.

Hooper, A.D.L. 1969 Soils of the Adelaide-Alligator area. In M.A.J. Williams, A.D.L. Hooper, R.E. O'Ferrall and J.R. McAlpine (eds) *Lands of the Adelaide-Alligator Area, Northern Territory Commonwealth Scientific and Industrial Research Organization, Land Research Series* No. 25, pp.95-113. Melbourne.

Hooper, J.N.A. 1988 Structural features of the benthic community of East Point reef fish reserve. A comparative study between oceanic, nearshore and inshore reefs of Northwest Australia. In H.K. Larson, M.G. Michie and J.R. Hanley (eds) *Proceedings of the Workshop on Research and Management held in Darwin 2-3 September, 1987, Mangrove Monograph* 4, pp.214-25. Australian National University, North Australian Research Unit, Darwin.

Hope, G., P.J. Hughes and J. Russel-Smith 1985 Geomorphological fieldwork and the evolution of the landscape of Kakadu National Park. In R. Jones (ed.) *Archaeological Research in Kakadu National Park. Special Publication* 13, pp.229-40. Australian National Parks and Wildlife Service, Canberra.

Hordern, M. 1989 *Mariners are warned!: John Lort Stokes and H.M.S. Beagle in Australia 1837-1843.* Melbourne University Press at the Miegunyah Press, Carlton, Victoria.

Horton, D.R. 1984 Minimum numbers: a consideration. *Journal of Archaeological Science* 11:255-71.

Horton, D.R. 1993 Here be dragons: a view of Australian archaeology. In M.A. Smith, M. Spriggs and B. Fankhauser (eds) *Sahul In Review*, pp.11-16. Department of Prehistory, Research School of Pacific Studies, The Australian National University, Canberra.

Huchet, B.M.J. 1991 The nature of analogies in Australian archaeology. *The Artefact* 14:3-12.

Hughes, P.J 1977 A Geomorphological Interpretation of Selected Archaeological Sites in Southern Coastal New South Wales. Unpublished PhD thesis, School of geography, University of New South Wales, Sydney.

Hughes, P. 1980 I dig dirt: geomorphological methods in Australian archaeology. In I. Johnson (ed.) *Holier Than Thou: Proceedings of the 1978 Kioloa conference on Australian Prehistory* pp.3-17. Department of Prehistory, Research School of Pacific Studies, The Australian National University, Canberra.

Hughes, P.J 1983 Geoarchaeology in Australia. In G. Connah (ed.) *Australian Field Archaeology: A Guide to Techniques*, pp.109-17. Australian Institute of Aboriginal Studies, Canberra.

Hughes, P. and P. Hiscock 1981 Olympic Dam Project: Baseline Archaeology Study. Unpublished report to Roxby Management Services Pty Ltd through Kinhill Pty Ltd, Adelaide. June 1981.

Hughes, P.J. and R.J. Lampert 1977 Occupational disturbance and types of archaeological deposits. *Journal of Archaeological Science* 4:135-40.

Hughes, P.J. and R.J. Lampert 1982 Prehistoric population change in southern coastal New South Wales. In S. Bowdler (ed.) *Coastal Archaeology in Eastern Australia: Proceedings of the 1980 Valla Conference on Australian Prehistory, Occasional Papers in Prehistory* 11, pp.16-28. Department of Prehistory, Research School of Pacific Studies, Australian National University, Canberra.

Hughes, P.J. and M.E. Sullivan 1974 The re-deposition of midden material by storm waves. *Journal and Proceedings of the Royal Society of New South Wales* 107:6-10.

Hughes, P.J. and M.E. Sullivan 1984 Environmental approaches to assessing archaeological significance. In S. Sullivan and S. Bowdler (eds) *Site Surveys and Significance Assessment in Australian Archaeology* pp.34-47. Australian National University, Canberra.

Hughes, P.J. and M.E. Sullivan 1999 A 5000 year old shell midden near Tea Gardens, New South Wales. In Hall, J. and I. McNiven (eds) *Australian Coastal Archaeology*, pp.219-27. ANH Publications, Department of Archaeology and Natural History, RSPAS, The Australian National University, Canberra.

Hutchings, P.A. and H.F. Recher 1983 The faunal communities of Australian mangroves. In H.J. Teas (ed.) *Biology and Ecology of Mangroves*. W. Junk Publishers, The Hague.

Hutchings, P. and P. Saenger 1987 *Ecology of Mangroves*. University of Queensland Press, St. Lucia, Queensland.

Hynes, R. and A.K. Chase 1982 Plants, sites and domiculture: Aboriginal influence on plant communities, Cape York Peninsula. *Archaeology in Oceania* 17:138-50.

Irvine, F.R. 1970 Evidence of change in the vegetable diet of Australian Aborigines. In A.R. Pilling and R.A. Waterman (eds) *Diprotodon to Detribalization: Studies of Change among Australian Aborigines*, pp.278-84. Michigan State University Press, East Lansing.

James, S.R. 1997 Methodological issues concerning screen size recovery rates and their effects on archaeofaunal interpretations. *Journal Of Archaeological Science* 24: 385-97.

Jerardino, A. 1995 The problem with density values in archaeological analyses: a case study from Tortoise Cave, Western Cape, South Africa. *South African Archaeologist Bulletin* 50:21-7.

Jerardino, A. 1997 Changes in shellfish species composition and mean shell size from a late-Holocene record of the west coast of Southern Africa. *Journal Of Archaeological Science* 24:1031-44.

Jerardino, A. 1998 Excavations at Poncho's Kitchen Midden, Western Cape Coast, South Africa: further observations into the Megamidden Period. *South African Archaeologist Bulletin* 53:16-25.

Jerardino, A. and R. Yates 1997 Excavations at Mike Taylor's Midden: a summary report and implications for a re-characterisation of megamiddens. *South African Archaeologist Bulletin* 52:43-51.

Johnson, I. 1980 Bytes from sites: the design of an excavation data recording system. In I. Johnson (ed.) *Holier Than Thou: Proceedings of the 1978 Kioloa conference on Australian Prehistory*, pp.91-118. Department of Prehistory, research School of Pacific Studies, Australian National University, Canberra.

Johnston, H. 1993 Pleistocene shell middens of the Willandra Lakes. In M.A. Smith, M. Spriggs and B. Fankhauser (eds) *Sahul In Review*, pp.197-203. Department of Prehistory, Research School of Pacific Studies, The Australian National University, Canberra.

Johnston, H. and P. Clark 1998 Willandra Lakes archaeological investigations. *Archaeology in Oceania* 33(3):105-19.

Jones, R. 1969 Fire stick farming. *Australian Natural History* 16:224-28.

Jones, R. 1977 The Tasmanian paradox. In R. Wright (ed.) *Stone Tools as Cultural Markers: change, evolution and complexity*, pp.189-204. Australian Institute of Aboriginal Studies, Canberra.

Jones, R. 1980a Different strokes for different folks: sites, scale and strategy. In I. Johnson (ed.) *Holier Than Thou: Proceedings of the 1978 Kioloa conference on Australian Prehistory*, pp.151-71. Department of Prehistory, Research School of Pacific Studies, The Australian National University, Canberra.

Jones, R. 1980b Hunters in the Australian coastal savannah. In D.R. Harris (ed.) *Human Ecology in Savannah Environments*, pp.107-46. Academic Press, London.

Jones, R. (ed.) 1985a *Archaeological Research in Kakadu National Park, Special Publication* 13. Australian National Parks and Wildlife Service, Canberra.

Jones, R. 1985b Archaeological conclusions. In R. Jones (ed.) *Archaeological Research in Kakadu National Park, Special Publication* 13, pp.291-8. Australian National Parks and Wildlife Service, Canberra.

Jones, R. 1990 Hunters of the dreaming: some ideational, economic and ecological parameters of the Australian Aboriginal productive system. In D.E. Yen and J.M.J. Mummery (eds) *Pacific Production Systems: Approaches to Economic Prehistory*, pp.25-53. Canberra: Department of Prehistory, Research School of Pacific Studies, Australian National University, Canberra.

Jones, R. and J. Bowler 1980 Struggle for the Savannah: Northern Australia in ecological and prehistoric perspective. In R. Jones (ed.) *Northern Australia: Options and Implications*, pp.3-31. Research School of Pacific Studies, Australian National University, Canberra.

Jones, R. and I. Johnson 1985a Rockshelter excavations: Nourlangie and Mt Brockman massifs. In R. Jones (ed.) *Archaeological Research in Kakadu National Park, Special Publication* 13, pp.39-76. Australian National Parks and Wildlife Service, Canberra.

Jones, R. and I. Johnson 1985b Deaf Adder Gorge: Lindner site, Nauwalabila 1. In R. Jones (ed.) *Archaeological Research in Kakadu National Park, Special Publication* 13, pp.165-227. Australian National Parks and Wildlife Service, Canberra.

Jones, R. and B. Meehan 1989 Plant foods of the Gidjingali: ethnographic and archaeological perspectives from northern Australia on tuber and seed exploitation. In D.R. Harris and G.C. Hillman (eds) *Foraging And Farming: The Evolution of Plant Exploitation*, pp.120-35. Unwin Hyman, London.

Jones, R. and N. White 1988 Point blank: stone tool manufacture at the Ngiliptji quarry, Arnhem Land. In B. Meehan and R. Jones (eds) *Archaeology with Ethnography: An Australian Perspective*, pp.51-87. Department of Prehistory, Research School of Pacific Studies, Australian National University, Canberra.

Jones, T.L., G.M. Brown, L.M. Raab, J.L. McVickar, W.G. Spaulding, D.J. Kennett, A. York and P.L. Walker 1999 Environmental imperatives reconsidered. *Current Anthropology* 40(2):137-69.

Jorde, L.B. 1977 Precipitation cycles and cultural buffering in the prehistoric southwest. In L.R. Binford (ed.) *For Theory Building in Archaeology*, pp.385-96. Academic Press, New York.

Kaberry, P.M. 1939 *Aboriginal Woman: Sacred and Profane*, George Routledge, London.

Kamminga, J. 1981 Comment on B. Hayden 1981 Research and development in the Stone Age: technological transitions among hunter-gatherers. *Current Anthropology* 22(5):535-6.

Kamminga, J. 1982 *Over the Edge: Functional analysis of Australian stone tools. Occasional Papers in Anthropology* 12, University of Queensland.

Kamminga, J. and H. Allen 1973 *Report of the archaeological survey. Alligator Rivers environmental fact-finding study*. Government Printer, Darwin.

Karkanas P., N. Kyparissi-Apostolika, O. Bar-Yosef and S. Weiner 1999 Mineral assemblages in Theopetra, Greece: a framework for understanding diagenesis in a prehistoric cave. *Journal of Archaeological Science* 26:1171-80.

Kefous, K. 1977 We have a fish with ears and wonder if it is valuable. Unpublished B.A. Hons thesis, Australian National University, Canberra.

Kelly, R.L. 1983 Hunter-gatherer mobility strategies. *Journal of Anthropological Research* 39:277-306.

Kelly, R.L. 1992 Mobility/Sedentism: concepts, archaeological measures, and effects. *Annual Review of Anthropology* 21:43-66.

Kench, P.S. 1999 Geomorphology of Australian estuaries: review and prospect. *Australian Journal of Ecology* 24:367-80.

Kennett, D.J. and B. Voorhies 1996 Oxygen isotope analysis of archaeological shells to direct seasonal use of wetlands on the southern Pacific coast of Mexico. *Journal of Archaeological Science* 23:689-704.

Kent, S. 1987 Parts as wholes: a critique of theory in archaeology. In S. Kent (ed.) *Method and Theory for Activity Area Research: An Ethnoarchaeological Approach*, pp.513-46. Columbia University Press, New York.

Kent, S. 1993 Models of abandonment and material culture frequencies. In C.M. Cameron and S.A. Tomka (eds) *Abandonment of settlements and regions: Ethnoarchaeological and archaeological approaches*, pp.54-73. Cambridge University Press, Cambridge.

Kerr, M.G. 1971 *The Surveyors: The story of the founding of Darwin*, Rigby: Adelaide.

Kimber, R.G. 1988 Smallpox in central Australia: evidence for epidemics, and postulations about the impact. *Australian Archaeology* 27:63-8.

King, P.P. 1969 [1827] *Narrative of a Survey of the Intertropical and Western Coasts of Australia, 1818-1822.* Vol.1, John Murray, London.

Kinhill Engineers 1992 Aboriginal and historic archaeological sites. Appendix C, *Haycock Reach Aquaculture Development, Preliminary Environmental Report*. Australian Frontier Holidays, Darwin.

Knight, J.G. 1874 Report on government buildings and works at Palmerston, 1874. *South Australian Parliamentary Papers,* No.137.

Knight, J.G. 1876 Supplementary Chapter. The Northern Territory. In W. Harcus (ed.) *South Australia, its History, Resources and Productions*, pp.148-86. Sampson Low, Marston, Searle and Rivington, London.

Knox, J.C. 1993 Large increases in flood magnitude in response to modest changes in climate. *Nature* 361:430-2.

Knuckey, I. 1999 Mud crab (*Scylla serrata*) population dynamics in the Northern Territory, Australia and their relationship to the commercial fishery. Unpublished PhD Thesis, Northern Territory University, Darwin.

Knudson, J. 1997 Observations on some intertidal marine prosobranch molluscs from Darwin Harbour, Northern Territory. In J.R. Hanley, G. Caswell, D. Megirian and H.K. Larson (eds) *The Marine Flora and Fauna of Darwin Harbour, Northern Australia, Proceedings of the Sixth International Marine Biological Workshop*, pp.199-212. Museum and Art Galleries of the Northern Territory and the Marine Sciences Association, Darwin.

Koepping, K.P. 1979 Artifacts and theory building in Anthropology. *Readings in Material Culture, Occasional Papers in Anthropology* 9:1-8.

Koike, H. 1986 Prehistoric hunting pressure and palaeobiomass: an environmental reconstruction and archaeozoological analysis of a Jomon shellmound area. In T. Akazawa. and C.M. Aikens (eds) *Prehistoric Hunter-Gatherers in Japan*, pp.27-53. University of Tokyo Press, Tokyo.

Koike, H. and M. Okamura 1994 Time-scaling of successively accumulated shell layers for exploitation diary analysis. *Archaeozoologia* VI/2:23-36.

Kolig, E. 1988 Australian Aboriginal totemic systems: structures of power. *Oceania* 58(3):212-30.

Larson, H.K. 1988 The fishes of Darwin Harbour. In H.K. Larson, M.G. Michie and J.R. Hanley (eds) *Proceedings of the Workshop on Research and Management held in Darwin 2-3 September, 1987, Mangrove Monograph* 4, pp.153-64. Australian National University, North Australian Research Unit, Darwin.

Larson, H.K. and K.C. Martin 1989 *Freshwater Fishes of the Northern Territory*. Handbook Series No.1., Museum and Arts Galleries of the Northern Territory.

Larson, H.K. and R.S. Williams 1997 Darwin Harbour fishes: a survey and annotated checklist. In J.R. Hanley, G. Caswell, D. Megirian and H.K. Larson (eds) *The Marine Flora and Fauna of Darwin Harbour, Northern Australia, Proceedings of the Sixth International Marine Biological Workshop*, pp.339-80. Museum and Art Galleries of the Northern Territory and the Marine Sciences Association, Darwin.

Lee, H.J., S.S. Chun, J.H. Chang and S.J. Han 1994 Landward migration of isolated shelly sand ridge (chenier) on the macrotidal flat of Gomso Bay, west coast of Korea: controls of storms and typhoon. *Journal of Sedimentary Research* A64(4):886-893.

Lees, B.G. 1985 Geomorphology and development of the Victoria delta. In K.N. Bardsley, J.D.S. Davie and C.D. Woodroffe (eds) *Coasts and Tidal Wetlands of the Australian Monsoon Region, Mangrove Monograph* 1, pp.31-42. North Australian Research Unit, Darwin.

Lees, B.G. 1987 Age structure of the Point Stuart chenier plain: a reassessment. *Search* 18(5):257-9.

Lees, B.G. 1991 Radiocarbon dating of cheniers. In R. Gillespie (ed.) *Quaternary Dating Workshop 1990*, pp.11-12. Department of Biogeography and Geomorphology, Research School of Pacific and Asian Studies, Australian National University, Canberra.

Lees, B.G. 1992a The development of a chenier sequence on the Victoria Delta, Joseph Bonaparte Gulf, northern Australia. *Marine Geology* 103:215-24.

Lees, B.G. 1992b Geomorphological evidence for late Holocene climatic change in northern Australia. *Australian Geographer* 23(1):1-10.

Lees, B.G. and Clements, A. 1987 Climatic implications of chenier dates in northern Australia. *Radiocarbon* 25:311-7.

Lees; B.G., Y. Lu and J. Head 1990 Reconnaissance thermoluminescence dating of northern Australian coastal dune systems. *Quaternary Research* 34(2):169-85.

Levins, R. and R. Lewontin 1985 *The Dialectical Biologist* Harvard University Press, Cambridge, Massachusetts.

Lewis, D. 1988 *The rock paintings of Arnhem Land: social, ecological, and material culture change in the post-glacial period. BAR International Series* 415, Oxford.

Lieberman, L.S. 1987 Biocultural consequences of animals versus plants as sources of fats, proteins and other nutrients. In M. Harris and E.B. Ross (eds) *Food and Evolution: Toward a Theory of Human Food Habits*, pp.225-58. Temple University Press, Philadelphia.

Lilley, I. 1993 Recent research in southwestern Western Australia: A summary of initial findings. *Australian Archaeology* 36:34-41.

Ling, Shao-Wen 1977 *Aquaculture in Southeast Asia, A Historical Overview*. Edited by L. Mumaw, University of Washington Press, Seattle.

Linse, A.R. 1992 Is bone safe in a shell midden? In Stein, J.K. (ed.) *Deciphering a Shell Midden*, pp.327-45. Academic Press, Inc., San Diego, California.

Lourandos, H. 1980 Forces of change: Aboriginal technology and population in south-western Victoria. Unpublished PhD thesis. University of Sydney, Sydney.

Lourandos, H. 1983 Intensification: a Late Pleistocene-Holocene archaeological sequence from Southwestern Victoria. *Archaeology in Oceania* 18(2):81-92.

Lourandos, H. 1984 Changing perspectives in Australian prehistory: a reply to Beaton. *Archaeology in Oceania* 19:29-33.

Lourandos, H. 1985a Problems with the interpretation of late Holocene changes in Australian prehistory. *Archaeology in Oceania* 20:37-9.

Lourandos, H. 1985b Intensification and Australian Prehistory. In T.D. Price and J.A. Brown (eds) *Prehistoric Hunter-Gatherers: The Emergence of Cultural Complexity*, pp.385-423. Academic Press, Orlando, Florida.

Lourandos, H. 1993 Hunter-gatherer cultural dynamics: long and short term trends in Australian prehistory. *Journal of Archaeological Research* 1:67-88.

Lourandos, H. and A. Ross 1994 The great "Intensification Debate": its history and place in Australian archaeology. *Australian Archaeology* 39:54-63.

Luby, E.M. and M.F. Gruber 1999 The dead must be fed: symbolic meanings of the shellmounds of the San Francisco bay area. *Cambridge Archaeological Journal* 9(1):95-108.

MacDonald, K. 1991 Hundreds and Thousands: The investigation of small sized lithic debitage from Bone Cave, Tasmania. Unpublished Honours thesis, Department of Archaeology, La Trobe University, Tasmania.

Macintosh, D.J. 1982 Fisheries and aquaculture significance of mangrove swamps, with special reference to the Indo-West Pacific region. In J.F. Muir and R.J. Roberts (eds) *Recent Advances in Aquaculture*, pp.5-85. Croom Helm, Sydney.

Macknight, C.C. 1976 *The Voyage to Marege*. Melbourne University Press, Melbourne.

Macknight, C.C. 1986 Macassans and the Aboriginal past. *Archaeology in Oceania* 21:69-75.

Macnae, W. 1968 A general account of the fauna and flora of mangrove swamps and forests in the Indo-West-Pacific region. *Advances in Marine Biology* 6:73-270.

Madsen, T. and R. Shine 1996 Seasonal migration of predators and prey - a study of pythons and rats in tropical Australia. *Ecology* 77(1):149-56.

Manton, J.T. 1866 Northern Territory Correspondence. *South Australian Parliamentary Papers,* No.80.

Martin, K. 1978 The reef dwelling fishes of the Darwin area. *Northern Territory Naturalist* 1:15-20.

Mason, G.M. 1991 The molluscan remains. In C.F.W. Higham, and R. Bannanurag (eds) *The Excavation of Khok Phanom Di: a Prehistoric Site in Central Thailand*, Vol. II, *The Biological Remains*, pp.249-319. The Society of Antiquaries of London, London.

Mason, R.D., M.L. Peterson and J.A. Tiffany 1998 Weighing vs. counting: measurement reliability and the California school of midden analysis. *American Antiquity* 63(2):303-24.

McAlpine, J.R. 1969 Climate of the Adelaide-Alligator area. In M.A.J. Williams, A.D.L. Hooper, R.E. O'Ferrall and J.R. McAlpine, *Lands of the Adelaide-Alligator Area, Northern Territory Commonwealth Scientific and Industrial Research Organization, Land Research Series* No. 25, pp.49-55. Melbourne.

McBryde, I. 1962 Archaeological field survey work in northern New South Wales. *Oceania* 33:12-17.

McBryde, I. 1974 *Aboriginal prehistory in New England: An archaeological survey of northeastern New South Wales*. Sydney University Press, Sydney.

McBryde, I. 1978 Wil-im-ee Moor-ring: or where do the axes come from? *Mankind*. 10:354-82.

McBryde, I. 1979 Ethnohistory in an Australian context: independent discipline or convenient data quarry? *Aboriginal History* 3(1):128-50.

McBryde, I. 1984 Kulin greenstone quarries: the social contexts of production and distribution for the Mt William site. *World Archaeology* 16(3):267-85.

McBurney, C.B.M. 1967 *The Haua Fteah (Cyrenaica) and the Stone Age of the south-east Mediterranean*. Cambridge University Press, Cambridge.

McCarthy, F. and M. McArthur 1960 The food quest and the time factor in Aboriginal economic life. In C.P. Mountford (ed.) *Records of the American-Australian Expedition to Arnhem Land* 2, pp.145-214. Melbourne University Press, Melbourne.

McCarthy, F. and F.M. Setzler 1960 The archaeology of Arnhem Land. In C.P. Mountford (ed.) *Records of the American-Australian Expedition to Arnhem Land* 2, pp.215-59. Melbourne University Press, Melbourne.

McConnell, U. 1957 *Myths of the Mungkan*. Melbourne University Press, Melbourne.

McConvell, P. 1985 The origin of subsections in northern Australia. *Oceania* 56(1):1-33.

McConvell, P. 1990 The linguistic prehistory of Australia: opportunities for dialogue with archaeology. *Australian Archaeology* 31:3-27.

McConvell, P. 1997 Long lost relations: Pama-Nyungan and northern kinship. In P. McConvell and N. Evans (eds) *Archaeology and Linguistics: Aboriginal Australia in Global Perspective*, pp.207-36. Oxford University Press, Melbourne.

McDonald, N.S. and J. McAlpine 1991 Floods and droughts: the northern climate. In C.D. Haynes, M.G. Ridpath and M.A.J. Williams (eds) *Monsoonal Australia: Landscape, Ecology and Man in the Northern Lowlands*, pp.19-29. A.A. Balkema, Rotterdam.

McGlade, J. 1994 Archaeology and ecodynamics. *Antiquity* 69:113-32.

McGlone, M.S., A.P. Kershaw and V. Markgraf 1992 El Nino/Southern Oscillation climatic variability in Australasian and South American palaeoenvironmental records. In H.F. Diaz and V. Markgraf (eds) *El Nino. Historical and Palaeoclimatic Aspects of the Southern Oscillation*, pp.435-62. Cambridge University Press, Cambridge.

McIntyre, K.G. 1984 *The Secret Discovery of Australia*. Pan Books, Sydney.

McMichael, D.F. 1960 Notes on some Australian volutidae. *Journal of the Malacological Society of Australia* 4:4-13.

McNiven, I. 1989 Aboriginal shell middens at the mouth of the Maroochy River, southeast Queensland. *Queensland Archaeological Research* 6:28-52.

McNiven, I. 1991 Teewah Beach: new evidence for Holocene coastal occupation in southeast Queensland. *Australian Archaeology* 33:14-27.

McNiven, I.J. 1999 Fissioning and regionalisation: the social dimensions of changes in Aboriginal use of the Great Sandy Region, southeast Queensland. In Hall, J. and I. McNiven (eds) *Australian Coastal Archaeology*, pp.157-68. ANH Publications, Department of Archaeology and Natural History, RSPAS, The Australian National University, Canberra.

Meehan, B. 1982a *Shell Bed to Shell Midden*. Australian Institute of Aboriginal Studies, Canberra.

Meehan, B. 1982b Ten fish for one man: some Anbarra attitudes towards food and health. In J. Reid (ed.) *Body, Land and Spirit: Health and Healing in Aboriginal Society*, pp.96-120. University of Queensland Press, St Lucia.

Meehan, B. 1983 A matter of choice? Some thoughts on shell gathering strategies in northern Australia. In C. Grigson and J. Clutto-Brock (eds*) Animals and Archaeology: 2. Shell Middens, Fishes and Birds, BAR International Series* 183, pp.3-17.

Meehan, B. 1988 Changes in Aboriginal exploitation of wetlands in northern Australia. In D. Wade-Marshall and P. Loveday (eds) *Northern Australia: Progress and Prospect*.2, Appendix 2, pp.1-23. North Australia Research Unit, Australian National University Press, Darwin.

Meehan, B. 1991 Wetland hunters: some reflections. In C.D. Haynes, M.G. Ridpath and M.A.J. Williams (eds) *Monsoonal Australia: Landscape, Ecology and Man in the Northern Lowlands*, pp.197-206. A.A. Balkema, Rotterdam.

Meehan, B., S. Brockwell, J. Allen and R. Jones 1985 The wetland sites. In R. Jones, (ed.) *Archaeological Research in Kakadu National Park, Special Publication* 13, pp.103-153. Australian National Parks and Wildlife Service, Canberra.

Meese, D.A., A.J. Gow, P. Grootes, P.A. Mayewski, M. Ram, M. Stuiver, K.C. Taylor, E.D. Waddington and G.A. Zielinski 1994 The accumulation record from the GISP2 core as an indicator of climate change throughout the Holocene. *Science* 266:1680-2.

Meighan, C.W., D.M. Pendergast, B.K. Swartz, Jr., and M.D. Wissler 1958 Ecological interpretation in archaeology. *American Antiquity* 24:1-23, 131-50.

Messel, H., G.C. Vorlicek, M. Elliot, A.G. Wells and W.J. Green 1981 *Darwin and Bynoe Harbours and their tidal waterways. Surveys of tidal river systems in the Northern Territory of Australia and their crocodile populations, Monograph* 17, Pergamon Press, Sydney.

Micha, F.J. 1970 Trade and change in Australian Aboriginal cultures: Australian Aboriginal trade as an expression of close culture contact and as a mediator of culture change. In A.R. Pilling and R.A. Waterman (eds) *Diprotodon to Detribalization: Studies of Change among Australian Aborigines*, pp.285-313. Michigan State University Press, East Lansing.

Michie, M.G. 1984 Holocene foraminifera from the Camerons Beach chenier plain, Shoal Bay, Northern Territory. Addendum 1 to S.H. Hickey 1981 Preliminary investigation of stranded beach ridges, Shoal Bay, Northern Territory: a small chenier plain? Technical Report GS81/1, Northern Territory Geological Survey, Darwin.

Michie, M.G. 1988 Sediments, sedimentary environments and palaeoenvironments in Port Darwin. In H.K. Larson, M.G. Michie and J.R. Hanley (eds) *Proceedings of the Workshop on Research and Management held in Darwin 2-3 September, 1987, Mangrove Monograph* 4, pp.32-41. Australian National University, North Australian Research Unit, Darwin.

Michie, M.G., D.L. Grey and R.K. Griffin 1991 Sea water salinity and temperature measurements in the Darwin Harbour area 1972-1978. Unpublished Fisheries report No.24, Department of Primary Industry and Fisheries, Darwin.

Mitchell, S. 1993 Shell mound formation in northern Australia: a case study from Croker Island, Northwestern Arnhem Land. *The Beagle* 10(1):179-92.

Mitchell, S. 1994a Foreign Contact and Indigenous Economies on the Cobourg Peninsula, Northwestern Arnhem Land. Unpublished PhD thesis, Northern Territory University, Darwin.

Mitchell, S. 1994b Stone exchange network in north-western Arnhem Land: evidence for recent chronological change. In M. Sullivan, S. Brockwell and A. Webb (eds) *Archaeology In The North*, pp.188-200. North Australia Research Unit, The Australian National University, Darwin.

Mitchell, S. 1995 Foreign contact and indigenous exchange networks on the Cobourg Peninsula, north-western Arnhem Land. *Australian Aboriginal Studies* 2:44-8.

Moore, J. 1983 Carrying capacity, cycles and culture. *Journal of Human Evolution* 12:505-14.

Moore, J.D. 1991 Cultural responses to environmental catastrophes: post-El Nino subsistence on the prehistoric north coast of Peru. *Latin American Antiquity* 2(1):27-47.

Moratto, M.J. 1984 *California Archaeology*. Academic Press Inc., London.

Moreno, C.A., J.P. Sutherland and H.F. Jara 1984 Man as predator in the intertidal zone of southern Chile. *Oikos* 42:155-160.

Morse, K. 1993 New radiocarbon dates from North West Cape, Western Australia: a preliminary report. In M.A. Smith, M. Spriggs and B. Fankhauser (eds) *Sahul In Review*, pp.155-63. Department of Prehistory, Research School of Pacific Studies, The Australian National University, Canberra.

Morton, S.R., K.G. Brennan and M.D. Armstrong 1990 Distribution and abundance of magpie geese, *Anseranas semipalmata*, in the Alligator Rivers region, Northern Territory. *Australian Journal of Ecology* 15:307-20.

Morwood, M.J. 1987 The archaeology of social complexity in South-east Queensland. *Proceedings of the Prehistoric Society* 53:337-50.

Morwood, M.J. and D.R. Hobbs 1995 Themes in the prehistory of tropical Australia. *Antiquity* 69:747-68.

Morwood, M.J. and S. L'Oste-Brown 1995 Excavations at Hann River 1, Central Cape York Peninsula. *Australian Archaeology* 40:21-8.

Moseley, M. 1975 *The Maritime Foundations of Andean Civilization*. Cummings, Menlo Park, California.

Moss, M.L. 1993 Shellfish, gender, and status on the Northwest Coast: reconciling archaeological, ethnographic, and historical records of the Tlingit. *American Anthropologist* 95(3):631-52.

Mountain, M. 1990 Taphonomic aspects of faunal analysis from Nombe Rockshelter, Highlands of Papua New Guinea. In S. Solomon, I. Davidson and D. Watson (eds) *Problem Solving in Taphonomy*, *Tempus* 2, pp.207-18. The University of Queensland, St. Lucia, Queensland.

Mowat, F. M. 1994 Size really does matter: factors affecting shell fragmentation. In M. Sullivan, S. Brockwell and A. Webb (eds) *Archaeology In The North*, pp.201-10. North Australia Research Unit, The Australian National University, Darwin.

Mowat, F. M. 1995 Variability in Western Arnhem Land Shell Midden Deposits. Unpublished (MA) thesis. Department of Anthropology, Northern Territory University, Darwin.

Muckle, R.J. 1985 Archaeological Considerations of Bivalve Shell Taphonomy. Unpublished (MA) thesis, Department of Archaeology, Simon Fraser University.

Mulvaney, J. 1975 *The Prehistory of Australia*. (rev. ed.), Penguin, Melbourne.

Mulvaney, J. 1981 Salvaging salvage archaeology. *Australian Archaeology* 12:61-2.

Mulvaney, J. 1990 The chain of connection: the material evidence. In J.M.J. Mummery (ed.) *Prehistory and Heritage*: *The writings of John Mulvaney*, pp.82-95. Department of Prehistory, Research School of Pacific Studies, Australian National University, Canberra.

Mulvaney, K. 1993 Aboriginal Archaeological Investigation, East Arm Port development, Darwin Harbour. Unpublished report to Acer Vaughan.

Murray, P.F. and J.R. Hanley 1986 Unmuddling the mudlobster; observations on the age and taphonomy of fossil *Thalassina*. *The Beagle* 3(1):59-70.

Murray, T. 1988 Ethnoarchaeology or Palaeoethnology? In B. Meehan and R. Jones (eds) *Archaeology with Ethnography*: *An Australian Perspective*, pp.1-16. Department of Prehistory, Research School of Pacific Studies, Australian National University, Canberra.

Nash, D.T. and M.D. Petraglia (eds) 1987 *Natural Formation Processes and the Archaeological Record, BAR International Series* 352. Osney Mead, Oxford, England.

Navin, K. and K. Officer 1995 Archaeological Survey Marrara Swamp – Rapid Creek RAAF Base Darwin. Unpublished report to GHD Pty Ltd for the Department of Defence.

Nichol, R.K. and L.J. Williams 1981 Quantifying shell midden: weights or numbers? *New Zealand Archaeological Association* Newsletter 24(2):87-91.

Nicholson, R.A. 1993 A morphological investigation of burnt animal bone and an evaluation of its utility in archaeology. *Journal of Archaeological Science* 20:411-428.

Nicholson, A. and S. Cane 1994 Pre-European coastal settlement and use of the sea. *Australian Archaeology* 39:108-17.

Nix, H.A. and J.D. Kalma 1972 Climate as a dominant control in the biogeography of Northern Australia and New Guinea. In D. Walker (ed.) *Bridge and Barrier: The Natural and Cultural History of Torres Strait*, pp.61-91. Research School of Pacific Studies, Department of Biogeography and Geomorphology, Publication BG/3, Australian National University, Canberra.

Nolan, A. 1986 Sandstone Point: temporal and spatial patterns of Aboriginal site use at a midden complex, S.E. Queensland. Unpublished B.A.(Hons) thesis, University of Queensland, Brisbane.

Noli, D. and G. Avery 1988 Protein poisoning and coastal subsistence. *Journal Of Archaeological Science* 15:395-401.

Nott, J. 1994 Long-term landscape evolution in the Darwin region and its implications for the origin of landsurfaces in the north of the Northern Territory. *Australian Journal of Earth Sciences* 41:407-15.

NTRS 828 Government Resident of the Northern Territory. Inwards correspondence, 1870-1903 Item A102. Letter of application from Burlow Perkins & Parr requesting shell from Government Paddock. Northern Territory Archives Service, Darwin.

Nunn, P.D. 1998 Sea-level changes over the past 1,000 years in the Pacific. *Journal of Coastal Research* 14(1):23-30.

O'Connell, J.F. and Hawkes, K. 1981 Alyawara plant use and optimal foraging theory. In B. Winterhalder and E.A. Smith (eds) *Hunter-gatherer foraging strategies*, pp.99-125. University of Chicago Press, Chicago.

O'Connor, S. 1989 Contemporary island use in the west Kimberley, Western Australia, and its implications for archaeological site survival. *Australian Institute of Aboriginal Studies* 2:25-31.

O'Connor, S. 1990 30,000 Years in the Kimberley: A prehistory of the islands of the Buccaneer Archipelago and adjacent mainland, west Kimberley, Western Australia. Unpublished PhD thesis, Department of Archaeology, The University of Western Australia.

O'Connor, S. 1996 Where are the middens? – an overview of the archaeological evidence for shellfish exploitation along the northwestern Australian coastline. In I.C. Glover and P. Bellwood (eds) *Indo-Pacific Prehistory: The Chiang Mai Papers 2, Bulletin of the Indo-Pacific Prehistory Association* 15:165-80.

O'Connor, S. 1999a A diversity of coastal economies: shell mounds in the Kimberley region in the Holocene. In Hall, J. and I. McNiven (eds) *Australian Coastal Archaeology*, pp.37-50. ANH Publications, Department of Archaeology and Natural History, RSPAS, The Australian National University, Canberra.

O'Connor, S. 1999b *30,000 Years of Aboriginal Occupation: Kimberley, North West Australia. Terra Australis 14*. Department of Archaeology and Natural History and Centre for Archaeological Research, The Australian National University, Canberra.

O'Connor, S. and M. Sullivan 1994a Coastal archaeology in Australia; developments and new directions. *Australian Archaeology* 39:87-96.

O'Connor, S. and M. Sullivan 1994b Distinguishing middens and cheniers: a case study from the Southern Kimberley, W.A. *Archaeology in Oceania* 29, 16-28.

O'Connor, S. and P. Veth 1993 Where the desert meets the sea: a preliminary report of the archaeology of the Kimberley coast. *Australian Archaeology* 37:25-34.

O'Connor, S., P. Veth and A. Barham 1999 Cultural versus natural explanations for lacunae in Aboriginal occupation deposits in northern Australia. *Quaternary International* 59:61-70.

O'Connor, S., P. Veth and C. Campbell 1998 Serpent's Glen rockshelter: report of the first Pleistocene-aged occupation sequence from the Western Desert. *Australian Archaeology* 46:12-22.

O'Neil, D.H. 1993 Excavation sample size: a cautionary tale. *American Antiquity* 53(3):523-29.

Ostvos, E.G. and W.A. Price 1979 Problems of chenier genesis and terminology – an overview. *Marine Geology* 31:251-63.

Parkhouse, T.A. 1895 Native tribes of Port Darwin and its neighbourhood. *Proceedings of the Australian Association for the Advancement of Science,* Vol. vi, 638-47.

Parks and Wildlife Commission of the Northern Territory 1997 *A management program for cycads in the Northern Territory of Australia.* Parks and Wildlife Commission of the Northern Territory.

Parks and Wildlife Commission of the Northern Territory 1999 *Knuckey Lagoons Conservation Reserve: draft management plan.* Parks and Wildlife Commission of the Northern Territory, Darwin.

Pathansali, D. 1966 Notes on the biology of the cockle *Anadara granosa* L. *Proceedings of the Indo-Pacific Fisheries Council* 11:84-98.

Paton, R. 1994 Speaking through stones: a study from northern Australia. *World Archaeology* 26(2):172-84.

Peterson, C.H. and F.E. Wells 1998 Molluscs in marine and estuarine sediments. In P.L. Beesley, G.J.B. Ross and A. Wells (eds) *Mollusca: The Southern Synthesis. Fauna of Australia* Vol.5, Part A, pp.36-46. CSIRO Publishing, Melbourne.

Peterson, N. 1968 The pestle and mortar: an ethnographic analogy for archaeology in Arnhem Land. *Mankind* 6(11):567-70.

Peterson, N. 1972 Totemism yesterday: sentiment and local organisation among the Australian Aborigines. *MAN* 7(1):12-32.

Peterson, N. 1973 Camp site location amongst Australian hunter-gatherers: Archaeological and ethnographic evidence for a key determinate. *Archaeology and Physical Anthropology in Oceania* 8:173-93.

Peterson, N. 1983 *Donald Thompson in Arnhem Land.* Currey O'Neil, Victoria.

Pianka, E.R 1970 On *r*- and *k*- selection. *American Naturalist* 104:592-7.

Pickering, M. 1994 The physical landscape as a social landscape: a Garawa example. *Archaeology in Oceania* 29:149-61.

Pickup, G., G. Allan and V.R. Baker 1988 History, palaeochannels and palaeofloods of the Finke River, central Australia. In R.F. Warner (ed.) *Fluvial Geomorphology of Australia*, pp.177-200. Academic Press, Sydney.

Pietsch, B. A. 1983 *Darwin 5073, 1:100 000 Geological Map Series, Explanatory Notes*. Government Printer of the Northern Territory, Darwin.

Pietsch, B.A. 1985 *Koolpinyah 5173, 1:100 000 Geological Map Series, Explanatory notes.* Department of Mines and Energy, Northern Territory Geological Survey, Government Printer of the Northern Territory, Darwin.

Pietsch, B.A. 1986 *Bynoe 5072, 1:100 000 Geological Map Series, Explanatory notes.* Department of Mines and Energy, Northern Territory Geological Survey, Government Printer of the Northern Territory, Darwin.

Pietsch, B. A. and P.G. Stuart-Smith 1987 *Darwin SD52-4, 1:250 000 Geological Map Series, Explanatory Notes.* Government Printer of the Northern Territory, Darwin.

Povinelli, E.A. 1992 "Where we gana go now": foraging practices and their meanings among the Belyuen Australian Aborigines. *Human Ecology* 20(2):169-202.

Price, O. and D.M.J.S. Bowman 1994 Fire-stick forestry: a matrix model in support of skilful fire management of *Callitris intratropica* R.T. Baker by north Australian Aborigines. *Journal of Biogeography* 21:573-80.

Quitmyer, I.R. and D.S. Jones 1997 The sclerochronology of hard clams, Mercenaria spp., from the south-eastern U.S.A.: A method of elucidating the zooarchaeological records of seasonal resource procurement and seasonality in prehistoric shell middens. *Journal of Archaeological Science* 24: 825-40.

Ranson, D. 1980 Open area excavation in Australia: a plea for bigger holes. In I. Johnson (ed.) *Holier Than Thou. Proceedings of the 1978 Kioloa conference on Australian Prehistory*, pp.77-90. Department of Prehistory, research School of Pacific Studies, Australian National University, Canberra.

Rasmussan, E.M. and Carpenter, T.H. 1982 Variations in tropical sea surface wind fields associated with the southern oscillation/El Nino. *Monthly Weather Review* 110:354-84.

Read, D.W. 1979 Regional sampling. In J.W. Mueller (ed.) *Sampling in Archaeology*, pp.45-60. The University of Arizona Press, Tuscon, Arizona.

Read, P. 1991 Where have all the people gone? In A. Roberts (ed.) *Writing Aboriginal History: Proceedings of the Federation of Australian Historical Societies Conference in Canberra 1988*, pp.19-31. Federation of Australian Historical Societies, Canberra.

Redman, C.L. 1987 Surface collection, sampling, and research design: A retrospective. *American Antiquity* 52:249-56.

Registrar-General's Office, NT. Document No. 308909 Surrender - Pastoral Lease Koolpinyah Station, NT Portions 1625 & 2990.

Reid, D.G. 1985 Habitat and zonation patterns of Littoraria species (Gastropoda: Littorinidae) in Indo-Pacific mangrove forests. *Biological Journal of the Linnean Society* 26:39-68.

Reid, G. 1990 *A Picnic with the Natives: Aboriginal-European Relations in the Northern Territory to 1910*. Melbourne University Press, Carlton, Victoria.

Reynolds, T. 1873 The report of the Commissioner of Crown Lands and Immigration of his Official Visit to the Northern territory, via Galle, Singapore, Java and Timor. *Report on Northern Territory July 30 1873*, G. & N. Ingleton.

Rhode, D. 1988 Measurement of archaeological diversity and the sample-size effect. *American Antiquity* 53(4):708-16.

Rhodes, E.G., H.A. Polach, B.G. Thom and S.R. Wilson 1980 Age structure of Holocene sediments: Gulf of Carpentaria, Australia. *Radiocarbon* 22(3):718-27.

Richardson, N. 1988 The Archaeological Component of Environmental Impact Studies in Queensland. Unpublished report for the Archaeology Branch, Department of Community Services, Queensland. March 1988.

Richardson, N. 1996a Archaeological site survey and documentation - Darwin Region. Unpublished report to the Australian Heritage Commission. Museums and Art Galleries of the Northern Territory.

Richardson, N. 1996b Seeing is believing: a graphical illustration of the vertical and horizontal distribution of conjoined artefacts using Design CAD 3D. In S. Ulm, I. Lilley and A. Ross (eds) *Australian Archaeology '95, Tempus* 6, pp.81-95. Anthropology Museum, The University of Queensland, St. Lucia, Queensland.

Richman, J.R. 1999 East Gippsland coastal archaeology: Past, present, and future. In Hall, J. and I. McNiven (eds) *Australian Coastal Archaeology*, pp.255-61. ANH Publications, Department of Archaeology and Natural History, RSPAS, The Australian National University, Canberra.

Rideout, B.L. 1982 [Reprinted] The Buffalo meat industry. In Ford, B.D. and D.G. Tulloch (eds) *The Australian Buffalo - A Collection of Papers*, pp.35-8. Technical Bulletin No.62, Department of Primary Production, Division of Agriculture and Stock, Darwin.

Ridpath, M.G., M.A.J. Williams and C.D. Haynes 1991 A torrid land. In C.D. Haynes, M.G. Ridpath and M.A.J. Williams (eds) *Monsoonal Australia: Landscape, Ecology and Man in the Northern Lowlands*, pp.207-21. A.A. Balkema, Rotterdam.

Rimmer, M.A. and J.R. Merrick 1983 A review of reproduction and development in the fork-tailed catfishes (Ariidae). *Proceedings of the Linnean Society of New South Wales* 107(1):41-50.

Ringrose, T.J. 1993 Bone counts and statistics: a critique. *Journal of Archaeological Science* 20(2):121-57.

Ritchie, D.J. 1998 A Painful Wrench for a European Mind: Land Claims and Representations of History and Aboriginal Tradition in The Darwin Hinterland. Unpublished PhD thesis, La Trobe University, Melbourne.

Roberts, A. 1991 An Analysis of Mound Formation at Milingimbi, N.T. Unpublished Master of Letters Thesis, U.N.E. Armidale.

Roberts, A. 1994 Cultural landmarks: The Milingimbi mounds. In M. Sullivan, S. Brockwell and A. Webb (eds), *Archaeology in the North*, pp.176-87. North Australian Research Unit, Australian National University, Darwin.

Robins, R.P. and E.C. Stock 1990 The burning question: a study of molluscan remains from a midden on Moreton island. In S. Solomon, I. Davidson and D. Watson (eds) *Problem Solving in Taphonomy, Tempus* 2, pp.80-100. The University of Queensland, St Lucia, Queensland.

Roddom, W. 1997 Like, - But Oh How Different: Stone Point Variability in the Top End, N.T. Unpublished Honours thesis, Department of Anthropology, Northern Territory University, Darwin.

Rollins, B., J.B. Richardson III and D.H. Sandweiss 1986 The birth of El Nino: Geoarchaeological evidence and implications. *Geoarchaeology* 1(1):3-15.

Ross, A. 1985 Archaeological evidence for population change in the middle to late Holocene in southeastern Australia. *Archaeology in Oceania* 20:81-9.

Roth, W.E. 1904 Domestic implements, arts and manufactures. *North Queensland Ethnography Bulletin No.7*. Government Printer, Brisbane.

Roth, W.E. 1984a [1901] Food, its capture and preparation. In K.F. MacIntyre (ed.) *The Queensland Aborigines* Vol.I, Hesperian Press, Perth.

Roth, W.E. 1984b [1907-1910] Decoration, deformation and clothing. In K.F. MacIntyre (ed.) *The Queensland Aborigines* Vol.III. No.15. Hesperian Press, Perth.

Rowland, M.J. 1983 Aborigines and environment in Holocene Australia: changing paradigms. *Australian Aboriginal Studies* 2:62-77.
Rowland, M.J. 1989 Population increase, intensification or a result of preservation? Explaining site distribution patterns on the coast of Queensland. *Australian Aboriginal Studies* 2:32-41.
Rowland, M. 1994a Mounds around the world: A contribution to understanding Australian shell middens and mounds. In M. Sullivan S. Brockwell and A. Webb (eds) *Archaeology In The North*, pp.107-29. North Australian Research Unit, Australian National University, Darwin.
Rowland, M. 1994b Size isn't everything. Shells in mounds, middens and natural deposits. *Australian Archaeology* 39:118-24.
Rowland, M.J. 1999 Holocene environmental variability: have its impacts been underestimated in Australian pre-History? *The Artefact* 22:11-48.
Rowley, C.D. 1970 *The destruction of Aboriginal society*. Australian National University Press, Canberra.
Rowley-Conwy, P. 1983 Sedentary hunters: the Ertebolle example. In G. Bailey (ed.) *Hunter-Gatherer Economy in Prehistory: A European Perspective*, pp.111-26. Cambridge University Press, Cambridge.
Russel-Smith, J. 1985 A record of change: studies of Holocene vegetation history in the South Alligator River region, Northern Territory. *Proceedings of the Ecological Society of Australia* 13:191-202.
Russel-Smith, J., D. Lucas, M. Gapindi, B. Gunbunuka, N. Kapirigi, G. Namingum, K. Lucas, P. Giuliana and G. Chaloupka 1997 Aboriginal resource utilization and fire management practice in western Arnhem Land, monsoonal northern Australia: notes for prehistory, lessons for the future. *Human Ecology* 25(2):159-95.
Sandweiss, D.H. 1996 Environmental change and its consequences for human society on the Central Andean Coast: A malacological perspective. In E.J. Reitz, L.A. Newsome and S.J. Scudder (eds) *Case Studies in Environmental Archaeology*, pp.127-46. Plenum Press, New York.
Sanger, D. 1981 Unscrambling messages in the midden. *Archaeology of Eastern North America* 9:37-42.
Sawbridge, D.F. and M.A.M. Bell 1972 Vegetation and soils of shell middens on the coast of British Columbia. *Ecology* 53(5):840-9.
Schiegl, S., P. Goldberg, O. Bar-Yosef and S. Weiner 1996 Ash deposits in Hayonim and Kebara caves, Israel: macroscopic, microscopic and mineralogical observations, and their archaeological implications. *Journal of Archaeological Science* 23:763-81.
Schiffer, M.B. 1987 *Formation Processes of the Archaeological Record*. University of New Mexico Press: Albuquerque.
Schiffer, M.B., A.P. Sullivan and T.C. Klinger 1978 The design of archaeological surveys. *World Archaeology* 10:1-28.
Schimmelmann, A., M. Zhao, C.C. Harvey and C.B. Lange 1998 A large California flood and correlative global climatic events 400 years ago. *Quaternary Research* 49(1):51-61.
Schrire, C. 1982 *The Alligator Rivers: Prehistory and Ecology in Western Arnhem Land. Terra Australis 7*, Department of Prehistory, Research School of Pacific Studies, Australian National University, Canberra.
Schrire, C. 1984 Interactions of past and present in Arnhem Land, North Australia. In C. Schrire (ed.) *Past and Present in Hunter Gatherer Studies*, pp.67-93. Academic Press, Inc., London.
Schrire, C. 1984 Wild surmises on savage thoughts. In C. Schrire (ed.) *Past and Present in Hunter Gatherer Studies*, pp.1-25. Academic Press, Inc., London.
Semeniuk, V. 1985 Mangrove environments of Port Darwin, Northern Territory: the physical framework and habitats. *Journal of the Royal Society of Western Australia* 67 (3 & 4):81-97.
Semeniuk, V. 1996 The Holocene record of climatic, eustatic and tectonic events along the coastal zone of Western Australia – a review. *Journal of Coastal research Special Issue* 17:247-59.
Shackleton, N.J. 1982 Stratigraphy and chronology of the KRM deposits: oxygen isotope evidence. In R. Singer and J. Wymer (eds) *The Middle Stone Age at Klasies River Mouth in South Africa*, pp.194-9. University of Chicago Press, Chicago.
Shaffer, B.S. and J.L. Sanchez 1994 Comparison of 1/8"- and 1/4"- mesh recovery of controlled samples of small-to-medium-sized mammals. *American Antiquity* 59(3):525-530.
Sharp, R.L. 1939 Tribes and totemism in North-East Australia. *Oceania* 9(3):254-75, (4):439-61.
Sheaves, M. 1998 Spatial patterns in estuarine fish faunas in tropical Queensland: a reflection of interaction between long-term physical and biological processes? *Marine and Freshwater Research* 49:31-40.
Shenkel, J.R. 1971 Cultural Adaptation to the Mollusk: A Methodological Survey of Shellmound Archaeology and a Consideration of the Shellmounds of The Marismas Nacionales, West Mexico. Unpublished PhD thesis, University of New York at Buffalo.
Shipman, P., G. Foster and M. Schoeninger 1984 Burnt bones and teeth: an experimental study of colour, morphology, crystal structure and shrinkage. *Journal of Archaeological Science* 11:307-25.
Shoocongdej, R. 1996 Rethinking the development of sedentary villages in western Thailand. In P. Bellwood (ed.), *Indo-Pacific Prehistory: The Chiang Mai Papers, Bulletin of the Indo-Pacific Prehistory Association* 14:203-15.
Shulmeister, J. 1992 A Holocene record from lowland tropical Australia. *The Holocene* 2:107-16.
Shulmeister, J. and B.G. Lees 1992 Morphology and chronostratigraphy of a coastal dunefield, Groote Eylandt, northern Australia. *Geomorphology* 5:521-34.

Shulmeister, J. and B.G. Lees 1995 Pollen evidence from tropical Australia for the onset of an ENSO-dominated climate at c. 4000 BP. *The Holocene* 5(1):10-18.

Shurcliff, K.S. 1993 Seasonal patterns of Wader populations in Darwin, Northern Territory 1974-1987. *Northern Territory Naturalist* 14:10-18.

Siegfried, W.R., P.A.R. Hockey and A.A. Crowe 1985 Exploitation and conservation of brown mussel stocks by coastal people of Transkei. *Environmental Conservation* 12(4):303-7.

Singh, B., S. O'Connor, P. Veth, and R. Gilkes 1991 Detection of amorphous alumino-silicate by X-ray diffraction and chemical analysis to detect firing in archaeological sediments. *Archaeology in Oceania* 26:17-20.

Singh, G. 1983 Holocene palaeoclimates and the seasonality of rainfall at Lake Frome, South Australia. In *Proceedings of the first CLIMANZ Conference*, p.118. Department of Biogeography and Geomorphology, Research School of Pacific Studies, Australian National University, Canberra.

Slack-Smith, S.M. 1998 Order Ostreoida. In P.L. Beesley, G.J.B. Ross and A. Wells (eds) *Mollusca: The Southern Synthesis. Fauna of Australia* Vol.5, Part A, pp.268-82. CSIRO Publishing, Melbourne.

Smith, A.N., J.R. Hanley and B. Love 1997 Comparisons of different quadrat sizes for measuring the densities of crabs and macromolluscs of the mangrove forest floor. In J.R. Hanley, G. Caswell, D. Megirian and H.K. Larson (eds) *The Marine Flora and Fauna of Darwin Harbour, Northern Australia, Proceedings of the Sixth International Marine Biological Workshop*, pp.461-6. Museum and Art Galleries of the Northern Territory and the Marine Sciences Association, Darwin.

Smith, M. 1981 Field Archaeologist's Report for 1980, Museum and Art Gallery of the Northern Territory, Darwin.

Smith, M.A. 1986 The antiquity of seedgrinding in arid Australia. *Archaeology in Oceania* 21:29-39.

Smith, M.A 1987 Pleistocene occupation in arid central Australia. *Nature* 328:710-11.

Smith, M.A. 1988a Central Australian seed grinding implements and Pleistocene grindstones. In B. Meehan and R. Jones (eds) *Archaeology with Ethnography: An Australian Perspective*, pp.94-108. Department of Prehistory, Research School of Pacific Studies, Australian National University, Canberra.

Smith, M.A. 1988b The pattern and timing of prehistoric settlement in Central Australia. Unpublished PhD thesis, University of New England, Armidale.

Smith, M.A. 1993 Biogeography, human ecology and prehistory in the Sandridge Deserts. *Australian Archaeology* 37:35-50.

Smith, M. and S. Brockwell 1994 Archaeological investigations at a stratified open site near Humpty Doo, northern Australia. In M. Sullivan, S. Brockwell and A. Webb (eds) *Archaeology In The North*, pp.84-104. North Australia Research Unit, Australian National University, Darwin.

Solomon, S., I. Davidson and D. Watson (eds) 1990 *Problem Solving in Taphonomy*, *Tempus* 2. The University of Queensland, St Lucia, Queensland.

Sorant, P.E. and Shenkel J.R. 1984 The calculation of volumes of middens, mounds, and strata having irregular shapes. *American Antiquity* 49(3):599-603.

Specht, J. 1985 Crabs as disturbance factors in tropical archaeological sites. *Australian Archaeology* 21:11-18.

Spencer, W. B. 1912 *An introduction to the study of certain native tribes of the Northern Territory. Bulletin of the Northern Territory of Australia*, No.2, McCarron, Bird & Co., Collins St. Melbourne.

Spencer, W. B. 1914 *Native tribes of the Northern Territory of Australia*. Macmillan, London.

Spencer, W.B. and F.J. Gillen 1904 *Northern Tribes of Central Australia*. Macmillan, London.

Spennemann, D.H.R. 1987 Availability of shellfish resources on prehistoric Tongatapu, Tonga: effects of human predation and changing environment *Archaeology in Oceania* 22(3):81-96.

Spennemann, D.H.R. and S.M. Colley 1989 Fire in a pit: the effects of burning on faunal remains. *Archaeozoologia* 111(1,2):51-64.

Speth, J.D. 1987 Early hominid subsistence strategies in seasonal habitats. *Journal of Archaeological Science* 14:13-29.

Speth, J.D. 1990 Seasonality, resource stress, and food sharing in so-called "egalitarian" foraging societies. *Journal of Anthropological Archaeology* 9:148-88.

Speth, J.D. and K.A. Spielmann 1983 Energy source, protein metabolism and hunter-gatherer subsistence strategies. *Journal of Anthropological Archaeology* 2:1-31.

Spiess, A.E. 1988 On New England shell middens: response to Sanger's cautionary tale. *American Antiquity* 53(1):174-77.

Spriggs, M. 1999 Archaeological dates and linguistic sub-groups in the settlement of the Island Southeast Asia-Pacific region. In P. Bellwood, D. Bowdery, D. Bulbeck, M. Fiskesjo, R. Green, I. Lilley and B. Maloney (eds) *Indo-Pacific Prehistory: The Melaka Papers* 2, *Bulletin of the Indo-Pacific Prehistory Association* 18:17-24.

Stahl, A.B. 1993 Concepts of time and approaches to analogical reasoning in historical perspective. *American Antiquity* 58(2):235-60.

Stahl, P.W. 1996 Holocene biodiversity: an archaeological perspective from the Americas. *Annual Review of Anthropology* 25:105-26.

Stanner, W.E.H. 1933 The Daly River Tribes – A report of fieldwork in north Australia. *Oceania* 3:377-405.

Stanner, W.E.H. 1962-3 On Aboriginal religion. VI. Cosmos and society made correlative. *Oceania* 33:239-73.

Stein, J.K. (ed.) 1992 *Deciphering a Shell Midden,* Academic Press Inc., San Diego, California.

Stein, J.K. 1992a The analysis of shell middens. In J.K. Stein (ed.) *Deciphering a Shell Midden*, pp.1-15. Academic Press Inc., San Diego, California.

Stein, J.K. 1992b Interpreting stratification of a shell midden. In J.K. Stein (ed.) *Deciphering a Shell Midden*, pp.71-95. Academic Press Inc., San Diego, California.

Stein, J.K. 1992c Sediment analysis of the British Camp shell midden. In J.K. Stein (ed.) *Deciphering a Shell Midden*, pp.135-62. Academic Press Inc., San Diego, California.

Stein, J.K. 1992d Organic matter in archaeological contexts. In V.T. Holliday (ed.) *Soils in Archaeology: Landscape Evolution and Human Occupation*, pp.193-216. Smithsonian Institution Press, Washington.

Stiner, M.C. and S.L. Kuhn 1995 Differential burning, recrystallization, and fragmentation of archaeological bone. *Journal of Archaeological Science* 22:223-37.

Stocker, G.C. 1966 Effects of fires on vegetation in the Northern Territory. *Australian Forestry* 30(3):223-30.

Stockton, E.D. 1973 Shaw's Creek shelter: human displacement of artefacts and its significance. *Mankind* 9:112-7.

Stockton, E.D. 1977 Review of early Bondaian dates. *Mankind* 11:48-51.

Stockton, E. 1980 The shifting sands of stratigraphic excavations. In I. Johnson (ed.) *Holier Than Thou: Proceedings of the 1978 Kioloa conference on Australian Prehistory*, pp.127-31. Department of Prehistory, Research School of Pacific Studies, The Australian National University, Canberra.

Stokes, J.L. 1846 *Discoveries in Australia*. T&W Boone, London.

Stone, T. 1987 Geomorphology of Channel Island near Darwin. In A. Fox and Associates (eds) Interpretation Plan for the Channel Island Environment with Recommendations on Environmental Management. Vol.2. Unpublished report to the Northern Territory Electricity Commission, Darwin.

Stone, T. 1989a An Archaeological Survey of the Proposed Second Cavalry Regiment Base near Palmerston, Northern Territory. Unpublished report to Australian Construction Services, National Heritage Studies, Adelaide.

Stone, T. 1989b Origins and environmental significance of shell and earth mounds in Northern Australia. *Archaeology in Oceania* 24:59-64.

Stone, T. 1991 Two birds with one Stone: a reply. *Archaeology in Oceania* 26:26-28.

Stone, T. 1992 Origins of the Weipa Shell Mounds. Unpublished MA(Sc) thesis, Australian National University: Canberra.

Storrs, M.J. and M. Finlayson 1997 *Overview of the conservation status of wetlands of the Northern Territory. Supervising scientist report*: 116, Barton, A.C.T.

Story, R. 1969 Grasses of the Adelaide-Alligator area and their utilization. In M.A.J. Williams, A.D.L. Hooper, R.E. O'Ferrall and J.R. McAlpine (eds) *Lands of the Adelaide-Alligator Area, Northern Territory Commonwealth Scientific and Industrial Research Organization, Land Research Series* No. 25, pp.131-39. Melbourne.

Story, R. 1969 Vegetation of the Adelaide-Alligator area. In M.A.J. Williams, A.D.L. Hooper, R.E. O'Ferrall and J.R. McAlpine (eds) *Lands of the Adelaide-Alligator Area, Northern Territory Commonwealth Scientific and Industrial Research Organization, Land Research Series* No. 25, pp.114-30. Melbourne.

Story, R., M.A.J. Williams, A.D.L. Hooper, R.E. O'Ferrall and J.R. McAlpine 1969 Summary description of the Adelaide-Alligator area. In *Lands of the Adelaide-Alligator Area, Northern Territory Commonwealth Scientific and Industrial Research Organization, Land Research Series* No. 25, pp.16-23. Melbourne.

Stuiver, M., and T. F. Braziunas, 1993 Modeling atmospheric ^{14}C influences and ^{14}C ages of marine samples back to 10,000 BC. *Radiocarbon* 35:137-89.

Stuiver, M. and P.J. Reimer 1993 Extended ^{14}C data base and revised CALIB radiocarbon calibration program. *Radiocarbon* 35:215-30.

Suguio, K., L. Martin and J-M. Flexor 1992 Paleoshorelines and the sambaquis of Brazil. In L.L. Johnson (ed.) *Paleoshorelines and Prehistory: An Investigation of Method*, pp.83-99. CRC Press, London.

Sullivan, M.E. 1982 Aboriginal Shell Middens in the Coastal Landscape of New South Wales. Unpublished PhD thesis, Australian National University, Canberra.

Sullivan, M.E. 1984 A shell midden excavation at Pambula Lake on the far south coast of New South Wales. *Archaeology in Oceania* 19:1-15.

Sullivan, M. 1996 Northern Australian landscapes. In P. Veth and P. Hiscock (eds) *Archaeology of Northern Australia, Tempus* 4, pp.1-8. Anthropology Museum, University of Queensland, St. Lucia, Queensland.

Sullivan, S. and S. Bowdler 1984 (eds) *Site Surveys and Significance Assessment in Australian Archaeology*. Australian National University, Canberra.

Sullivan, M. and S. O'Connor 1993 Middens and cheniers: implications of Australian research. *Antiquity* 67, 776-88.

Sullivan M., S. Brockwell and A. Webb (eds) 1994 *Archaeology In The North*. North Australian Research Unit, Australian National University, Darwin.

Sundstrom, L. 1993 A simple mathematical procedure for estimating the adequacy of site survey strategies. *Journal of Field Archaeology* 20:91-6.

Suzuki, K. 1986 Volumetry and nutritional analysis of a Jomon shell-midden. In T. Akazawa and C.M. Aikens (eds) *Prehistoric Hunter-Gatherers in Japan*, pp.55-71. University of Tokyo Press, Tokyo.

Swadling, P. 1976 Changes induced by human exploitation in prehistoric shellfish populations. *Mankind* 10:156-62.

Swadling, P. 1997 Changing shorelines and cultural orientations in the Sepik-Ramu, Papua New Guinea: implications for Pacific prehistory. *World Archaeology* 29(1):1-14.

Tacon, P.S. 1993 Regionalism in the recent rock art of western Arnhem Land, Northern Territory. *Archaeology in Oceania* 28:112-20.

Tacon, P.S.C. 1994 Socialising landscapes: the long-term implications of signs, symbols and marks on the land. *Archaeology in Oceania* 29:117-29.

Taylor, J.A. and D.G. Tulloch 1985 Rainfall in the wet-dry tropics: extreme events at Darwin and similarities between years during the period 1870-1983 inclusive. *Australian Journal of Ecology* 10:281-95.

Testart, A. 1981 On the anteriority of matrilineal descent in Australia: logical mathematical evidences. *Mankind* 13:15-24.

Thomas, D.H. 1975 Non-site sampling in Archaeology: up the creek without a site? In J.W. Mueller (ed.) *Sampling in Archaeology*, pp.61-81. University of Arizona Press, Tuscon.

Thomas, D.H. 1989 Diversity in hunter-gatherer cultural geography. In R.D Leonard and G.T. Jones (eds) *Quantifying Diversity in Archaeology*, pp. 85-91. Cambridge University Press.

Thomson, D.F. 1939 The seasonal factor in human culture –illustrated from the life of a contemporary nomadic group. *Proceedings of the Prehistoric Society* 5:209-21.

Thomson, D.F. 1949 *Economic Structure and the Ceremonial Exchange Cycle in Arnhem Land.* Macmillan, Melbourne.

Thomson, D. 1983 *Donald Thomson in Arnhem Land.* Compiled and introduced by Nicolas Peterson. Currey O'Neil, Melbourne.

Thomson, J.M. 1954 The genera of oysters and the Australian species. *Australian Journal of Marine and Freshwater Research* 5:132-68.

Tindale, N.B. 1925-28 *Natives of Groote Eylandt and of the west coast of the Gulf of Carpentaria.* Hassell Press, Adelaide.

Tindale, N. 1974 *Aboriginal Tribes of Australia.* University of California Press, Berkeley.

Todd, T.W. 1968 Dynamic diversion: influence of longshore current-tidal flow interaction on chenier and barrier island plains. *Journal of Sedimentary Petrology* 38(3):734-46.

Tomka, S.A. and M.C. Stevenson 1993 Understanding abandonment processes: summary and remaining concerns. In C.M. Cameron and S.A. Tomka (eds) *Abandonment of settlements and regions: Ethnoarchaeological and archaeological approaches*, pp.191-5. Cambridge University Press, Cambridge.

Treganza, A.E. and S.F. Cook 1948 The quantitative investigation of Aboriginal sites: Complete excavation with physical and archaeological analysis of a single mound. *American Antiquity* 13:287-97.

Triggs, B. 1996 *Tracks, Scats and Other Traces.* Oxford University Press, Melbourne.

Tulloch, D.G. 1981 Previous attempts at the development of a buffalo industry. In Best, F.G. (ed.) *Proceedings of the Buffalo Industry Symposium* pp.6-16. The Department of Primary Production and the Northern Territory Buffalo Industry Council, Darwin.

Tulloch, D.G. 1985 The Magpie Goose (Anseranas semipalmata) and the black soil plains: a biological case history of waterbirds in coastal North Australia. In K.N. Bardsley, J.D.S. Davie and C.D. Woodroffe (eds) *Coasts and Tidal Wetlands of the Australian Monsoon Region, Mangrove Monograph* 1, pp.285-94. North Australia Research Unit, Australian National University, Darwin.

Turner, D.H. 1976 Levels of organisation and communication in Aboriginal Australia. In N. Peterson (ed.) *Tribes and Boundaries in Australia*, pp.180-91. Australian Institute of Aboriginal Studies, Canberra.

Underwood, A.J. 1977 *Barnacles.* Reed Education, Sydney.

Urry, J. 1979 Beyond the frontier: European influence, Aborigines and the concept of traditional culture. *Journal of Australian Studies* 5:2-16.

Van Geel, B. J. Buurman and H.T. Waterbolk 1996 Archaeological and palaeoecological indications of an abrupt climate change in the Netherlands, and evidence for climatological teleconnections around 2650 BP. *Journal of Quaternary Science* 11(6):451-60.

Vanderlely, R. 1996 Nesting Ecology of Flatback Sea Turtles of West Alligator Head and Field Island, Kakadu National Park, MASc thesis, Northern Territory University.

Vardon, M.J. and C.R. Tidemann 1999 Flying-foxes (*Pteropus alecto* and *P. scapulatus*) in the Darwin region, north Australia: patterns in camp size and structure. *Australian Journal of Zoology* 47:411-23.

Veitch, B. 1994 Hearth stones in the mound. In M Sullivan, S. Brockwell and A. Webb (eds) *Archaeology In The North*, pp.167-75. North Australia Research Unit, Australian National University, Darwin.

Veitch, B. 1996 Evidence for mid-Holocene change in the Mitchell Plateau, Northwest Kimberley, Western Australia. In P. Veth and P. Hiscock (eds) *Archaeology of Northern Australia, Tempus* 4, pp.66-89. Archaeology and Material Culture Studies in Anthropology, Anthropology Museum, University of Queensland, St. Lucia, Queensland.

Veitch, B. 1999 Shell middens on the Mitchell Plateau: a reflection of a wider phenomenon? In Hall, J. and I. McNiven 1999 (eds) *Australian Coastal Archaeology*, pp.51-64. ANH Publications, Department of Archaeology and Natural History, RSPAS, The Australian National University, Canberra.

Veth, P. 1989 Islands in the interior: a model for the colonisation of Australia's arid zone. *Archaeology in Oceania* 24:81-92.

Veth, P. 1995 Aridity and settlement in northwest Australia. *Antiquity* 69:733-46.

Veth, P. and P. Hiscock (eds) 1996 *Archaeology of Northern Australia. Tempus* 4, pp.90-105. Anthropology Museum, The University of Queensland, St. Lucia, Queensland.

Veth, P. and B. O'Brien 1986 Middens on the Abydos Plain, northwest Australia. *Australian Archaeology* 22:45-59.

Villa, P. 1982 Conjoinable pieces and site formation processes. *American Antiquity* 47:276-90.

Villa, P. and J. Courtin 1983 The interpretation of stratified sites: a view from underground. *Journal of Archaeological Science* 10:267-81.

Walshe, K. 1998 Taphonomy of Mungo B assemblage: indicators for subsistence and occupation of Lake Mungo. *Archaeology in Oceania* 33:200-6.

Walters, I. 1979 Information loss in zooarchaeology: a study of vertebrate remains from two middens, Moreton Island, Queensland. Unpublished B.A.(Hons) thesis, University of Queensland, Brisbane.

Walters, I. 1992 Farmers and their fires, fishers and their fish: production and productivity in pre-European South-East Queensland. *Dialectical Anthropology* 17:167-82.

Walters, I., P. Lauer, A. Nolan, G. Dillon and M. Aird 1987 Hope Island: salvage excavation of a Kombumerri site. *Queensland Archaeological Research* 4:80-95.

Ward, G.K. 1994 On the use of radiometric determinations to 'date' archaeological events. *Australian Aboriginal Studies* 2:106-9.

Warner, W.L. 1969 [1958] *A Black Civilization: A Social Study of an Australian Tribe,* Revised Edition, Harper and Row, New York.

Waselkov, G.A. 1987 Shellfish gathering and shell midden archaeology. In M.B. Schiffer (ed.) *Advances in Archaeological Method and Theory* 10:93-210.

Waters, M.R. and D.D. Kuehn 1996 The geoarchaeology of place: the effect of geological processes on the preservation and interpretation of the archaeological record. *American Antiquity* 61(3):483-97.

Wedd, R. 1993 Report on the partial processing of the Bayview Haven Estate shell midden excavation material. In N. Richardson (ed.) Archaeological Site Survey and Documentation – Darwin Region. Unpublished report to the Australian Heritage Commission, Museum and Art Galleries of the Northern Territory, Darwin.

Weisler, M.I. 1993 The importance of fish otoliths in Pacific Island archaeofaunal analysis. *New Zealand Journal of Archaeology* 15:131-59.

Wells, F.E. 1980 A comparative study of distributions of the mudwhelks *Terebralia sulcata* and *T. palustris* in a mangrove swamp in northwestern Australia. *Malacological Review* 13:1-5.

Wells, F.E. 1997 A review of the Northern Australian species of the genera *Cassidula* and *Ellobium* (Gastropoda:Ellobiidae). In J.R. Hanley, G. Caswell, D. Megirian and H.K. Larson (eds) *The Marine Flora and Fauna of Darwin Harbour, Northern Australia, Proceedings of the Sixth International Marine Biological Workshop*, pp.213-29. Museum and Art Galleries of the Northern Territory and the Marine Sciences Association, Darwin.

Wells, F.E. and C.W. Bryce 1988 *Seashells of Western Australia.* Western Australian Museum, Perth.

Wells, F.E. and S.M. Slack-Smith 1981 Zonation of molluscs in a mangrove swamp in the Kimberley, Western Australia. In *Biological Survey of Mitchell Plateau and Admiralty Gulf, Kimberley, Western Australia*, pp.265-74. Western Australian Museum, Perth.

Wells, S.J. 1995 Town camp or homeland?: a history of the Kulaluk Aboriginal community. A report prepared for the Australian Heritage Commission, Northern Territory University, Darwin.

Wenke, R.J. 1981 Explaining the evolution of cultural complexity: a review. In M.B. Schiffer (ed.) *Advances in Archaeological Method and Theory* 4:79-127.

White, C. [Schrire] 1967 Plateau and Plain: Prehistoric Investigations in Arnhem Land, Northern Territory. Unpublished PhD thesis, Australian National University, Canberra.

White, J.P. and J.F. O'Connell 1982 *A Prehistory of Australia, New Guinea and Sahul.* Academic Press, Sydney.

White, N. 1976 A preliminary account of the correspondence among genetic, linguistic, social and topographical divisions in Arnhem Land. *Mankind* 10:240-7.

White, N. 1997 Genes, languages and landscapes in Australia. In P. McConvell and N. Evans (eds) *Archaeology and Linguistics: Aboriginal Australia in Global Perspective*, pp.45-81. Oxford University Press, Melbourne.

White, N., B. Meehan, L. Hiatt, and R. Jones 1990 Demography of contemporary hunter-gatherers: lessons from Arnhem Land. In B. Meehan and N. White (eds) *Hunter-Gatherer Demography, Past and Present*, *Oceania Monograph* 39, pp.171-85. University of Sydney.

Whitehead, P.J. and K. Tschirner 1991 Lead shot ingestion and lead poisoning of Magpie Geese *Anseranas semipalmata* foraging in a northern Australian hunting reserve. *Biological Conservation* 58:99-118.

Whitehead, P.J., P. Bayliss and R.E. Fox 1988 Recreational waterfowl hunting activity and harvests in the Northern Territory, Australia. *Australian Wildlife Research* 15:625-31.

Whitehead, P.J., B.A. Wilson and D.M.J.S. Bowman 1990 Conservation of coastal wetlands of the Northern Territory of Australia: the Mary River floodplain. *Biological Conservation* 52:85-111.

Wightman, G.M. 1989 *Mangroves of the Northern Territory. Northern Territory Botanical Bulletin* No.7. Conservation Commission of the Northern Territory.

Williams, C.K., and A.E. Newsome 1991 Adaptation in native animals. In C.D. Haynes, M.G. Ridpath and M.A.J. Williams (eds) *Monsoonal Australia: Landscape, Ecology and Man in the Northern Lowlands*, pp.151-67. A.A. Balkema, Rotterdam.

Williams, E. 1988 *Complex hunter-gatherers. A late Holocene example from temperate Australia. BAR International Series* 423, Oxford.

Williams, M.A.J. 1969 Geomorphology of the Adelaide-Alligator area. In M.A.J. Williams, A.D.L. Hooper, R.E. O'Ferrall and J.R. McAlpine (eds) *Lands of the Adelaide-Alligator Area, Northern Territory Commonwealth Scientific and Industrial Research Organization, Land Research Series* No. 25, pp.71-94. Melbourne.

Williams, M.A.J. 1994 Some implications of past climatic changes in Australia. *Transactions of the Royal Society of South Australia* 118(1):17-25.

Wilson, B. A. and D. M. J. S. Bowman 1986 A map and description of the vegetation of the Howards Peninsula/ Shoal Bay area near Darwin, Northern Territory. Unpublished report, Wildlife Research Unit, Conservation Commission of the Northern Territory, Darwin.

Wilson, B. A. and D. M. J. S. Bowman 1987 Fire, storm, flood and drought: the vegetation ecology of Howards Peninsula, Northern Territory, Australia, *Australian Journal of Ecology* 12, 165-74.

Wilson, B. A., P.S. Brocklehurst, M.J. Clark and K.J.M. Dickinson 1990 *Vegetation Survey of the Northern Territory, Australia.* Technical Report No. 49. Conservation Commission of the Northern Territory, Darwin.

Withnell, J.G. 1965 [1901] *The customs and traditions of the Aboriginal natives of North Western Australia.* Libraries Board of South Australia, Adelaide. Facsimile edition, Originally published, Roebourne, WA.

Wohl, E.E. 1992 Bedrock benches and boulder bars: floods in the Burdekin Gorge of Australia. *Geological Society of America Bulletin* 104:770-8.

Wohl, E.E., S.J. Fuertsch and V.R. Baker 1994 Sedimentary records of late Holocene floods along the Fitzroy and Margaret Rivers, Western Australia. *Australian Journal of Earth Sciences* 41:273-80.

Woinarski, J.C.Z. 1992 Biogeography and conservation of reptiles, mammals and birds across north-western Australia: an inventory and base for planning an ecological reserve system. *Wildlife Research* 19:665-705.

Woinarski, J.C.Z., S.C. Tidemann and S. Kerin 1988 Birds in a tropical mosaic: the distribution of bird species in relation to vegetation patterns. *Australian Wildlife Research* 15:171-96.

Wolstenholme J. Z.D. Dinesen and P. Alderslade 1997 Hard corals of the Darwin region, Northern Territory, Australia. In J.R. Hanley, G. Caswell, D. Megirian and H.K. Larson (eds) *The Marine Flora and Fauna of Darwin Harbour, Northern Australia, Proceedings of the Sixth International Marine Biological Workshop*, pp.381-98. Museum and Art Galleries of the Northern Territory and the Marine Sciences Association, Darwin.

Wood, B.G., P.J. Fogarty and K.J. Day 1985 *The Land Systems of the Darwin Region.* Technical Report No. 24. Conservation Commission of the Northern Territory, Darwin.

Wood, R.W. and D.L. Johnson 1978 A survey of disturbance processes in archaeological site formation. In M.B. Schiffer (ed.) *Advances in Archaeological Method and Theory* 1:315-81.

Woodroffe, C.D. 1993 Late Quaternary evolution of coastal and lowland riverine plains of Southeast Asia and northern Australia: an overview. *Sedimentary Geology* 83:163-75.

Woodroffe, C.D 1995 Response of tide-dominated mangrove shorelines in northern Australia to anticipated sea-level rise. *Earth Surface Processes and Landforms* 20:65-85.

Woodroffe, C.D. and K.N. Bardsley 1988 The distribution and productivity of mangroves in creek H, Darwin Harbour. In H.K. Larson, M.G. Michie and J.R. Hanley (eds) *Proceedings of the Workshop on Research and Management held in Darwin 2-3 September, 1987, Mangrove Monograph* 4, pp.82-121. Australian National University, North Australian Research Unit, Darwin.

Woodroffe, C.D. and D. Grime 1999 Storm impact and evolution of a mangrove-fringed chenier plain, Shoal Bay, Darwin, Australia. *Marine Geology* 159:303-21.

Woodroffe, C.D. and J. Grindrod 1991 Mangrove biogeography: the role of Quaternary environmental and sea-level change. *Journal of Biogeography* 18:479-92.

Woodroffe, C.D. and M.E. Mulrennan 1993 *Geomorphology of the Lower Mary River Plains, Northern Territory.* North Australia Research Unit, Australian National University, Darwin.

Woodroffe, C.D., J.M.A. Chappell and B.G. Thom 1988 Shell middens in the context of estuarine development, South Alligator River, Northern Territory. *Archaeology in Oceania* 23:95-103.

Woodroffe, C.D., M.E. Mulrennan and J. Chappell 1993 Estuarine infill and coastal progradation, southern van Dieman Gulf, northern Australia. *Sedimentary Geology* 83:257-75.

Woodroffe, C.D., B.G. Thom and J. Chappell 1985 Development of widespread mangrove swamps in mid-Holocene times in northern Australia. *Nature* 317:711-3.

Woodroffe, C.D., E.A. Bryant, D.M. Price and S.A. Short 1992 Quaternary inheritance of coastal landforms, Cobourg Peninsula, Northern Territory. *Australian Geographer* 23(2):101-9.

Woodroffe, C.D., B.G. Thom, J. Chappell, E. Wallensky, J. Grindrod and J. Head 1987 Relative sea level in the South Alligator River region, north Australia, during the Holocene. *Search* 18(4):198-200.

Wright, R.V.S. 1963 Report on archaeological reconnaissance work in Cape York. Document No.63/83, Australian Institute of Aboriginal Studies, Canberra.

Wright, R.V.S. 1971 Prehistory in the Cape York Peninsula. In D.J. Mulvaney and J. Golson (eds) *Aboriginal Man and Environment in Australia*, pp.133-40. Australian National University Press, Canberra.

Wylie, A. 1985 The reaction against analogy. In M.B. Schiffer (ed.) *Advances in Archaeological Method and Theory* 8:63-111.

Yellen, J. 1977 Cultural patterning in faunal remains: evidence from the !Kung Bushmen. In D. Ingersoll, J.E. Yellen and W. Macdonald (eds) *Experimental Archaeology*, pp.271-331. Columbia University Press, New xYork.

Yengoyan, A.A. 1976 Structure, event and ecology in Aboriginal Australia: a comparative viewpoint. In N. Peterson (ed.) *Tribes and Boundaries in Australia*, pp.121-32. Australian Institute of Aboriginal Studies, Canberra.

Yesner, D.R. 1980 Maritime hunter-gatherers: ecology and prehistory. *Current Anthropology* 21(6):727-50.

Young, E. 1866 Journal.-Escape Cliffs, Northern Territory. *South Australian Parliamentary Papers*, No.80.

Zobel, D.E., R.L. Vanderwal and D. Frankel 1984 The Moonlight Head rockshelter. *Proceedings of the Royal Society of Victoria* 96:1-24.

www.ingramcontent.com/pod-product-compliance
Ingram Content Group UK Ltd.
Pitfield, Milton Keynes, MK11 3LW, UK
UKHW061848041026
15591UKWH00015B/965

9 781407 309231